Geoenvironmental Disaster Reduction

The International Consortium on Geo-disaster Reduction (ICGdR) decided to create a series of books which could help to reduce the overall risk from natural hazards of different origin in concordance with the main strategy and objectives of ICGdR. These are:

1. Promote geo-disaster reduction for the benefit of human society and the natural environment, and capacity development, including education;
2. Combine international expertise and coordinate their efforts in geo-disaster reduction, thereby resulting in an effective international organization which will act as a partner in various projects; and
3. Promote regional and global, multidisciplinary activity on geo-disaster reduction.

The idea is to bring the attention of authors and research teams to the subject of mitigation of natural risks irrespective the detailed topic of hazards or region of occurrence. The International Consortium on Geo-disaster Reduction in cooperation with Springer Verlag with its broad variety of scientific interests and large international operation is a suitable platform to manage such important mission. The International Consortium on Geo-disaster Reduction offer to the international scientific community to take the patronage over selected book publications printed in the Springer Verlag regardless the authors are members of ICGdR or non-members. The submitted books will go through the standard editorial process and will be visible labelled by ICGdR. The ICGdR will help to promote these books. We welcome your proposals.

Rui Carrilho Gomes · Vítor Correia ·
Balazs Bodó
Editors

Citizens' Observatories on Geohazards

Lessons from Five Pilots

Springer

Editors
Rui Carrilho Gomes
CERIS—Civil Engineering Research
and Innovation for Sustainability,
Instituto Superior Técnico
Universidade de Lisboa
Lisbon, Portugal

Vítor Correia
International Raw Materials
Observatory
Brussels, Belgium

Balazs Bodó
La Palma Research Centre
La Palma, Spain

ISSN 2946-5710 ISSN 2946-5729 (electronic)
Geoenvironmental Disaster Reduction
ISBN 978-3-031-53370-9 ISBN 978-3-031-53371-6 (eBook)
https://doi.org/10.1007/978-3-031-53371-6

This Springer imprint is published by the registered company Springer Nature Switzerland AG
The registered company address is: Gewerbestrasse 11, 6330 Cham, Switzerland

QR codes'

AGEO Citizens' Observatory: https://ageoplatform.eu/

AGEO Citizens' Observatory, users' manual: https://ageo-manual.web.app/

AGEO Mobile app (iOS)

AGEO Mobile app (android)

Acknowledgments

This book results from the implementation of project AGEO—Platform for Atlantic Geohazard Risk Management funded by the Interreg Atlantic Area Programme, financed by the European Regional Development Fund (ERDF) under the European Territorial Cooperation objective of the European Union Cohesion Policy for the programming period 2014–2020.

About This Book

This open-source e-book demonstrates the effective use of Citizens' Observatories for geohazards monitoring under operational conditions and its possible connectivity with regional/national risk management systems, with better capacity in disaster risk reduction based on the principle of public engagement: "together we are strong."

This book compiles the tools, experience, and information accumulated throughout project AGEO—Platform for Atlantic Geohazard Risk Management—implemented in the Atlantic Area between 2019 and 2023.

Five diversified Citizens' Observatory pilots were deployed during the course of this project, each covering different geohazards, exposures, and environments, with the goal of engaging local communities to actively participate in risk preparedness and monitoring and incorporate local capacities into risk management systems, aiming at serving as the foundation for further efforts following the project's conclusion.

More efficient uptake of Copernicus data on a regional level, new methodologies, a new open-source app, and a new open-source platform developed for risk reduction raising citizens' interest for the topic, increased public confidence in the regional authorities' abilities to manage natural hazards as a result of increased transparency, empowerment of community leaders to take up active roles in risk prevention, and recovery and regeneration actions are described in this book.

Introduction

All Atlantic regions are affected by natural hazards (geohazards) and various cascading hazard risk scenarios, such as rockfalls, landslides, earthquakes, tsunamis, geotechnical hazards, floods including flash floods, and coastal hazards.

These geohazards have the potential to cause extensive damage to economic, social, and natural assets. This is partly due to the likelihood of their occurrence, the high costs associated with mitigating their effects, and the lack of preparedness to effectively monitor and respond to them.

Active participation from local communities in risk preparedness and monitoring is essential to creating a stronger public awareness of the importance of geohazard risk prevention and a strong social license for the implementation of new measures.

This shift will empower community leaders to take on more active roles in risk prevention, recovery, and regeneration efforts.

Strengthening the territory's resilience to risks of natural, climate, and human origin through the strengthening of risks management systems is a priority for the Interreg Atlantic Area European program. The AGEO—Platform for Atlantic Geohazard Risk Management project (2019/23) is a direct response to this priority, aimed at strengthening risk management systems.

The AGEO project has formulated a set of recommendations for the creation of future observatories in response to a wide range of hazards, including both natural and human-induced threats in the Atlantic region. The recommendations include:

- creating a cooperation and resource platform focused on Atlantic geohazards risk assessment, preparedness, mitigation, and prevention.
- providing concrete case studies to demonstrate the capacities of Citizens' Observatories in improving risk management systems.
- encouraging the regional-level uptake and use of products and services provided by European spatial data infrastructures, such as Copernicus or EGDI.

By engaging citizens, the AGEO project has been able to gather valuable data streams of local information complementing state and EU-level systems and data sources. During the course of the AGEO pilots, citizens provided their own observations and utilized mobile devices to collect and transmit valuable data for the project. The project thus demonstrated that the active engagement with citizens and the local uptake of innovative Earth Observation-based products and services improves risk preparedness and management in a cost-efficient manner.

Citizens' Observatories can serve as the interface between individuals and risk management systems, as they can be defined as "innovative earth observation technologies, in particular those based on use of mobile telephony, and community-based environmental monitoring, data collection, interpretation and information delivery systems; empower communities with the capability to monitor and report on their environment; and enable communities to access the information they need to make decisions in an understandable and readily usable form" (Rubio Iglesias, 2013).

Rubio-Iglesias (2013) described Citizens' Observatories as having at least four distinctive features:

1. Bidirectional information flows, i.e., "citizens are recipients of information but also important providers."
2. New citizen functions, e.g., "the public should be given the means to aggregate, combine and generally reuse information according to their various needs."
3. Support for multi-scalar governance, e.g., "participation in assessing the success of European Union (EU) environment policies."
4. Complementarity, e.g., "the potential to enormously expand in situ monitoring capability, and ...limit the charge on the public purse..."

The incorporation of Citizens' Observatories into existing risk management systems has the potential to significantly improve the resilience and planning capacity at local and regional level for the management of natural and climatic risks as well as those derived from human activities.

The AGEO project deployed five unique Citizens' Observatory pilots, each covering different geohazards, exposures, and environments, according to regional priorities, actively engaging local communities in risk preparedness and monitoring, incorporating local capacities into risk management systems.

Table 1 outlines the Citizens' Observatories deployed during AGEO project execution and the geohazards addressed by each pilot.

The development of harmonized solutions for disaster risk reduction in the Atlantic region required a transnational approach that involved

Table 1 Citizens' observatories deployed by the AGEO project

Citizens' Observatories Pilots	Geohazards
1. Citizens' observatory on rockfalls in the Canary Islands, Spain	Rockfall
2. OSIRISC: co-design of an integrated observatory to monitor vulnerability to coastal risks of erosion and flooding in Brittany, France	Coastal erosion and floods
3. Multihazard Citizens' Observatory, Lisbon, Portugal	Earthquakes, floods, landslides, geotechnical risks
4. Causeway Coast, Northern Ireland, UK	Rockfalls
5. Citizens' observatory on rockfalls in Madeira, Portugal	Rockfalls

key stakeholders from all participating countries: Portugal, Spain, France, Ireland, and Northern Ireland/UK. These stakeholders were directly involved in the development work, while representatives of European Union data infrastructures participated in consultation and scoping processes via dedicated workshops and conferences.

In addition to the benefits of exchanging experience and best practices in disaster risk reduction along the Atlantic, the project also promoted cooperation on low probability high impact events, which are inherently transboundary in nature.

In addition to the five Citizens' Observatories pilots, the AGEO project has delivered:

(i) A comprehensive toolbox that facilitates the deployment of the Citizens' Observatories to be also used in follow-up projects, interoperable, and compatible with national data systems;
(ii) Educational contents, information leaflets, and dissemination materials;
(iii) A project website and social media channels for direct engagement with citizens and the professional community;
(iv) Several high-impact conferences and workshops organized on a national and international level with the aim of stakeholder engagement and the mobilization of external expertise;
(v) An app and an IT platform dedicated to long-term cooperation and information sharing on geohazards;
(vi) An evaluation of the AGEO Citizens' Observatories and recommendations for the future;
(vii) An evaluation of the status of regional use of Copernicus products and recommendations for the future;
(viii) Overarching recommendations for the development of new policies and measures in view of improving the resilience of land and maritime areas to climate and nature changes.

With the fulfillment of the above achievements, AGEO serves as an example of the benefits of engaging civil society and local authorities in disaster risk reduction, building on local capacity combined with innovative Earth Observation products and services provided by European data infrastructures.

AGEO demonstrated the effective use of Citizens' Observatories for geohazards monitoring under operational conditions as integral part of regional/national risk management systems, with better capacity in disaster risk reduction based on the principle of public engagement: "together we are strong."

With this book, the AGEO consortium aims to promote the following ambition:

- Citizens' Observatories to become a regular component of risk management systems at local, regional, and national levels.
- Stronger public awareness to be reached for the importance of geohazard risk prevention and a strong social license to be developed for the implementation of new measures.
- More efficient uptake of Copernicus data on a regional level and a new range of innovative EO services and products seamlessly integrated with Citizens Observatories.

- New methodologies and technologies, for example, new apps, to be developed for risk reduction raising young people's interest for the topic.
- Increased public confidence in the regional authorities' abilities to manage natural hazards as a result of increased transparency.
- Community leaders to be empowered to take up active roles in risk prevention, and recovery and regeneration actions.

Reference

Rubio Iglesias, J.M., 2013. Citizens' observatories for monitoring the environment: A commission perspective. In Proceedings of Workshop on Citizen's Involvement in Environmental Governance, Arlon, Belgium, 7 October 2013; Directorate General Research and Innovation, European Commission: Brussels, Belgium, 2013.

AGEO Consortium

Partners	Team
Instituto Superior Técnico, Lisbon, Portugal (Lead Partner)	Rui Carrilho Gomes Mónica Amaral Ferreira Carlos Sousa Oliveira Hugo O'Neill Ana Paula Flor Ana Carolina Antunes Paulo Fonseca Teixeira
Associação Portuguesa de Geologia, Lisbon, Portugal	Vitor Correia José Romão Pavlos Tyrologou
La Palma Research Centre, Canary Islands, Spain	Adrienn Cseko Ariadna Ortega Carlos Garcia Gabriella Foti Balazs Bodó
Instituto Geológico y Minero de España, Agencia Estatal Consejo Superior de Investigaciones Científicas, Canary Islands, Spain	Inés Galindo Mª Nieves Sánchez Isabel Montoya Juan Carlos García Inmaculada García Moreno
Université de Bretagne Occidentale, Brest, France	Nicolas Le Dantec Marion Jaud Quentin Ruaud Caroline Lummert Thibaud Lami Mathias Rouan Julie Pierson Alain Henaff Iwan Le Berre Catherine Meur Ferec Elisabeth Guillou Manuelle Philippe Pauline Letortu Christophe Delacourt

Partners	Team
Geological Survey of Northern Ireland, British Geological Survey, United Kingdom Research and Innovation, UK	Kirstin Lemon Kieran Parker
University College Dublin, Dublin, Ireland	Eleni Mangina Eoghan Holohan Levent Gorgu Alexis Hrysiewicz Xiaowen Wang Gregory O'Hare Michael O'Grady Maria O'Brien
Centre d'Etudes et d'Expertise sur les Risques Environnement Mobilité et Aménagement, Brest, France	Roche Amélie Sylvain lendre Marie Caouissin Perherin Céline
Laboratório Nacional de Energia e Geologia, Lisbon, Portugal	Lídia Quental Elsa Ramalho Pedro Gonçalves Susana Machado Teresa Pena Silva
Universidad de La Laguna, Canary Islands, Spain	Juan Carlos Santamarta Cerezal Noelia Cruz Perez Joselin Sarai Rodríguez Alcántara
Laboratório Nacional de Engenharia Civil, Lisbon, Portugal	Laura Caldeira Filipe Telmo Jeremias João Paulo Serra António Roque Maria João Coelho Delta Sousa e Silva Joana Carreto Rute Ramos
Universidade da Madeira, Funchal, Portugal	Domingos Rodrigues Fátima Gouveia Paulo Silva Lobo João Martins
Câmara Municipal de Lisboa, Lisbon, Portugal	Cláudia Pinto Carlos Sousa Ferreira Paula Duarte Carla Pousada Miguel Inácio Ana Luísa Seabra Maria Manuel Pinto Inês Ascenso Ana Runa Pedro Leonardo Alexandra Frias Mário Fontes Maria José Cerqueira Carolina Almeida Rita Dórdio Vanessa Fernandes Joana Fernandes Filipa Vilela

Contents

About the Editors

Rui Carrilho Gomes is a Professor of geotechnics at Instituto Superior Técnico of the University of Lisbon, Portugal, Department of Civil Engineering. From 2019 to 2023, he was the coordinator of project AGEO—Platform for Atlantic Geohazard Risk Management—funded by the Interrg Atlantic Area. From 2000 to 2013, he was member of the Board of the Portuguese Society for Earthquake Engineering. His research interests focus on risk assessment, earthquake engineering, soils dynamics, and seismic surface waves using computational methods and field testing. He has published more than 50 papers in refereed journals and conference proceedings.

Rui has a 5-year Diploma in Civil Engineering (1996), Master in Structural Engineering (2000), and received his Ph.D. in Civil Engineering (2009) from Instituto Superior Técnico, Portugal.

Vitor Correia currently serves as Secretary-General of the International Raw Materials Observatory, an independent, international association that supports international cooperation in the mineral raw materials' field, and he works as an independent consultant providing his expertise to EU and development banks projects. From 2019 to 2023, he led the exploitation work package of the AGEO project, using his experience in management, innovation and organizational effectiveness to support the project's success.

Vitor began his career as a mine geologist, and he worked in mining, geological engineering and environmental geology projects in Europe, Africa and South America. Vitor is a member of the Expert Group on Resource Classification of the United Nations Economic Commission for Europe and he is Past President of the European Federation of Geologists. He was member of the Board of the Portuguese Association of Geologists and he's an Honorary Fellow of the Geological Society of London. Vitor holds a B.Sc. in Geology (1992) and an M.B.A. (1994), both from the University of Lisbon, and he holds the EuroGeologist professional title (since 2012).

Balazs Bodó is the Head of the Technology Foresight Unit at La Palma Research Centre S.L., an internationally active geoscience think-tank headquartered on La Palma Island, Spain. Balazs has an M.Sc. in Geological Engineering (1996) from the University of Miskolc, Hungary and an M.Sc. in Environmental Engineering (1998) from the Royal Institute of Technology, Sweden. He has a research interest in natural hazards and societal resilience.

He was the Project Coordinator of the EU FP7-funded project SECURENV (Assessment of environmental accidents from a security perspective, Grant Agreement No. 218152). He has been a Project Manager in AGEO, developing public outreach strategies for the project.

Enhancing Risk Management Systems with Citizens' Observatories for Geohazards

1

Rui Carrilho Gomes and Ana Carolina Antunes

Abstract

Risk management is the use of disaster risk reduction policies and strategies to prevent the occurrence of future disasters, reduce current disaster risk, and manage residual risk, thereby enhancing resilience and reducing catastrophic losses, despite the fact that citizens may take an active role in geohazard risk management. Citizens' Observatories are community-based information and monitoring systems that invite citizens to submit observations, generally by mobile phone or the internet (WeObserve, 2018) and directly link volunteers and scientists for the bidirectional exchange of data. This chapter examines how Citizens' Observatories can improve risk management systems by promoting citizen involvement, educating and making citizens aware of the possible geological hazards that may occur in their area of residence and surroundings, and thereby encouraging them to exert pressure on decision-makers and local and government authorities to implement preventative and protective measures.

R. C. Gomes (✉) · A. C. Antunes
CERIS, Instituto Superior Técnico, Universidade de Lisboa, Lisbon, Portugale-mail: rui.carrilho.gomes@tecnico.ulisboa.pt

1.1 Introduction

The increasing urbanization is leading to the concentration of population in geohazard-prone areas, increasing losses and fatalities.

Risk management is the application of disaster risk reduction policies and strategies to avoid new disaster risk, reduce current disaster risk, and manage residual risk, thereby enhancing resilience and reducing catastrophic losses, aiming to create a prevention ethic. Citizens are generally only seen as victims, disregarding that they can be active contributors in the management of geohazards, something that is intended to change.

We live in a digital era, where the population has access to technology that eases monitoring data on geohazards, such as earthquakes, landslides, flash floods, and coastal erosion. For example, the citizens can fill up a short survey, send images and videos of a disaster, or simply allow the mobile phone to share the phone accelerometer data. This process can be conducted through an app and an associated platform linking citizens with the established risk management system, named Citizens' Observatory.

The implementation of Citizens' Observatories for geohazards have so far not reached the necessary maturity, unlike in other areas, e.g. birdwatching, galaxy classification, despite citizens may have an active role.

R. C. Gomes et al. (eds.), *Citizens' Observatories on Geohazards*, Geoenvironmental Disaster Reduction,
https://doi.org/10.1007/978-3-031-53371-6_1

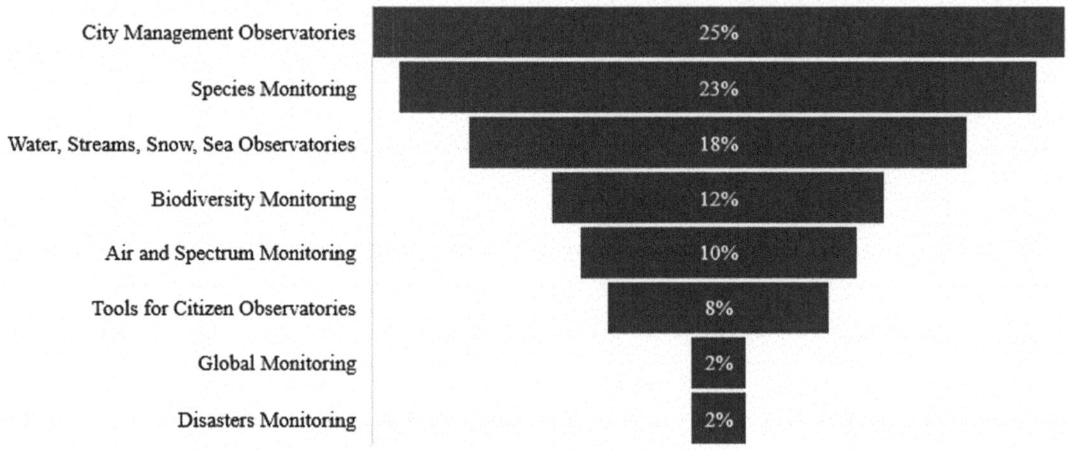

Fig. 1.1 Types of citizen observatories around the world (adapted from Palacin-Silva et al., 2016)

Figure 1.1 shows the existing Citizens' Observatories by type. It can be noticed that only a small number of the existing Citizens' Observatories is dedicated to monitoring geohazards, more specifically monitoring earthquakes and their early detection (Palacin-Silva et al., 2016).

The goal of a Citizens' Observatory for geohazards is promoting citizen involvement in the detection and/or characterization of the geohazard, but also to raise citizen awareness on geohazards in their area of residence, of work or of tourism, establishing a bidirectional communication.

Engaged citizens can better exert their citizenship, promoting better decision-making and pressuring authorities to implement preventive and protective measures.

This chapter contributes to define a conceptual framework on the use of Citizens' Observatories for geohazard, enhancing the risk management system. The link between Citizens' Observatories and the risk management cycle is established, and intervention measures are identified.

1.2 Geohazards

A geohazard is a natural hazard with potential to cause losses and/or casualties. The characteristics of geohazards are diverse, and they can occur on a wide range of spatial and temporal scales.

A brief overview of the geohazards covered in this book is presented, namely earthquakes, landslides, rockfalls, floods, and coastal erosion, covering their definitions and temporal e spatial scale.

Earthquakes are characterized by the rapid release of energy in the form of mechanical waves induced by tectonic and volcanic activity. The shaking duration ranges from seconds to tens of seconds, and the effects can be felt up to tens to thousands of kilometers away from the epicentre. The vulnerability and exposure of the built environment to hazards are critical elements for risk assessment.

Mass movements is the downslope movement of a mass of surface materials, such as soil or rock, that can also be called as rockfall or landslide, respectively.

In a rockfall, the block falls, rolls, or bounces rapidly, a few meters per second, and may occurs multiple times per year covering an area proportional to the slope height. Landslides move a few meters per minute or hour, covering from small to very large areas, depending on the volume of the unstable mass.

Floods are overflows of water that can occur due to heavy and/or persistent precipitation, thunderstorms, cyclones, or melting snow.

Flood event duration can range from flash floods, which last from minutes to hours, to major regional floods that may last weeks. The extent is determined by the stage of overflowing water and the morphology of the area. Water can flood and occupy kilometre-sized areas in flat lands for many days, whereas in mountain areas, impacts are related to water erosion power and flooded areas are less common.

Coastal erosion is the gradual disintegration of rocks along the shore generated by intense wave action, sea level rise or coastal flooding. Storm surges, hurricanes and typhoons, tsunamis, windstorms, and flooding are all known to cause coastal erosion, affecting a large percentage of the world's shorelines. Thus, the spatial scale of coastal erosion can be of a few to thousands of kilometres of coastline exposed to water and wind. Coastal erosion may be due to episodic event, as a tsunami or a single very strong storm, short-term over few years, or it may be long-term, e.g. over tens to thousands of years.

This brief overview on the geohazards characteristics covered in this book supports Fig. 1.2, that plots the temporal scale versus the spatial scale of each geohazard.

The specific characteristics of each geohazard has an impact on how the Citizens' Observatory can be linked to the risk management cycle, that is analysed in the next sections.

1.3 Risk Management Cycle

Risk is the human, material or economic loss probability in a period of time. Natural risk consists of the combination of geohazard with exposure, vulnerability, and capacity.

Geohazard is the probability of occurrence of a specified level of hazard intensity in a specified period of time, including eventual cascading effects as landslides, liquefaction, tsunamis.

Vulnerability is the degree of damage caused by various levels of loading. The vulnerability may be calculated in a probabilistic or deterministic way.

Exposure includes the value elements, e.g. population, assets, and environment, exposed to the geohazard.

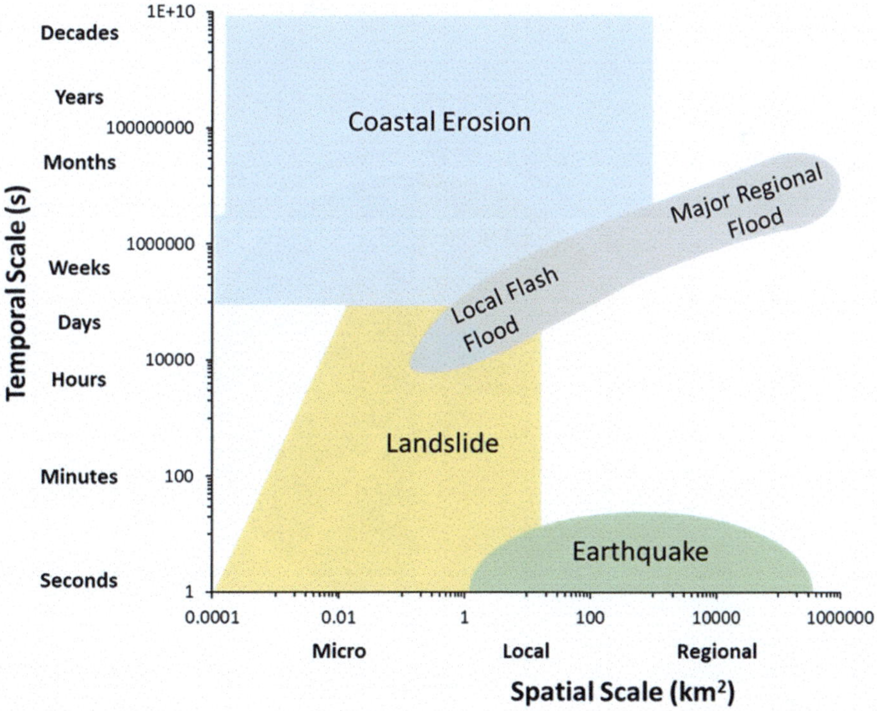

Fig. 1.2 Geohazards: temporal scale versus spatial scale (adapted from Eurosion (2004) and Gill and Malamud (2014))

Capacity is the combination of all resources and strengths to manage and mitigate the risks and strengthen resilience.

The risk management is the process of identifying potential threats, minimizing the impact of disasters on society and the environment, assure the safety during critical events, and craft disaster recovery plans to quickly remedy the effects of a geohazard event on community.

Risk management can be applied across many sectors including, government, healthcare, higher education, IT operations and more. Regardless of industry sector, the risk management system must quickly inform the population on crisis situations.

Managing disasters effectively requires risk management systems to:

- Identify, evaluate and prioritize potential threats.
- Establish procedures and safety measures.
- Immediately notify people of distructive events.
- Deploy disaster recovery programs and business continuity strategies.

The risk management is usually divided in stages that describe a cycle before, during and post-event (e.g. UNESCO, 2007; UN-ISDR, 2017) (Fig. 1.3).

The pre-event stage aims to minimise community vulnerability to the impact of natural events and to reduce community exposure to dangers or even reduce the geohazard itself. Pre-event stage can be subdivided in:

- prevention and mitigation, aiming to minimize impact of future disaster;
- preparedness, aiming to prepare for managing crisis, including knowledge and skills required to effectively foresee, respond to, and recover from the effects of potential, imminent, or ongoing geohazard;
- early warning, consisting of an integrated system of hazard monitoring, forecasting and prediction, disaster risk assessment, communication and preparedness activities systems and processes that enables individuals, communities, governments, businesses, and others to take prompt action to reduce disaster risks before hazardous events.

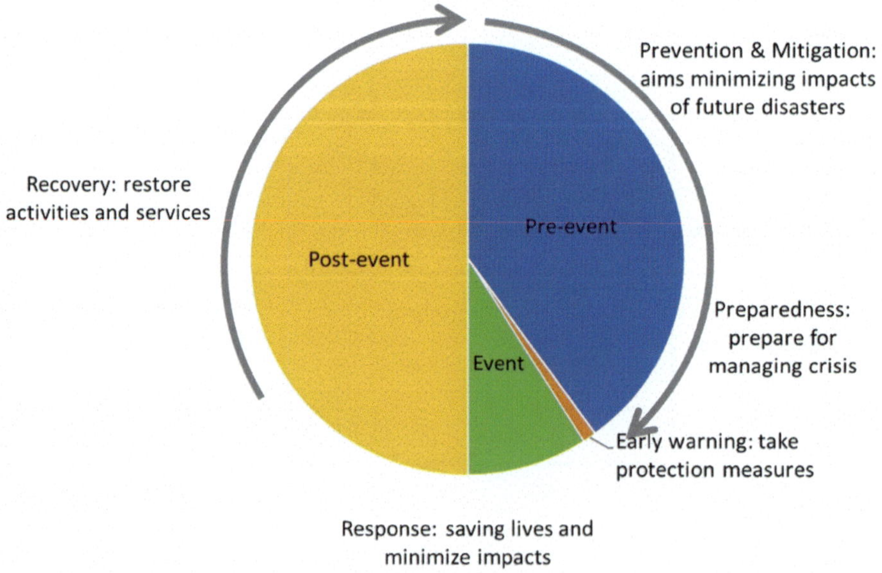

Fig. 1.3 Risk management cycle

During the event phase is primarily concerned with immediate and short-term requirements, including rescue, damage assessment, and response procedures are required. Depending on the events duration, this phase can last from a few seconds to a few days.

Finally, the post-event includes the recovery phase, in which communities restore normal operations and disaster lessons are incorporated into development policies, and a new cycle is restarted.

1.4 Citizens' Observatory Concept

1.4.1 Overview

Citizens' Observatories are community-based information and monitoring systems that encourage citizens to contribute observations through cell phone or the Internet (WeObserve, 2018), and directly linking volunteers and scientists in a bidirectional sharing of information. This bidirectional sharing (Fig. 1.4), is based on two types of approaches (Liu et al., 2014):

- top-down, led by scientists and experts;
- bottom-up, citizen initiative to solve problems or situations.

Disaster risk management is usually carried out using a top-down approach, giving decision-making power and strategy development to government bodies, researchers, and disaster management bodies, while not encouraging the participation of affected communities (Solinska-Nowak et al., 2018).

The adoption of a more participatory strategy by the population in disaster risk management strategies has the advantage that "the end users of information [are treated] not merely as a target audience but as partners in co-learning through processes and products that reflect their own contributions" (Solinska-Nowak et al., 2018).

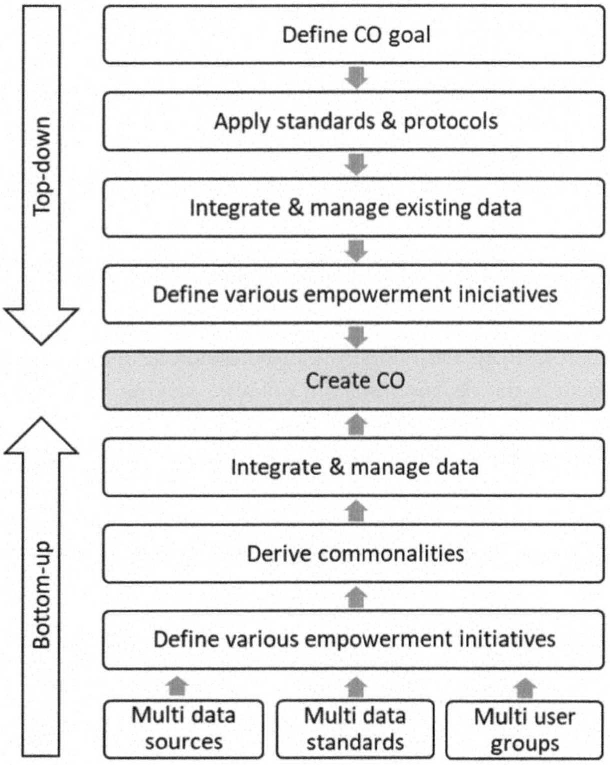

Fig. 1.4 Citizens' observatory concept (adapted from Liu et al., 2014)

The use of new technologies has facilitated the bidirectional communication, because allowed to quickly acquire and transmit in situ data (Haklay, 2015). Data sent by citizens comes in various formats, such as XML, text, CSV, images, and videos that can be collected through web platforms, smartphone applications, or low-sensors (Palacin-Silva et al., 2016).

Social networks can be used for data collection since a large number of users upload disaster related information shortly after the disaster, but its data quality data needs to be confirmed which requires time and resources.

Citizens' Observatories can be viewed as an excellent opportunity; nonetheless, the following concerns must be properly addressed (Palacin-Silva et al., 2016):

- Minimal citizens' involvement without providing sufficient data quality and quantity.
- Concerns of citizens over the privacy of their data and how it will be utilised.
- The collected data are available in a variety of forms, and the extraction procedure can be time-consuming and error-prone.
- Mobile devices have several restrictions that may affect the quality of collected data.
- The formation and growth of an observatory may be negatively affected by a lack of scientific expertise and financing.
- Citizens' Observatories must acknowledge the contributions of their volunteers and make the findings of data processing accessible to anyone who can benefit from them.

A Citizen Observatory can be a crucial element in the decision-making process, serving as a support system for authorities and scientists. These observatories are already present in several areas, including environment, astronomy, and medicine (Grossberndt & Liu, 2016), and there is a growing interest and development in the creation of Citizens' Observatories related to natural hazards.

1.4.2 Brief Overview of the Existing Citizens' Observatories for Geohazards

In this section are introduced existing Citizens' Observatories for geohazards, even if some of them were launched without having been named as such. In general, Citizens' Observatories aim to collect data and inform citizens on a specific geoan hazard.

The "Did you feel it?" project launched by USGS, USA, in the late 1990s aimed to collect a description of what was felt by citizens during an earthquake by filling out survey (Atkinson & Wald, 2007). The collected answers were used to plot a seismic intensities map that was published on the Community Internet Intensity Maps. The map was released and updated in nearly real time.

A similar "Did you feel it?" project was launched in Europe, led by EuroGeoSurveys consortium of 36 seismological organizations from 24 countries (Bossu et al., 2017). In this project, the citizens could fill up a survey on what they have felt during an earthquake available on a website and an app.

Google Inc. integrated the Android Earthquake Alerts System into Android operating system of smartphones. This freeware system detects earthquakes worldwide autonomously and can receive earthquake early-warnings. In California, Washington, and Oregon, this system works in partnership with ShakeAlert, which is an early warning system developed by the USGS that uses 1675 seismic sensors to detect and analyze seismic activity. ShakeAlert detects the earthquake and sends the warning to the Android Earthquake Alerts System. Outside these American states, data measured by the smartphone accelerometer is used to detect earthquakes. When an earthquake is detected, the data and the smartphone location is sent to the earthquake detection center. All data collected is processed first to cross-validate the data and, second, to issue a warning. The Android Earthquake Alerts System transform smartphones into passive data collection sensors, detecting the vibration and speed of an event.

NASA developed a worldwide open-access landslide platform, named Cooperative Open Online Landslide Repository, where citizens, authorities and scientists can report landslide events and see past-events locations and characteristics, aiming at enhancing landslide emergency response, scientific modeling, and public awareness.

The Citizens' Observatories for Situation Awareness in Flooding (WeSenseIt) (Lanfranchi, 2014) was launched to address flood hazard. This observatory collects data through traditional detection equipment and low-cost equipment. Traditional detection technology collects data such as air temperature, humidity, and precipitation automatically and remotely, but its high cost and continuing maintenance requirements limit their number. To increase the number of measurement points, low-cost equipment and a short training was given to residents, that although less precise complement the more accurate measurements. This observatory has also created a mobile application that enables the sharing of the sensor's acquired data and facilitates the communication in an emergency. A platform for large-scale real-time analysis and display, was also implemented (Lanfranchi, 2012), which works as a repository of all Citizens' Observatory information, accessible for citizens, professionals, stakeholders and authorities.

The USGS project "iCoast: Did the Coast Change?" operated between 1995 to 2000, collecting more than 140,000 photos related with the storms impact on the coastline, to monitor the coastal erosion. Volunteers were asked to compare coastal images before and after a storm on a web platform, to tag changes in the coastline. With this procedure, the volume of data increased enormously, permitting to automatize the assessment of the shoreline retreat.

These examples of citizens' observatories for geohazards only cover parts of the risk management cycle. In the next section it is discussed how citizens' observatory can be linked with the risk management cycle.

1.5 Linking Citizens' Observatories to the Risk Management Cycle

To analyze the potential contribution of Citizens' Observatories to the Risk Management Cycle, the next Tables present an exercise linking the Liu et al. (2014) Citizens' Observatory concept with the risk Management cycle, identifying the action that can be undertaken in each of them.

Table 1.1 addresses the potential use of Citizens' Observatories for prevention, mitigation and preparedness stages, while Table 1.2 is for early-warning stage and Table 1.3 for response and recovery stages.

1.6 Top-Down and Bottom-Up Actions

In continuation of the preceding part's discussion of the relationship between Citizens' Observatories and the Risk Management Cycle, this section focuses on actions that may be performed from top-down and bottom-up efforts.

The following Tables present a non-exhaustive list of measures for each stage of the Risk Management Cycle grouped by:

- Top-down, taken by decision-makers, scientists, among others.
- Bottom-up, citizens individually taking care of their property or in association taking care of their region.
- Engineering measures, mostly to be implemented top-down but also some to be implemented bottom-up.
- Planning, Management, Institutionalization Societal Measures, divided by both bottom-up and top-down initiatives (Tables 1.4, 1.5, 1.6, 1.7 and 1.8).

Table 1.1 Connecting citizens' observatories with risk management cycle: prevention, mitigation, and preparedness

Top-down		
Citizens' observatory goal	**Prevention/mitigation**	**Preparedness**
Data sources	– Ministries and government agencies – Non-governmental organizations – Academic institutions	– Satellite and/or drone imagery – In situ precision sensoring, e.g. hydrological stations, accelerometers network
Data	– Hazard database – Geographical settings – Physical vulnerability – Social & economic vulnerability – Inventories of assets	– Needs for assistance – Pre-disaster information on infrastructures network, land cover and use, critical structu-res, financial data – Hazard return period – Hazard-affected area
Apply standards & protocols	– Standards to assess risk – Standards and protocols for risk management – Standards for structural and non-structural prevention and mitigation measures – Standards and protocols for local, regional and national planning	– Protocols for response in medium to high-risk areas – Protocols to disseminate information, e.g. educational contents, areas affected, auto-protection measures – Protocols for evacuation drills in schools, offices, etc – Protocols to promote efficient responses
Integrate & manage existing data	– Interactive multihazard risk map – Prioritized action list for prevention/mitigation – Technical and financial implementation plan for risk prevention/ mitigation	– Disaster scenarios – Prioritized preparedness action list

(continued)

Table 1.1 (continued)

Top-down		
Define empowerment initiatives	– Inform the citizens in medium to high-risk areas – Disseminate mitigation measures – Insurance companies reflecting the risks in the prizes charged	– Spread information on how to be prepared, e.g. drill, safety measures, reaction, among others – Disseminate warning signs about hazards – Disseminate how the early warning system operated – Inform the population through conventional media on the reliability and advantages of the early warning system
Citizens' observatory (bi-directional exchange of information)		
Define empowerment initiatives	– Active civic participation in risk prevention and mitigation initiatives in cooperation with authorities and researchers – Training and social recognition – Gamification of prevention and mitigation measures – Direct participation in prevention and mitigation measures	– Form an emergency response team – Prepare an emergency kit – Prepare and practice a family emergency plan – Gamification of preparedness measures – Direct participation in prepa-redness measures
Derive commonalities + integrate & manage data	– Population exposure map in higher risk areas – Map areas/properties with higher risk – Inform citizens in the hazard-prone areas	
Multi data standards	– Standards to disseminate educational contents – Standards to process data provided by citizens – Standards for citizens to collect data	
Multisource data	– Characteristics of the citizens' properties exposed to hazard – User-friendly and/or low cost sensors	
Multiuser groups	– Citizens' initiatives, associations and cooperatives, e.g. cultural associations, community science, observatories, scouts – Local community leaders – Professional networks	
Bottom-up		

Table 1.2 Connecting citizens' observatories with risk management cycle: early-warning

Top-down	
Citizens' observatory goal	**Early warning**
Data sources	– Satellite and/or drone imagery – In situ sensoring, e.g. weather and hydrological stations, accelerometers network
Data	– Ground motion – Weather forecast
Apply standards & protocols	– Protocols to issue warning, with predefined level of reliability – Protocols to automatic protection measures, e.g. shut down electrical systems, activate emergency power supply systems
Integrate & manage existing data	– Estimated time to the event – Interactive map of estimated affected area
Define empowerment initiatives	– Inform the population through conventional media on protection measures to be implemented after receiving an early warning
Citizens' observatory (bi-directional exchange of information)	
Define empowerment initiatives	– Give support to elderly, disables and children – Implement protection measures
Derive commonalities + Integrate & manage data	– Sensoring standards – Estimated time to event – Estimated affected area
Multi data standards	– Map data retrieved by sensors
Multisource data	– Low-cost sensors and wearables managed by prepared citizens
Multiuser groups	– Citizens' initiatives, associations and cooperatives, e.g. cultural associations, community science, observatories, scouts – Local community leaders – Professional networks
Bottom-up	

1.7 Conclusions

Risk management is a crucial activity in modern societies to reduce, mitigate, prepare, respond and recover from geohazards impact.

Risk management systems can be enhanced if citizens play an active role. Citizens' Observatories can directly connect volunteers and scientists for the bidirectional exchange of data.

This chapter explored how Citizens' Observatories may provide a bottom-up component to the commonly used top-down strategy to risk management.

After a brief overview on the key characteristics of the geohazards covered by this book—earthquake, landslides, rockfall, floods and coastal erosion—the risk management cycle and the citizens' observatory concept were introduced.

An exercise liking Citizens' Observatory with the risk management cycle is shown to illustrate the complementarity of top-down and bottom-up approaches.

Table 1.3 Connecting citizens' observatories with risk management cycle: response and recovery

Top-down		
Citizens' observatory goal	**Response**	**Recovery**
Data sources	– Ministries and government agencies, as civil protection, and fire department – Non-governmental organizations – International organizations – In situ sensoring, e.g. weather and hydrological stations, accelerometers network – Satellite and/or drone imagery	– Ministries and government agencies – Non-governmental organizations – International organizations, as IMF and UN – Satellite and/or drone imagery
Data	– Post-disaster losses and damages maps – Affected citizens, e.g. requiring shelter, food, water, medical aid, etc.	– Variations in risk caused by differences in hazards, exposure, vulnerability, and capability
Apply standards & protocols	– Protocols for emergency response and rescue operations – Protocols to disseminate information on the event – Protocol to distribute relief	– Standards to build back better – Capacitation protocols
Integrate & manage existing data	– Affected area interactive map; – Prioritization of the emergency response, rescue operations, and relief distribution	– Prioritized actions list for recovery – Technical and financial implementation plan for recovery
Define empowerment initiatives	Inform the population on the: – Characteristics of the event and the affected area – Areas where relief is distributed	– Disseminate how to Build Back Better geohazards-resistant structures; – Involve the population in the community recovery plan
Citizens' observatory (bi-directional exchange of information)		
– Define empowerment initiatives	– Train citizens to be first responders – Train citizens to give support to the vulnerable persons	– Monitoring Building Back Better – Involvement in the community recovery plan
Derive commonalities + integrate & manage data	– Map areas with same level of intensity, damage, etc – Estimate losses and damages	– Map areas with problems in Building Back Better
Multi data standards	– Standard to analyse the coherence and process the data received	
Multisource data	– Photo, text, query form, georeferenced reporting event characteristics and/or damages – Location-aware mobile phone data – Social media publications with descriptions of event characteristics and/or damages – Information on location and number of affected citizens requiring shelter, food, water and medical aid	– Photo, text, query form, georeferenced reporting incorrect application of build back better
Multiuser groups	– Citizens' initiatives, associations and cooperatives, e.g. cultural associations, community science, observatories, scouts – Local community leaders – Professional networks	
Bottom-up		

Table 1.4 Actions for PREVENTION AND MITIGATION stages

Engineering measures	
Top-down	Bottom-up
– Assess hazard vulnerability of the building stock – Develop and implement strengthening plan for the building stock – Develop and implement codes and guidelines for hazard resistant structures – Ensuring the compliance of hazard-building codes, urban planning legislation and other safety regulations – Install devices to limit the possibility of gas leaks, such as a gas shut-off device and flexible connections – Build coastal protection structures, e.g. breakwaters – Build flood protection structures, e.g. dikes, floodways – Removing people and property from flooding areas, elevating structures, permanent relocation, among others	– Fixation of elevated furniture and items that may fall during an earthquake – Elevating utility infrastructure and expensive equipment, e.g. to an upper story to prevent flood damages – Waterproofing existing exterior walls and adding removable flood gates for door openings – Assessment of your property susceptible to ground movements – Use vegetation to stabilize slopes of your property
Planning, management, institutionalization societal measures	
Top-down	Bottom-up
– Preparing emergency action plans, warning and response systems – Characterize the potential extent and nature of risks – Regulatory measures for land use zoning – Ensuring the compliance of hazard-resistant building codes, town planning laws and other safety regulations	– Dissemination of safety do-it-yourself hazard-specific measures – Learn the history of geohazards past events in your area – Avoid building in areas where landslides and floods are likely to occur again

Table 1.5 Actions for PREPAREDNESS stage

Engineering measures	
Top-down	Bottom-up
– Prepare emergency strengthening measures for buildings and critical infrastructures – Prepare temporary housing – Prepare field hospitals – Install multi-hazard early warning systems	(–)
Planning, management, institutionalization societal measures	
Top-down	Bottom-up
– Organize mock drills in offices, schools, industrial units, among others – Educate and empower the population with response capabilities – Elaborate management plans with measures to facilitate efficient responses to hazards – Deploy warning signs for landslides, rockfalls and coastal erosion – Risk management plans should define the appropriate responses given the nature of the hazards and local site conditions, based on an assessment of all available options and their relative merits. Responsibilities and contacts should clearly specify a hierarchy of responsibilities for the emergency responses	– Know the geohazard risk level in your area – Keep critical items safe, e.g. critical documents – Learn self-protective measure – Prepare, review and practice regularly your family emergency and communication plan – In an high risk area prepare an evacuation plan for you, your family and your community – Build a disaster supply emergency kit – Properly trained and adequately motivated local communities and citizens can know the early warning signals of the geohazard and be attentive, e.g. heavy rain for floods and landslides – Training local communities to understand the nature of the hazards, their possible intensities and ranges, and react according to existing guidelines provided by the institutions identified by the authorities

Table 1.6 Actions for EARLY WARNINGS

Engineering measures	
Top-down	Bottom-up
– Sensors deployment and hazard detection protocol – Implement protection actions after receiving and early warning such as: (i) Hospitals: stop safely surgeries, opening automatic doors, secure radioactive sources; (ii) Schools: issue alerts. (iii) Transports: slow down trains. (iv) Factories: early warning system to safely shut down and reduce secondary hazards	– Send information on initial signs of a possible geohazard, e.g. landslide
Planning, management, institutionalization societal measures	
Top-down	Bottom-up
– Warning broadcast through media and mobile phones – Evacuation when risk appears imminent, such as evidence of downslope movements above a building, a rapidly eroding steep slope is approaching a building or adverse meteorological condition forecast	– Listen the warnings and instructions and take safety measures – Follow authorities' instruction for evacuation

Table 1.7 Actions for RESPONSE

Engineering measures	
Top-down	Bottom-up
– Emergency engineering responses to rescue and protect citizens	
Planning, management, institutionalization societal measures	
Top-down	Bottom-up
– Detecting missing persons and initiating search and rescue efforts – Planning and construction of emergency shelters and relief camps – Providing aid to the afflicted population, such as health care, water and sanitation, food, psychological assistance, etc. – Disseminate information on the event characteristics, event temporal evolution, safe areas, water and food supply spots, – Minimize damage to property through the removal of household contents or commercial stock and equipment from buildings	– Take care of yourself first; give your home first aid; dry out your flooded home; restore the utilities; check on financial assistance; rebuild – Take autoprotection measures, emergency supplies and evacuate carefully and calmly, try to assist children, elderly and neighbours – Examine your area for damage and report injuries to emergency personnel, damages or releases of hazardous materials; – Be aware and report cascade events, as tsunamis, fires, rockfalls, landslides and floods

The chapter concludes with a non-exhaustive list of actions for each stage of the risk management cycle that may be implemented by the top-down and bottom-up approaches, separated into engineering measures and planning, management, institutionalization societal measures.

The application of these measures is described in the book's chapters pertaining to the five pilots deployed during the AGEO project's execution.

Table 1.8 Actions for RECOVERY

Engineering measures	
Top-down	Bottom-up
– Build Back Better by improving the resilience of critical infrastructure, fixing problems that contributed to the disaster and rebuild with the next disaster in mind – Integrate resilience into the legal, policy and urban planning frameworks, including (i) mainstreaming disaster risk management, into the core government's urban planning documents and institutions to achieve 'risk-informed' development; (ii) reviewing disaster risk management legislation and policies and strengthening institutional capabilities of the DRM multi-sectoral system; (iii) provide basic first-responder training to community groups; (iv) creation of a network of community leaders who are involved in the early warning system and implementation of evacuation plans along with basic first aid and search and rescue sensitization – Collaboration and coordination between Public institutions, ONG's, private institutions, donors, and civil society, communities and individuals is crucial to successful early recovery efforts, to achieving the quickest and most effective recovery – Community recovery plan, minimizing relocation and resettlement, proximity to essential services, utilities and water, sanitation and hygiene needs, taking into account socio cultural preferences, being inclusive and sensitive to those with special requirements – Safety and security, environmental and ecological sustainability, energy efficiency and appropriateness to climatic conditions in new settlements, shelter, and infrastructure reconstruction – Recovery and rehabilitation plan specifying actions to be undertaken to mitigate any impact of emergency works – Replant the damaged area, since the loss of ground cover may lead to more soil erosion	– Informed citizens claim for hazard-resistant buildings and infrastructures
Spatial planning, management and institutionalization measures	
Top-down	Bottom-up
– Ensure coordination between shelter rehabilitation and reconstruction efforts – Community recovery programmes that effectively integrate civil society, the private sector, and social inclusion, strengthening vulnerable groups – Ensure that all early recovery interventions will better prepare the most socio-economically vulnerable to absorb the impact of future hazards and shocks	– Business continuity planning, e.g. recovery plan and mitigate possible risk of secondary effects as fire, gas release, water damage – Prioritize repairs and resumption of operations

References

Atkinson, G. M., & Wald, D. J. (2007). Did you feel it? Intensity data: A surprisingly good measure of earthquake ground motion. *Seismological Research Letters, 78*, 362–368.

Botzen, W. W., Kunreuther, H., & Michel-Kerjan, E. (2015). Divergence be-tween individual perceptions and objective indicators of tail risks: Evi-dence from floodplain residents in New York City. *Judgment and Decision Making, 10*(4), 365–385.

Bossu, R., Landès, M., Roussel, F., Steed, R., Mazet-Roux, G., Martin, S. S., & Hough, S. (2017). Thumbnail-based questionnaires for the rapid and efficient collection of macroseismic data from global earthquakes. *Seismo-Logical Research Letters, 88*, 72–81.

Eurosion. (2004). Living with coastal erosion in Europe: Sediment and space for sustainability. www.eurosion.org.

Gill, J. C., & Malamud, B. D. (2014). Reviewing and visualizing the inter-actions of natural hazards. *Reviews of Geophysics, 52*(4), 680–722.

Google Inc. Crisis Response. Retrieved from https://crisisresponse.google/android-alerts/ [Consult. 28 Fev. 2023].

Grossberndt, S., & Liu, H. Y. (2016). Citizen participation approaches in envi-ronmental health. In Environmental Determinants of Human Health (pp. 225–248). Springer, Cham.

Haklay, M. E. (2015). Citizen science and policy: A European perspective.

Hicks, A., Barclay, J., Chilvers, J., Armijos, M. T., Oven, K., Simmons, P., & Haklay, M. (2019). Global mapping of citizen science projects for disaster risk reduction. *Frontiers in Earth Science, 7*, 226. https://doi.org/10.3389/feart.2019.00226

Lanfranchi, V., Wrigley, S. N., Ireson, N., Wehn, U., & Ciravegna, F. (2014, January). Citizens' observatories for situation awareness in flooding. In ISCRAM 2014 Conference Proceedings-11th International Conference on Information Systems for Crisis Response and Management (pp. 145–154). Sheffield.

Liu, H. Y., Kobernus, M., Broday, D., & Bartonova, A. (2014). A conceptual ap-proach to a citizens' observatory–supporting community-based environ-mental governance. *Environmental Health, 13*(1), 107.

NASA. (2023). Cooperative Open Online Landslide Repository project. Retrieved form https://gpm.nasa.gov/landslides/about.html. [Consult. 03 Fev. 2023].

Palacin-Silva, M., Seffah, A., Heikkinen, K., Porras, J., Pyhälahti, T., Sucks-dorff, Y., ... & Junttila, S. (2016). State-of-the art study in citizen observa-tories: technological trends, development challenges and research avenues.

Solinska-Nowak, A., Magnuszewski, P., Curl, M., French, A., Keating, A., Mo-chizuki, J., & Jarzabek, L. (2018). An overview of serious games for disaster risk management–Prospects and limitations for informing actions to arrest increasing risk. *International Journal of Disaster Risk Reduction, 31*, 1013–1029.

UN-ISDR—United Nations-International Strategy for Disaster Reduction. Sendai Framework Terminology on Disaster Risk Reduction. (2017). Retrieved from https://www.undrr.org/terminology. [Consult. 05 Dez. 2021].

USGS. (2023). Did You Feel It? Retrieved from https://pubs.usgs.gov/fs/2005/3016/pdf/FS-2005-3016.pdf. [Consult. 03 Fev. 2023].

USGS. (2023). iCoast – Did the Coast Change? Retrieved from https://www.usgs.gov/centers/spcmsc/science/icoast-did-coast-change. [Consult. 28 Fev. 2023].

UNESCO, Integrated Global Observing Strategy—IGOS, Geohazards Theme Report. (2007). (https://unesdoc.unesco.org/ark:/48223/pf0000140532).

WeObserve. Citizen Observatories; International Institute for Applied Systems Analysis (IIASA): Luxembourg. (2018). Retrieved from https://www.weobserve.eu/about/citizen-observatories/. [Consult. 02 Jan. 2023].

Communication and Dissemination of AGEO

2

A. Ortega Rodríguez, A. Cseko, C. García, N. Cruz Pérez, J. C. Santamarta Cerezal, J. S. Rodríguez, I. Montoya-Montes, I. Galindo Jiménez, N. Sánchez Jiménez, J. C. García López-Davalillo and I. García-Moreno

2.1 Introduction

Communication was a vital component for AGEO project, as it enabled the Consortium to engage with the stakeholders and raise awareness about geohazards and how to react to them. Bidirectional communication was established with the public, receiving feedback on how to improve the citizens' preparedness towards geohazards.

The purpose of this chapter is to illustrate the actions that have been done in terms of communication for the AGEO project. In order to do so, a comprehensive analysis that served as the foundations for the communication plan will be presented, followed by the implementation strategy.

Project analysis from the communication standpoint

A. Ortega Rodríguez (✉) · A. Cseko · C. García
La Palma Research Centre, La Palma, Santa Cruz de Tenerife, Spain
e-mail: ariadna.ortega@lapalmacentre.eu

N. Cruz Pérez · J. C. Santamarta Cerezal · J. S. Rodríguez
Department of Agricultural and Environmental Engineering, University of La Laguna (ULL), La Laguna, Tenerife, Spain

I. Montoya-Montes · I. Galindo Jiménez · N. Sánchez Jiménez · J. C. G. López-Davalillo · I. García-Moreno
Department of GeoHazards and Climate Change, IGME-CSIC, Madrid, Spain

2.2 External Analysis

2.2.1 Microenvironment—Porter's 5 Forces

Threat of new entrants	Medium Recent events have highlighted the importance of geohazards and its prevention. This might result in an increase of projects and initiatives created to disseminate information about geohazards to citizens and other stakeholders, creating competition to take over the same communicational space
Threat of substitutes	High There is a high threat of substitutes as there are many sources of information about geohazards, such as universities, research institutions, researchers, independent experts, specialized journalists and geological surveys, among others Also, there are many citizen science projects that can attract the attention of users

R. C. Gomes et al. (eds.), *Citizens' Observatories on Geohazards*, Geoenvironmental Disaster Reduction,
https://doi.org/10.1007/978-3-031-53371-6_2

Bargaining power of customers	High In today's hyper-connected society, users have innumerable ways of obtaining information. This also applies to geohazards, where the audience can decide where to search for information between many different options (social media, Youtube, internet, podcasts, TV, radio, press, etc.) In this scenario, the strategies used by these information channels can determine what information will be accessed by the potential audience
Bargaining power of suppliers	Low Considering the AGEO consortium as suppliers of information, their bargaining power is low. There are many sources of information that citizens can access easily. Those channels, such as Google and Twitter, are top of mind in terms of information, being the primary source for citizens searching for knowledge
Competitive rivalry	High Taking into account the high number of information sources, as well as their positioning as top of the mind for users, the competitive rivalry will be high for AGEO in terms of being an information channel about geohazards It will have to compete with other information channels that are better established and have a big user network

2.2.2 Macroenviroment—Pestel Analysis

Political factors	Political changes or instabilities (COVID, elections, etc.) can make it difficult to communicate effectively about geohazards. This can include situations where the government is unable to provide reliable information or unwilling to openly discuss the risks Cooperation between different countries has been of the utmost importance for communicating AGEO as the project involves multiple regions. The willingness of different partners, stakeholders and governments to work together can impact how effective communication is in this regard
Economical factors	The availability of resources such as technology, personnel, and infrastructure can impact the ability to communicate about geohazards effectively The Atlantic Area has a large touristic infrastructure that can, and has been (La Palma volcano, 2021), significantly affected by geohazards, but, at the same time, communicating openly about said geohazards can be viewed as both an advantage and a disadvantage by the tourism industry in the area
Social factors	The target audiences for the project are very diverse. Not only are there public from different countries with different cultures and languages, but also differences inside each one of the countries. In this sense, there are populations with high access to information and technologies and others from more isolated places with limited resources Age also has to be considered, taking into account that different types of communication strategies will be needed to communicate effectively to people from different generations Reaching these various target audiences is a challenge, considering also the need to coordinate and adapt the communication strategy to each one of the countries' realities

Technological factors	Technology plays a fundamental role for AGEO, both in terms of project implementation and communication Regarding the implementation, it is essential to create the technological infrastructure needed for the recollection, analysis and visualization of the data Furthermore, the general population is technologically literate, being used to using apps, taking pictures and introducing data In terms of Internet access, "more than 188 million EU households (97.9%) had access to at least one of the main fixed broadband access technologies at the end of June 2021"[3] Plus, the smartphone penetration in Europe is estimated at 82% of the population[4] In terms of communication, the high penetration of smartphones and Internet coverage makes the information widely available to the population
Ecological factors	Populations from the pilot areas are already interested in the project and their outputs due to the impact that the geohazards studied have in their everyday life. Therefore, they can understand the importance of the project and what it would mean for their future, encouraging their participation and the use of the information provided by the project Plus, the occurrence of important geological events, such as La Palma volcano, the Portugal wildfires and the Funchal floods has put the informational focus on geohazards, creating the need of reliable information channels and preparedness programs Furthermore, there is a rising interest in environmental awareness from the citizens
Legal factors	EU Council Directive 96/82/EC defines hazard as "a condition with the potential for causing undesirable consequences" whilst risk is defined as "a measure of the probability and severity of an adverse effect to health, property of the environment" and vulnerability as "the degree of loss to a given element or set of elements within the area affected by the natural hazard"[1] EU Legislation on Human Rights together with articles 196 and 220 on civil protection of the Lisbon Treaty,[2] provide the right of all citizens to equal civil protection from natural hazards regardless of location As a part of the European Union, the data obtained from the citizens taking part in the pilots and those who live in the countries that conform once the app is launched, will be protected by GDPR. Therefore, it must be obtained and stored in concordance with this law

[1] https://eur-lex.europa.eu/legal-content/EN/TXT/PDF/?uri=CELEX:31996L0082&from=en—Accessed on the January 25th 2023.

[2] https://eur-lex.europa.eu/legal-content/EN/TXT/?uri=celex%3A12007L%2FTXT—Accessed on the January 25th 2023.

[3] https://digital-strategy.ec.europa.eu/en/library/broadband-coverage-europe-2021—Accessed on the January 25th 2023.

[4] https://www.statista.com/forecasts/1147144/smartphone-penetration-forecast-in-europe—Accessed on the January 25th 2023.

2.3 Internal Analysis

2.3.1 SWOT

Strengths	Opportunities
* Strong Consortium with top institutions in the field * Support from the EU * Access to information provided by the research centers and geological surveys participating in the project * Budget * Pilots cover a wide array of geohazards * Data generated by the project is open access	* Rising interest on topics related to geohazards due to recent events in Europe * Engaging in bidirectional communication with stakeholders using online and offline channels * Possibility of using the AGEO Platform to support the communication and educational efforts * Widespread use of smartphones between the target audiences * Collaboration with institutions and other projects to create synergies * Rising interest and awareness on sustainability and environmental preservation * Possibility of making a real impact by influencing policy making and preparedness * Potential to generate cooperation between civil protection, disaster response agencies and citizens
Weaknesses	**Threats**
* Different geographical areas, with different languages and characteristics, making it hard to create a cohesive communication strategy that can be adapted to everyone * Difficult to monitor how communication is managed by the pilots * Difficult to create engaging content	* Possibility of pilots doing their own communication * Other alternatives to get information * AGEO app would compete for screen time with other apps that are more popular * Potential danger to users using the app in hazardous conditions * Privacy concerns regarding the data gathered by the app

2.3.2 Target Groups

Different target groups were identified and grouped because of their relevance to the project objectives. For each one of the target groups, specific communication objectives and core messages were analyzed in order to deliver a tailored, case by case strategy.

Twelve groups were identified, classified in four different systems:

System	Groups of stakeholders
Political	1. Government bodies (e.g. ministries of environment, public works, planning, interior, tourism, education) 2. Regional and local authorities (e.g. regional councils, county councils, municipalities, cities) 3. Civil protection/disaster response agencies
Economic	4. Insurance companies 5. Industries and industry associations
Knowledge	6. Public research institutes (especially those active in public safety, territorial development and planning, economic development, transports and infrastructure, education) 7. Geological surveys 8. Weather forecast and seismological institutions/authorities 9. Related INTERREG and H2020 projects
Socio-cultural	10. Professional networks 11. Citizens' initiatives, associations and cooperatives (e.g. cultural associations, community science, observatories, scouts) 12. Local community leaders

Traditional typology	
Internal	AGEO Consortium, project coordinator, project officer, European Union
Mixed	Civil protection and disaster response agencies, insurance companies, public research institutes, geological surveys, industries and industry associations, weather forecast and seismological institutions

Traditional typology	
External	Citizens from the Atlantic Area, citizens from the pilot areas, local governments, national governments, cultural organizations, trekking clubs, children in schools, families of children in schools, school directors, media, community leaders
Matrat	
Decision	Project coordinator, European Union, government bodies, local and regional authorities
Consultation	AGEO Consortium, geological surveys, civil protection and disaster response agencies, weather forecast and seismological institutions, professional networks
Behaviour	Citizens from the pilot areas, cultural organizations, trekking clubs, children in schools, families of children in schools, school principals, community leaders, consortium members in charge of the pilots
Opinion	Citizens from the Atlantic Area, media, researchers, research institutes, specialized journalists
Grunig & Hunt	
Non-public	Citizens outside the Atlantic Area
Latent public	Citizens from the Atlantic Area, insurance companies, industries and industry associations
Aware public	Citizens living in the pilot areas, citizens participating in the activities and workshops, cultural organizations, trekking clubs, citizen organizations, community leaders, researchers, research institutes, geological surveys
Active public	Consortium, civil protection and disaster response agencies, government bodies, local and regional authorities, school principals

2.3.3 Targeting and Positioning

One of the more relevant aspects of the communication plan is to select the target audiences that will be reached by the communication actions, either because they are directly affected by the issues of the project, or because they play an essential role in its resolution or continuation (target groups, audiences or beneficiaries).

National and regional authorities were directly involved in the project and were in charge of the development of the main project outputs with the support of specific members of the Consortium. Public stakeholders (civil society and individuals) were directly engaged locally (via the AGEO pilots) and on a regional, interregional and national level (via social media).

At the beginning of the project, the focus was on understanding the requirements and priorities to design and deliver the outputs to meet stakeholder needs. During the final months of the project stakeholder satisfaction was assessed as a part of the evaluation of the pilots. This will help to anticipate future stakeholder needs of follow-up projects and replication of results.

2.3.3.1 Segmentation

After careful examination of the project and the objectives of the Consortium, the target audiences were divided in three general segments:

- Citizens in the Atlantic Area affected by geohazards.
- Companies and industries with an interest in the results from the project
- National, regional and local government bodies and authorities.

Each one of these segments was layered in sub-segments to better understand the idiosyncrasy of the audience and address them more efficiently.

The citizens living in the Atlantic Area who are susceptible to geohazards segment was divided into two sub-segments, using the criteria of age:

- Adults who live in areas affected by geohazards
- Children who live in areas affected by geohazards.

From the adults segment, another two sub segments were created based on their activity:

- Hobbyists who take part in outdoor activities that can be affected by geohazards (ex: hiking club members)
- Tourists.

In the same manner, companies and industries were divided in:

- Insurance companies
- Industries whose activity could be affected by geohazards (construction, logistics, etc.)

Finally, the government bodies and authorities were segmented as:

- City councils from the areas where the pilots were developed.
- Regional governments.
- National governments.
- Civil protection and disaster response agencies.

2.3.3.2 Targeting

During the project, communication efforts were targeted on the citizens' segment and its respective subsegments. As a citizen science project that nurtures from citizen participation, the Consortium understood that they were the primary focus for communication, as their role in the project was essential. Not only citizens can provide valuable data and input for the project, but also produce behavioral changes. Furthermore, citizen participation would be a lever to bring about changes in terms of legislation and prevention plans.

2.3.3.3 Positioning

For the communication strategy, the Consortium understood that a solution-based position strategy focused on the user would be the most optimal way to reach the objective. In this type of positioning strategy, the value is offering a service that can give a solution to a problem.

In the case of AGEO, the Consortium provides services to the users (workshops, training, data-driven information, AGEO App, AGEO Platform) to increase the risk preparedness and decrease the negative effects of geohazards in the affected areas. The actions are performed in the field, with direct contact with the final users of the product (citizens and civil protection/disaster response agencies) by experts in geohazards from the Consortium.

Positioning Statement:

To citizens in the Atlantic Area coexisting with geohazards, AGEO is the only Citizen Science project that engages citizens to prevent damages from geohazards because it gives them training on geohazard preparedness from experts on the field, analyzes data provided through the app, and elaborates action plans to produce changes through legislative action mechanisms.

2.4 Communication Strategy

2.4.1 Objective

The objective for the communication strategy was creating awareness of the outputs of AGEO project and the importance of citizen science between the relevant stakeholders. These outputs include the AGEO App, the AGEO Platform, the educational contents on geohazards created for schools and the use of Copernicus services for geohazard monitoring.

As analyzed previously, there is a vast number of information suppliers that are easily accessible in nowadays hyper connected society. The credibility of these can vary, often providing information that is not verified by experts. In this context, AGEO communication was envisioned as a trusted source of information in geohazards, developed by professionals working on the field and obtained through observation and data analysis.

2.4.2 Key Messages

AGEO's takeaway message for civil society is one of empowerment and inclusiveness in the scientific and policy process.

The project aims to prove the very important role citizens can play in territorial risks monitoring, mitigation and resilience by giving them usable tools and knowledge that will create an informed voice for the future. This is to be made through face-to-face and online workshops, through easy-to-follow instructions and specially through a bottom-up design and communication.

For the communication and dissemination strategy, five key messages were formulated:

1. AGEO stimulates the active participation of communities and groups of citizens in the prevention and response to natural hazards;
2. AGEO supports coalitions and networks for better risk assessment, preparedness, mitigation and prevention;
3. AGEO reinforces regional and local authorities (e.g. regional councils, county councils, municipalities, cities) capacity to design and implement better land use planning, civil protection and response to crisis requirements;
4. Citizens' Observatories increase knowledge and stimulate cooperation between civil protection/disaster response agencies and citizens;
5. Citizens' Observatories contribute to better adaptation and enhanced resilience of local communities to climate change.

The delivery of the key messages was phased through the project and divided in three stages:

Stages	Key messages	Actions
First stage (2019/2020)	What is AGEO and what are the outcomes? Stimulate the active participation of communities and groups of citizens in the prevention and response to natural hazards	Press conferences and workshops with stakeholders to present the project and inform about the outcomes, press releases, news, social media campaigns
Second stage (2020/2021)	Benefits and accomplishments of AGEO Support coalitions and networks for better risk assessment, preparedness, mitigation and prevention Reinforce regional and local authorities (e.g. regional councils, county councils, municipalities, cities) capacity to design and implement better land use planning, civil protection and response to crisis requirements	Workshops for stakeholders, technical visits, social media campaigns, participation in science fairs and conferences, press releases, news
Third stage (2022)	Roles and functions of the Citizens' Observatories Increase knowledge and stimulate cooperation between civil protection/disaster response agencies and citizens Contribute to better adaptation and enhanced resilience of local communities to climate change	Workshops for stakeholders, Social media campaigns, participation in science fairs, press releases, news

2.5 Implementation Strategy

The communication strategy was implemented through a variety of channels, using both physical and online resources to further convey the key messages to the target groups.

At the start of the project, a visual identity for AGEO was developed. This visual identity was used to support and give coherence to all the communication materials created during the project.

As a result of this effort, the AGEO Stylebook was created. The document comprises all the visual identity elements that will be used on the project: the brand elements (logo and colors), the rules for interaction of the AGEO logo and the INTERREG logo when presented together, typefaces and a set of presentation and document templates. The contents of the stylebook permeate all the content generated for the project in order to give a coherent vision of the brand through all of its communication materials.

Another pivotal part of the communication strategy for AGEO was the creation of a character, Riska, that was used to support the communication for the project. The role of Riska is being a nexus between AGEO and the general population, especially children, delivering the key messages in an easy and understandable manner.

A slogan ("Solution is prevention") was created to be included in AGEO communication. This slogan was selected because it summarizes the main idea behind the communication of AGEO for the general public. In consequence, it was used mainly in the informative videos about the project.

The different target groups were addressed using different channels in order to further extend the reach of the communication actions.

2.5.1 Online Channels

For the online communications, three channels were used: website, social media and the contact database.

The website was the main channel of information, serving as a repository for all project related news and materials. In this space, news about the meetings and workshops were shared, as well as the materials (*see Communication materials*) produced within the framework of AGEO.

The social media channels were used to spread information about AGEO activities, while also informing the public about relevant news related to geohazards and the use of Copernicus services. Facebook and Instagram were used to address the general public, sharing informational videos using the character Riska, while Twitter was more oriented to professionals and scientists, sharing updates on the progress of the project and related news and findings in geohazards and Copernicus services.

A synergy was developed between the website and the social media channels. These accounts were linked on the website to provide easier access to the visitors who wanted to learn more about the project. At the same time, news published on the website were linked on the social media channel to funnel visitors to the website.

2.5.2 Offline Channels

In the last decades, online channels have gained importance in the way projects are communicated. While they are useful resources to provide information to the public, offline channels still play a fundamental role in communication.

Despite the penetration of technology in society, face to face communication is still relevant as it provides a great opportunity of direct interaction with the stakeholders, as well as the chance to interact with audiences that are not digitally literate. Considering this, AGEO developed an offline strategy that complemented the online strategy.

This offline strategy was developed by each one of the pilots depending on their needs, being based on:

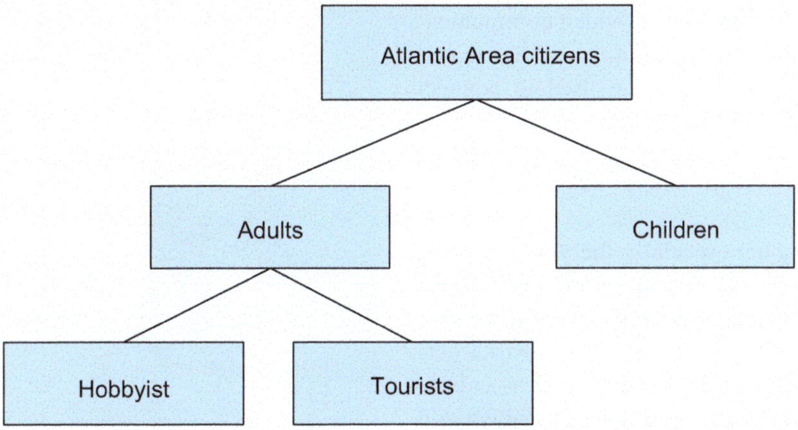

- Participation in news articles published on newspapers, radio and TV.
- Face-to-face interaction with stakeholders through:
- Meetings with authorities.
- Workshops on geohazards.

- Training sessions on data recollection for the pilots.
- Hands on data collection initiatives.
- Visits to schools.
- Presentations about AGEO at informative events.
- Leaflets.
- Scientific papers.

2.6 Communication Materials

Communication materials were created for AGEO. These materials comprise a roll up, brochures, visual materials for social media (images and videos), the character Riska, a press kit and press releases and news for the website.

Press releases and news on the website were published after the meetings as a summary of the main points discussed and the progress being made in the pilots and the development of the AGEO platform.

Riska was used in image and video communications for the project. This character was created to be the face of the project, used to connect with

the general public, especially the younger ones. The pilots made a series of informative videos on the actions taken to prevent the risks present in each of the locations, where Riska acts as a host. These videos were created as animations with voice overs in English, Spanish and Portuguese.

Furthermore, a campaign was developed by Câmara Municipal de Lisboa in collaboration with Delta Cafés to include Riska in the packaging of sugar packages to be distributed in local cafeterias. These sugar packages include information about AGEO as well as QR code to access the website.

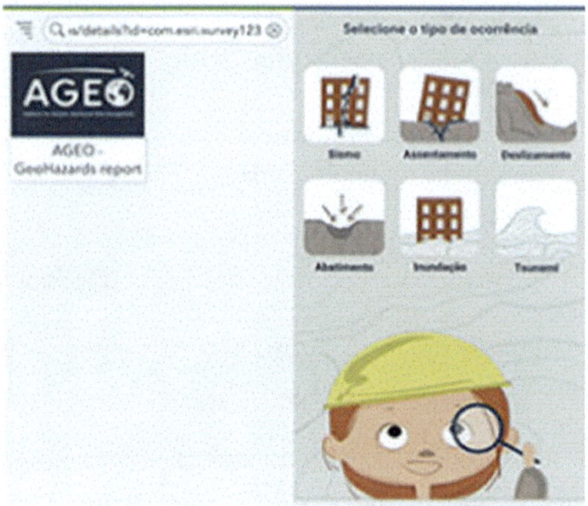

Empowering Citizen Science Through Stakeholder Engagement

3

Vitor Correia, José Romão and Pavlos Tyrologou

3.1 Introduction

According to the National Geographic Society (2020), 'citizen science is the practice of public participation and collaboration in scientific research to increase scientific knowledge'. Projects that involve citizen scientists are expanding, particularly in ecology and the environmental sciences, although the origins of citizen science have at least two centuries (Silvertown, 2009).

Perhaps the oldest example of citizen science is the survey that engaged a large number of clergymen throughout Portugal after the 1755 earthquake of Lisbon, ordered by the Marquis of Pombal, chief minister of the Portuguese King, to collect information about the degree of intensity and type of damage caused by the earthquake throughout the country.[1] Another well-known example of a citizen science project that is running for more than 100 years is the Christmas Bird Count sponsored by the National Audubon Society (National Geographic, 2020); since 1900, the organisation has sponsored a bird count that runs from December 14 through January 5 each year. An experienced birder leads a group (called a circle) of volunteers as they collect information about local populations of birds, and more than 2,000 such circles operate today across the United States and Canada.

Silvertown (2009) advances two main reasons for the expansion of citizen science: (a) scientists cannot always collect large amounts of data or cover big geographic areas for both data collection and documentation because of restricted time and limited monetary resources; and (b) to increase citizens' awareness of problems related to their immediate environment. This second reason is particularly relevant for AGEO, since the project aims to raise people's awareness of geohazards and risks, and to engage citizens in disaster prevention, preparedness, mitigation and warning.

Identifying and engaging relevant stakeholders is a crucial but demanding step in the

V. Correia (✉) · J. Romão · P. Tyrologou
International Raw Materials Observatory,
Brussels, Belgium
e-mail: vcorreia@intraw.eu

J. Romão
e-mail: manuel.romao@lneg.pt

P. Tyrologou
e-mail: pavlos.tyrologou@eurogeologists.eu

J. Romão
Laboratório Nacional de Energia e Geologia,
Lisbon, Portugal

P. Tyrologou
European Federation of Geologists,
Brussels, Belgium

[1] Interestingly, probably the oldest example of citizen science was designed to collect information related to a geohazard.

R. C. Gomes et al. (eds.), *Citizens' Observatories on Geohazards*, Geoenvironmental Disaster Reduction,
https://doi.org/10.1007/978-3-031-53371-6_3

development and implementation of citizen science projects. The following are some of the common challenges:

1. Reaching a diverse audience, citizen science projects often require the participation of a diverse range of stakeholders, including scientists, community leaders, educators, policy makers, and the general public. Reaching and engaging this diverse audience can be challenging, especially in underrepresented communities.
2. Building trust, many stakeholders may be sceptical of citizen science projects, particularly if they have limited understanding of the scientific process. Building trust and credibility with stakeholders is crucial for the success of a citizen science project.
3. Balancing scientific accuracy and community involvement, citizen science projects must strike a balance between scientific accuracy and community involvement. On one hand, the data collected by citizens must be of high quality, i.e., reliable and marked by rigorous observation to ensure that it can be used for scientific purposes. On the other hand, the data acquisition process must be accessible and engaging to encourage widespread participation.
4. Sustaining participation, maintaining stakeholder involvement over time can be challenging, especially if the process is time-consuming or if the results are not immediately apparent.
5. Managing data, citizen science projects often generate large amounts of data, which must be managed and analysed effectively. Ensuring that the data is secure, accessible, and usable can be a challenge, especially if the data is collected by a large number of volunteers.

Overcoming these challenges requires a collaborative, participatory approach that engages stakeholders at every stage of the project, from planning and design to data collection and analysis. Effective communication and transparency are also crucial for building trust and sustaining participation over time.

3.2 Engaging Stakeholders in Citizen Science

The field of citizen science has expanded rapidly with the development of smartphones in the last few years, allowing more information to be shared through digital media. By using phones that have built-in GPS receivers, volunteers can readily provide geo-location information about events in real time, and this is supporting the popularisation of science around the new concept of citizens' observatories. The term "citizens' observatories" was first used in 2014 in the EU FP7 Topic ENV.2012.6.5–1: 'Developing community-based environmental monitoring and information systems using innovative and novel earth observation applications' applied to a framework that combined participatory community monitoring with monitoring by policymakers, scientists and other stakeholders.

In practice, citizens' observatories typically share a similar model (Liu et al., 2017), that includes engaging the participation of citizens in data collection, data interpretation and information delivery. The model (Fig. 3.1.) normally combines (i) sequential aspects, (ii) interaction with citizens and other stakeholders, (iii) data collection tools, and (iv) an ICT infrastructure that underlies the citizens' observatories framework and supports effective citizen participation.

The ICT infrastructure (represented in the oval at the top of Fig. 3.1) is an essential part of the citizens' observatories model, and it normally encompasses boundary services with sensors and apps, data management and data storage support (Liu et al., 2017). In the case of AGEO, the ICT infrastructure (including the AGEO mobile app) enables a two-way communication between citizens and other stakeholders, fostering co-design and collaborative management processes. Moreover, the AGEO ICT infrastructure was designed to facilitate replication and the setting up of new citizens' observatories focused in geohazards across the European Atlantic region.

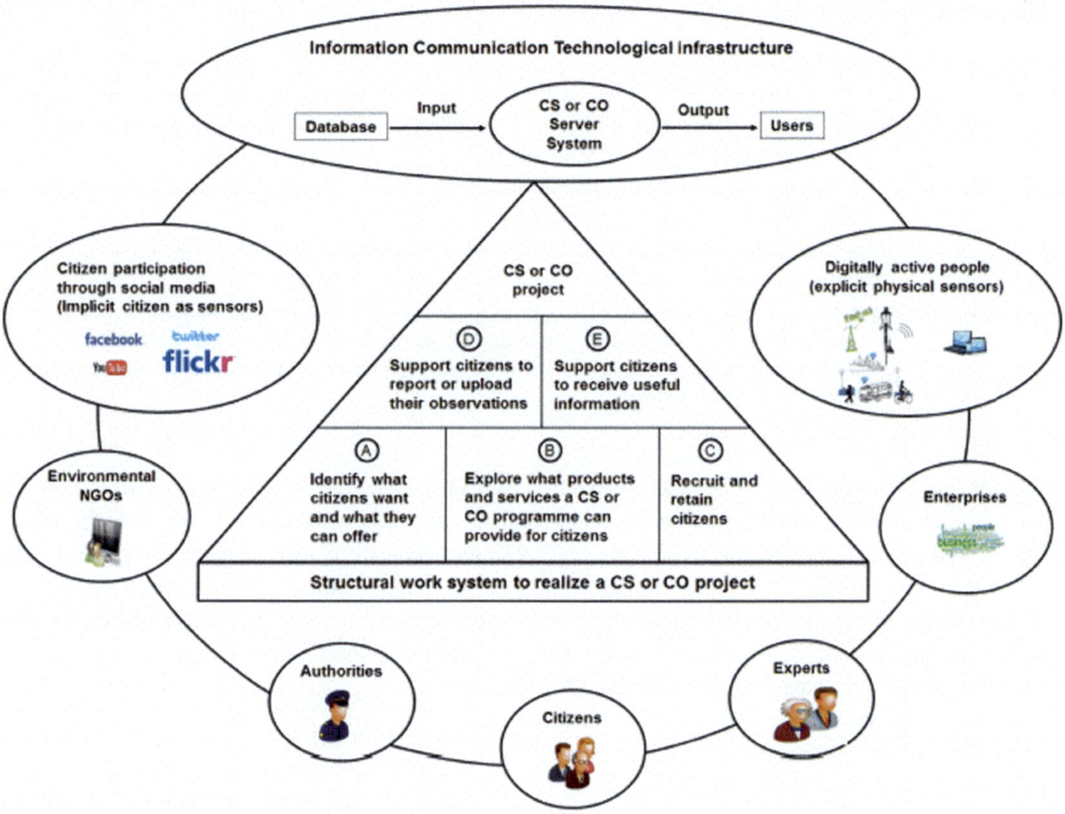

Fig. 3.1 A common model for citizens' observatories programmes in practice. *Source* Liu et al. (2017)

3.2.1 The AGEO Vision

In the vision of AGEO, local communities will have the ability to organise in Citizens Observatories and actively engage with local authorities for better risk assessment, preparedness, prevention and mitigation, using available data and data infrastructures (e.g., European spatial data infrastructures, such as Copernicus), hence becoming more resilient and capable of responding flexibly and adaptatively to climate change.

To advance this vision AGEO developed and implemented five Citizen Observatories pilots in France, Northern Ireland, Portugal and Spain, focused on the monitoring of particular geohazard risks (according to regional priorities—see Table 3.1).

Table 3.1 AGEO citizen pilots and lead organisations

Citizens' observatory pilot, focus	Responsible
Brittany, coastal risks	Université de Bretagne Occidentale
Canary Islands, rock falls	Instituto Geológico y Minero de España
Causeway Coast, complex landslides	Geological Survey of Northern Ireland
Lisbon city, multihazards	Câmara Municipal de Lisboa
Madeira, flash floods and landslides	Universidade da Madeira

Engaging stakeholders in AGEO's Citizen Observatories required a systematic approach, further detailed, encompassing four steps:

Table 3.2 Types of stakeholders relevant for AGEO

Systems	Groups of stakeholders
Political	1. Policymakers, as they can use AGEO outcomes to align broader policy goals and priorities, and to inform decision-making (e.g., government bodies, regional councils, municipalities, civil protection/disaster response agencies)
Economic	2. Industries and services, impacted by natural hazards or having assets that could be affected (e.g., insurance companies, energy providers, port operators)
Knowledge	3. Scientists and researchers, seeking to leverage the expertise, knowledge, and resources of citizens to address scientific questions and solve real-world problems
Socio-cultural	4. Educators, who could play a crucial role in citizen science projects by helping to raise awareness and engage students, teachers, and other members of the education community 5. The general public, as they are often the primary contributors of data and the ultimate beneficiaries of project's outcomes

1. Identifying stakeholders, identification of all potential stakeholders that could be impacted by AGEO.
2. Understanding their probable level of engagement, in particular the influence and potential interest of stakeholder groups in AGEO, vital to manage outreach and communication activities.
3. Assessing their interests and needs, once the stakeholders have been identified, it was necessary to assess their interests and needs through focus groups and interviews.
4. Engaging stakeholders, this involved the development of outreach and communication activities, the establishment of partnerships, and the provision of opportunities for hands-on involvement.

3.2.2 Stakeholders Identification

Citizen science projects can have a wide range of impacts, both positive and negative, on a variety of stakeholders. Determining who is affected by the project, and how, can be challenging and requires a systematic approach to stakeholder identification.

In the initial screening of stakeholders likely to be interested in AGEO outputs, it was possible to identify five relevant groups (Table 3.2), organised around four basic societal systems: political, economic, knowledge and socio-cultural.

In a subsequent step, and considering the diverse geographies, context and aims of the five AGEO pilots, the organisations working in the different pilots were asked to identify the stakeholders that were more relevant for them. The list produced was more specific than the general groups named in Table 3.2, and included the following stakeholders' groups:

1. Local authorities (e.g. local governments, municipalities).
2. Local communities (e.g. parishes, neighbours associations, recreational non-profit groups).
3. Landowners (e.g., farmers and tenant farmers).
4. Tourists (national and international visitors).
5. First responders (e.g., firefighters, medical emergency teams).
6. Public research institutes/universities (e.g., geological surveys, national laboratories, research departments and institutions).
7. Private agents (e.g., civil works companies, utilities' providers and consultants).
8. Professional associations (e.g., engineers, geographers, geologists professional and scientific associations).
9. Media (local, regional and national radio and newspapers).
10. Insurance companies (insurers of infrastructures and individuals).
11. Schools (from first grade to secondary, professional and tertiary levels).
12. Energy or utility providers/strategic infrastructures (e.g., electricity and gas providers, road authorities).

13. Federal/national government (e.g., ministries responsible for policy making, emergency preparedness and land use).

3.2.3 Stakeholders Classification

Assessing the level of interest and engagement among stakeholders can be difficult and requires a nuanced understanding of their motivations and expectations. To screen the potential engagement of the 13 stakeholders' groups in AGEO, the organisations in the Consortium directly linked to each of the five pilot Observatories classified the 13 groups using the Mendelow (1991) power-interest matrix, that considers two dimensions: (1) stakeholders power (or influence); and (2) and stakeholders interest (in the pilots or in the AGEO project). The perceptual map below (Fig. 3.2) represents the outcome of the general classification exercise. The size of the circles in the image represents the normalised number of individuals/entities in each group.

The perceptual map and the classification outlined are relevant to manage stakeholder groups. The management prescriptions arousing from the classification (Mendelow, 1991) is represented in Fig. 3.3. According to this matrix, there are four possible positions/prescriptions:

1. High power, highly interested groups (prescription *Manage Closely*): stakeholders in this position should be fully engaged, and meeting their needs should be a main concern;
2. High power, less interested groups (prescription *Keep Satisfied*): these stakeholders should be kept satisfied, but it is essential to avoid excessive interactions;
3. Low power, highly interested groups (prescription *Keep Informed*): this group of stakeholders should be adequately informed and engaged to ensure that no significant issues are raised. People in this category can easily be willing and helpful in implementation activities;
4. Low power, less interested groups (prescription *Monitor*): This group should be monitored, and excessive communication should be avoided.

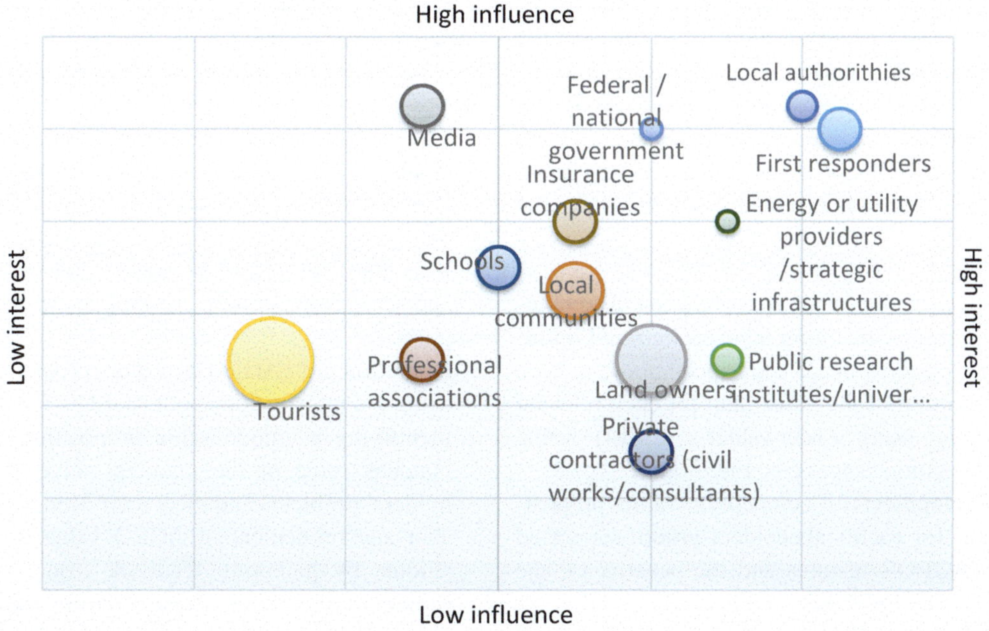

Fig. 3.2 Generic classification of relevant stakeholders for AGEO's pilots, according to their influence and interest

It must be emphasised that the generic perceptual map of stakeholders in Fig. 3.2 was later refined to each pilot. This adjustment was paramount to avoid simplifications that would mask over-representation or biases, and to consider existing stakeholder relationships, relevant to identify potential allies and influencers who could shore up support and participation. An evident example of the importance of the local refinement comes from the pilot on the Causeway Coast (Northern Ireland), where touristic activity is incredibly significant and where tourists would play a central role in the Citizens Observatory pilot (placing them in the "high interest" areas of the matrix). The adjustment to the specific features and groups of stakeholders pertinent to each pilot was made by the corresponding pilot leader.

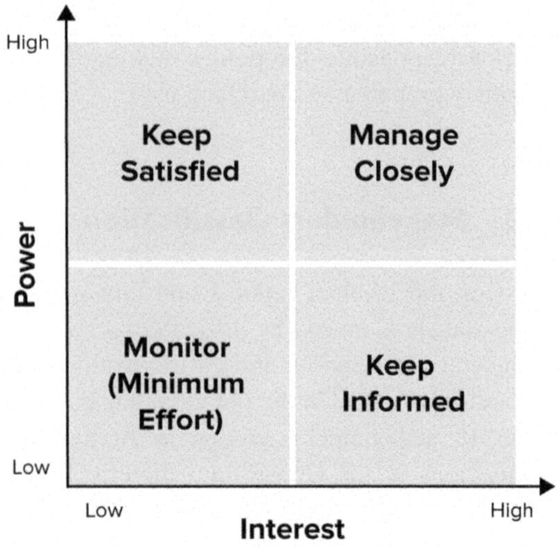

Fig. 3.3 Power/Interest (influence) grid for stakeholder prioritisation (adapted from Mindtools.com)

3.2.4 Assessing Stakeholders' Interests and Needs

A fundamental challenge for AGEO was to engage citizens in activities and dialogues with scientists and risk management authorities, hence setting the foundations for resilient, long-lasting Citizens Observatories. Instrumental to address this challenge was the development of compelling value propositions of AGEO and the Citizens Observatories' aligned with the interests and needs of the most relevant groups of stakeholders.

The definition of the appropriate value propositions followed a standardised method, that included:

- The selection of the most relevant stakeholders' groups, detailing their background, influence, specific needs and expectations;
- Definition of the value map for each relevant group, using Osterwalder's (2014) value proposition canvas (see Fig. 3.3);
- Development of compelling value propositions for each stakeholders group, supported by AGEO's features and the benefits for the pilot Citizens Observatories;
- Test and refinement of the value propositions (with the support of stakeholders classified in the low power, high interested quadrant of Mendelow's (1991) power-interest matrix);
- Fitting use of the tailored value propositions for each pilot to foster local engagement and boost the capitalisation and uptake of AGEO's results.

The value proposition canvas (Fig. 3.4) developed by Osterwalder (2014) was used to outline the value propositions for the most relevant groups of stakeholders, following three steps:

1. Analysis of stakeholders' jobs, pains, and gains; definition of what these stakeholders are trying to achieve (their jobs), what problems they are trying to solve (their pains), and what benefits they are looking for (their gains).
2. Definition of AGEO's value proposition; using the information about AGEO's features, identifying a unique value proposition, that matches some of the features, solve stakeholders' problems and meet their needs.
3. Test and refinement of AGEO value propositions; by gathering feedback from representatives of each stakeholders group and fine tuning the proposition to ensure that it effectively addresses specific needs.

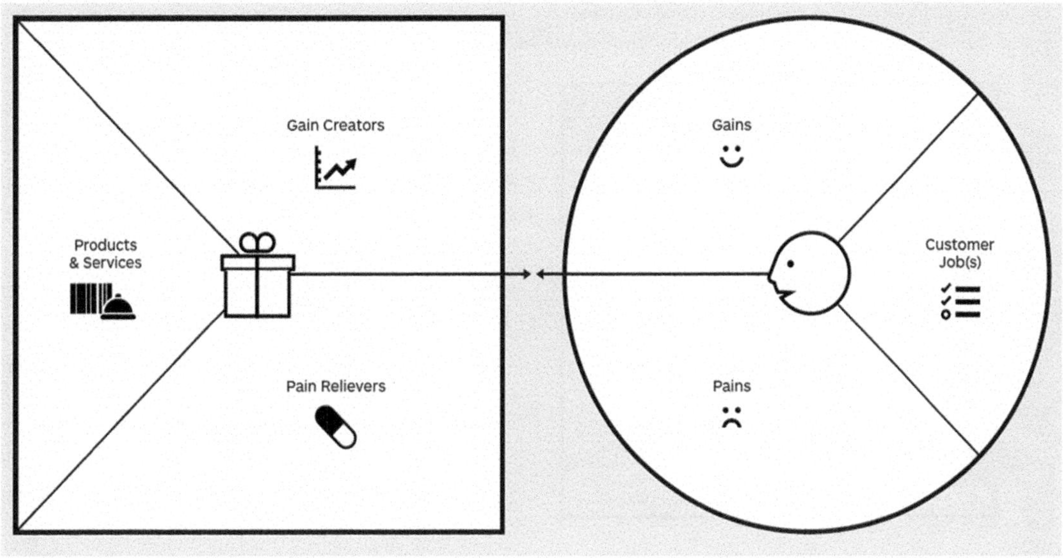

Fig. 3.4 Value proposition canvas to be used to define value for relevant stakeholder groups within each pilot (Osterwalder, 2014)

The development of value propositions was made in a workshop held with all Consortium members, and some of the generic value propositions canvas produced are shown below (Figs. 3.5, 3.6, 3.7, 3.8 and 3.9).

The value proposition outlined for local authorities says *AGEO reinforces regional and local authorities (e.g., regional councils, county councils, municipalities, cities) capacity to design and implement better land use planning, civil protection and response to crisis requirements*. This value proposition took in consideration the resources (financial and human) constraints that most local authorities have, as a result of tight budgeting, and the vital role local authorities play in land-use planning, civil protection and response to crisis.

The value proposition outlined for civil protection agencies says *Citizens' Observatories increase knowledge and stimulate cooperation between civil protection/disaster response agencies and citizens*. This value proposition builds on the importance of coordination between agencies and other disaster response players, and on the importance of public awareness to counter the lack of interest of the general public on risk prevention and mitigation.

The value proposition outlined for local communities says *Citizens' Observatories contribute to better adaptation and enhanced resilience of local communities to climate change*. This proposition builds on the increasing awareness of the effects of climate change (and the rollout of more frequent extreme climatic events) and on the public perception that is crucial to improve the resilience against the effects of climate change.

The value proposition outlined for landowners says *AGEO supports coalitions and networks for better risk assessment, preparedness, mitigation and prevention*. This value proposition underlines the importance of partnerships between scientists and local networks (such as associations of landowners) in defining and implementing actions that would limit the risks of natural hazards and the mitigation of their impact.

The value proposition outlined for tourists says *AGEO stimulates the active participation of tourists and groups of citizens in the prevention and response to natural hazards*. This value proposition points out the important role tourists and groups of citizens play in risk monitoring and response.

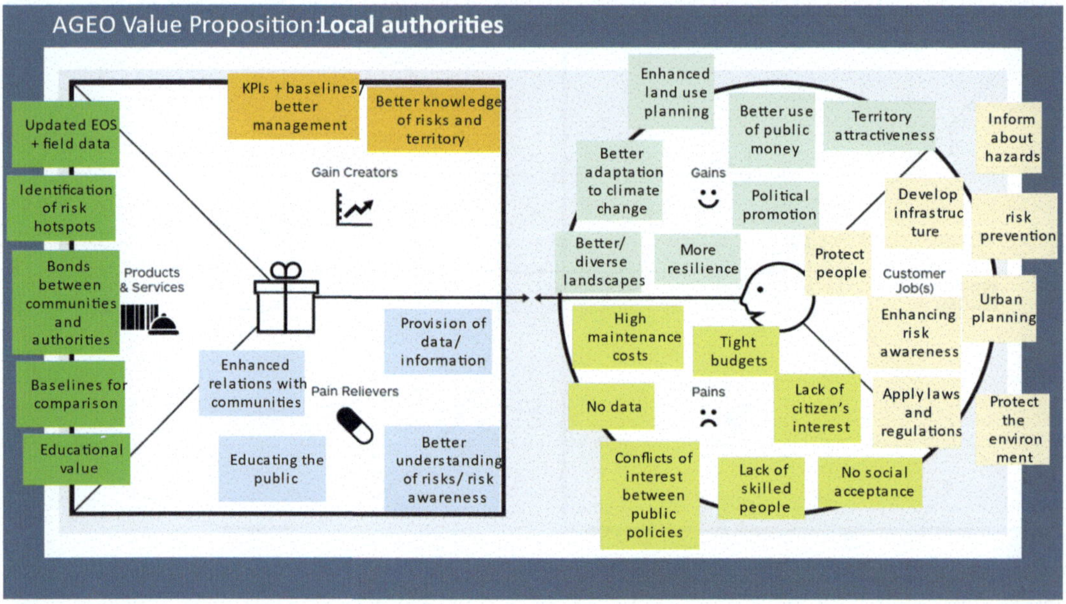

Fig. 3.5 AGEO's value proposition canvas used to define the value proposition for Local Authorities (canvas adapted from Osterwalder, 2014)

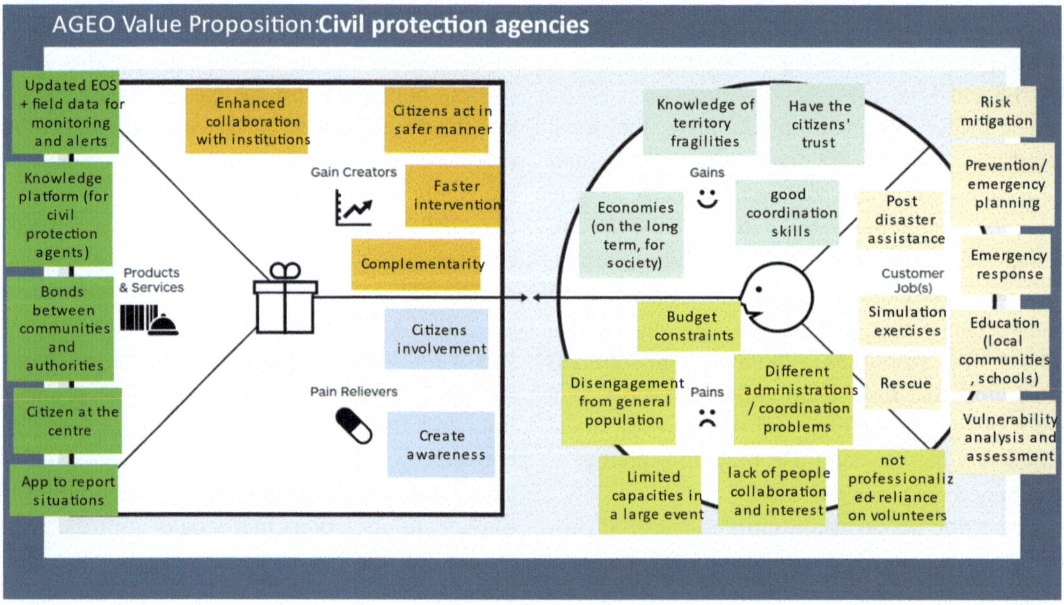

Fig. 3.6 AGEO's value proposition canvas used to define the value proposition for Civil Protection Agencies (canvas adapted from Osterwalder, 2014)

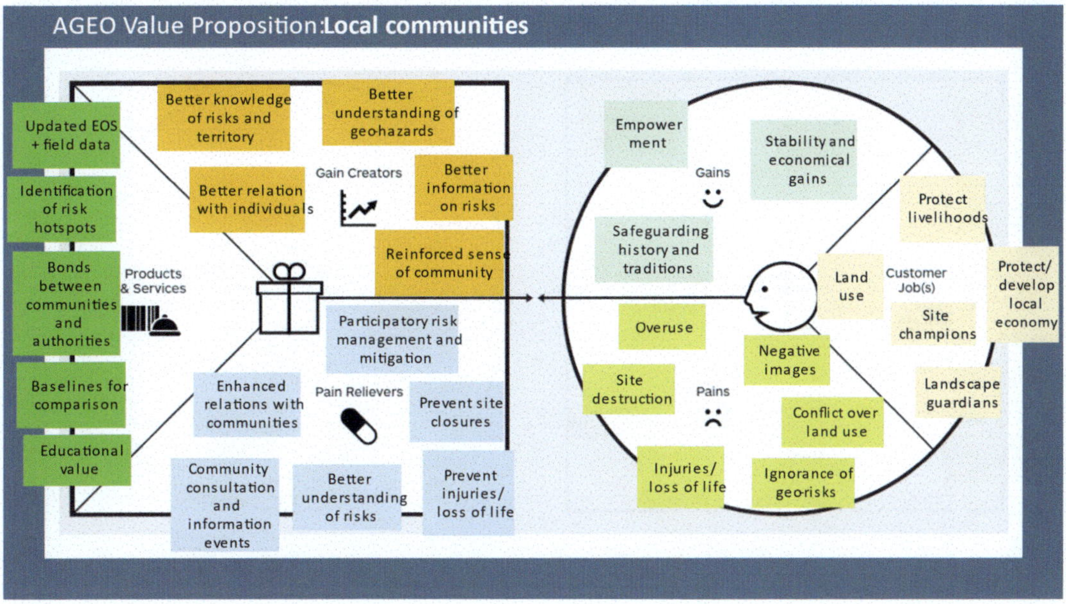

Fig. 3.7 AGEO's value proposition canvas used to define the value proposition for Local Communities (canvas adapted from Osterwalder, 2014)

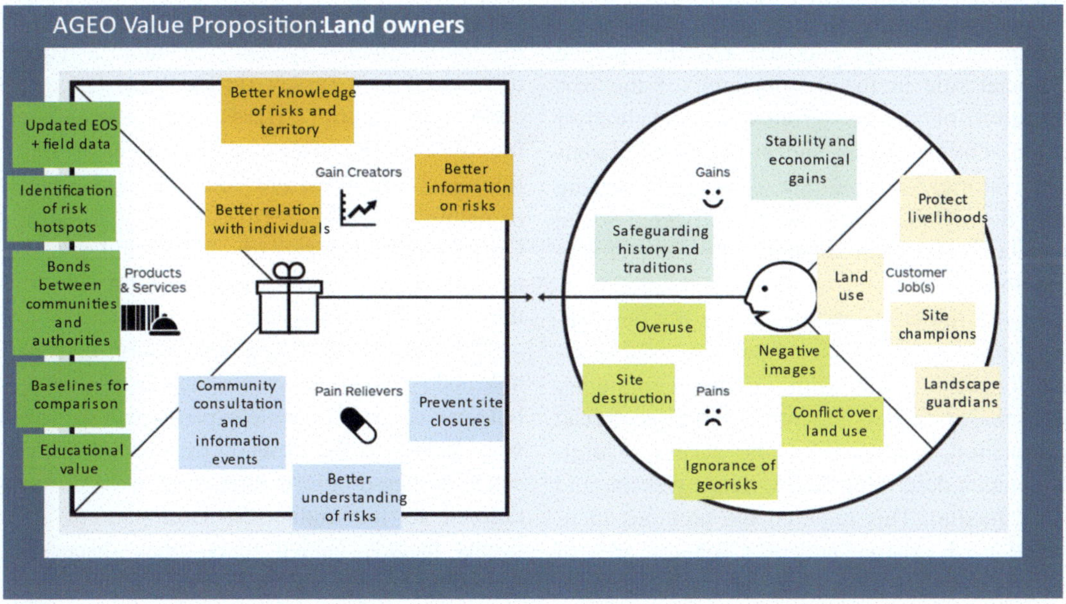

Fig. 3.8 AGEO's value proposition canvas used to define the value proposition for Landowners (canvas adapted from Osterwalder, 2014)

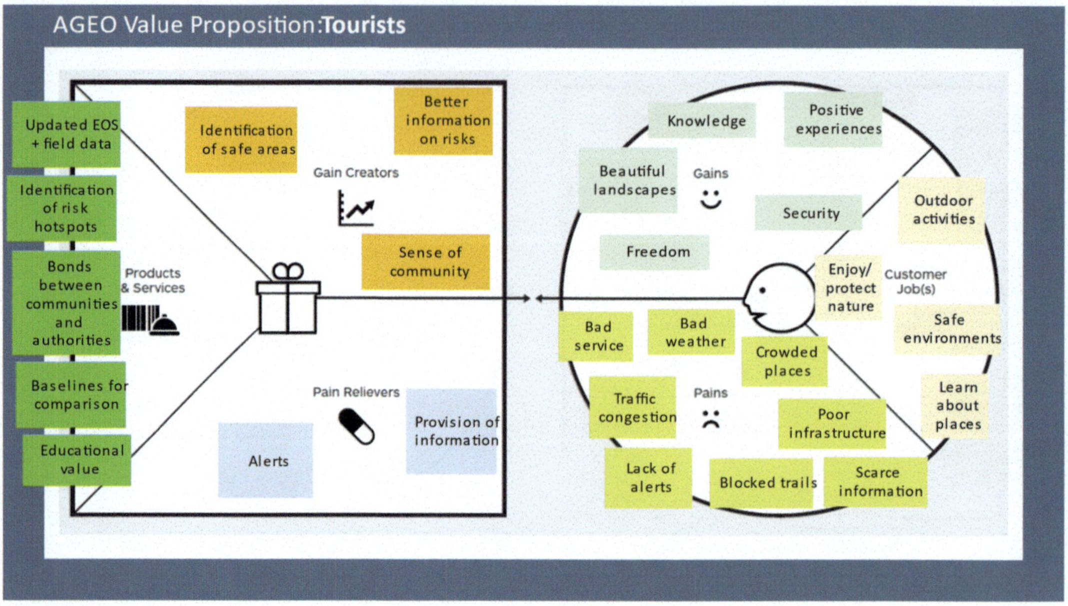

Fig. 3.9 AGEO's value proposition canvas used to define the value proposition for Tourists (canvas adapted from Osterwalder, 2014)

3.2.5 Engaging Stakeholders

Communication and outreach activities played a crucial role in the development of the participation of diverse groups of stakeholders in the activities of all and every pilot Citizens Observatories, despite the constraints arising from the COVID pandemic that affected social interactions for approximately half of the project duration.

In the initial stages of the project development pilot leaders and local partners used referrals to get in touch with existing community networks, and relied on word of mouth to spread information, facilitated by the access to digital sources describing AGEO (e.g., website and social media). This required the previous identification of existing networks (on the area of influence of each pilot site) that could be interested in receiving information on AGEO.

The approach was typically initiated with a face-to-face or online meeting with managers/responsible from the relevant stakeholders' groups (e.g., recreational non-profit groups, landowners' associations, fire squad/police stations and local/regional schools), where the goals of AGEO were explained, and digital/printed materials provided. The follow up on the initial contacts typically included a request to respond to a survey (in the local language) addressing natural hazards and generic community concerns. These surveys included questions such as "What are your concerns relating to natural hazards?", "How would you like the Citizens Observatory to be?", "Would you like to be involved?" and created the favourable conditions for further engagement.

The next step was the organisation of enjoyable and social events within each group of stakeholders, to boost people's participation. These were linked to the observatory topic, conveyed scientific-based information, and used formats tailored to the audiences. Examples of such events include guided tours, open discussion/workshop sessions and visits to and from schools.

The experience with AGEO has confirmed that the five crucial factors that drive stakeholder engagement are:

1. Relevance, citizens are more likely to engage in a citizen science project if they feel that the project is relevant to their interests and

concerns. Relevance can be enhanced by involving stakeholders in the design and implementation of the project and by clearly communicating the goals and benefits of the project.

2. Accessibility, citizen science projects must be designed to be accessible to all stakeholders, regardless of their level of scientific expertise or technology skills. This may involve providing training and support, developing user-friendly interfaces, and ensuring that the project is culturally and linguistically appropriate.
3. Trust, trust is a critical factor in stakeholder engagement, particularly for underrepresented communities. Building trust requires transparent communication, the use of reliable and trustworthy sources of information, and ensuring that the data collected is secure and protected.
4. Recognition and incentives, these play an important role in motivating citizens to participate in a citizen science project. This may involve providing opportunities for hands-on involvement, recognising contributions and achievements, and offering tangible benefits such as access to training and resources.
5. Sustainability, Maintaining stakeholder engagement over time can be challenging, especially if the process is time-consuming or if the results are not immediately apparent. Ensuring the sustainability of the project requires continuous engagement and communication, and providing ongoing opportunities for participation.

3.3 Sustaining the Participation of Stakeholders in Ageo

Broadly speaking, the likelihood of sustaining a citizen science project depends on: (1) initial project design; (2) project implementation; and (3) community support.

3.3.1 Initial Design

The AGEO project design considered, since its start, four assets (outcomes) that would be relevant to the uptake of project results:

(1)The AGEO Information and Communication Technology (ICT) infrastructure, including a mobile app, data management and information services, and data storage support, to be uptaken by civil protection entities and used by citizens and civil protection agents;

(2)Recommendations and examples of application of European spatial data infrastructures (such as Copernicus services and solutions) for a range of geohazards including landslides, floods, tsunamis, earthquakes, volcanic activity, cascading events that may include other types of natural hazards (such as wildfires) and human-induced geohazards (e.g., induced seismicity or surface subsidence);

(3)Recommendations to policymakers to increase the resilience of society against natural hazards by enhancing risk management structures and capacity in disaster risk reduction (DRR) via public engagement and better integration of Copernicus products and services;

(4)Recommendations for risk authorities on citizen engagement to enhancing geohazard risk preparedness in the Atlantic region through educational/capacity-building and communication tools and activities.

The development and fine-tuning of AGEO assets were made as the project progressed, with the active contribution of all consortium partners, and one can assert that these outcomes were fully achieved.

Also, at the start of AGEO, a group of sustainability factors were identified and further detailed into sustainability objectives (see Table 3.3), to harness the exploitation and replication potential of the five Citizen Science Observatories after the end of the project funding period.

3.3.2 AGEO's Implementation

During the project implementation, two supporting factors became very relevant to boost the uptake of AGEO results: (1) the perception

Table 3.3 AGEO's sustainability factors and objectives

Sustainability factors	Sustainability objectives
Project design and implementation	Define the value-added of AGEO Define what are the aspects of the Citizens' Observatories that are most effective Prioritise what to sustain from experience with the Citizens Observatories pilots Identify goals and action steps for long-term (after the AGEO funding period) sustainability
Project monitoring and evaluation	Define what activities result in a significant positive impact Design and implement a monitoring dashboard to assess the impact and outcomes of Citizens' Observatories
Financial resources	Define which Citizens' Observatories components can be sustained with minimal financial resources Evaluate the feasibility of different sources of funding Establish budget/resources needed Develop fundraising activities
Organisational structure	Define which Citizens' Observatories components can be sustained with minimal human resources Provide staff and community stakeholders training in creative and flexible problem solving to consolidate and expand the Citizens' Observatories Recruit and engage volunteers
Project Champions	Recruit, train and engage community champions who facilitate community activities Uphold the endorsement and support of project champions
Managerial support and flexibility	Disseminate AGEO impacts to leaders of Citizens' Observatories and educate them on the importance of adapting to external changes and using opportunities to meet AGEO's goals Define which aspects of AGEO can be integrated into other existing/ongoing programs or partnerships
Community support for the project	Define the aspects of AGEO that are highly respected or supported by the broader community Hold regular dissemination meetings with the stakeholder community and professional conferences to disseminate the results of the project and build recognition for success

of climatic extremes and the increasing risk of geohazards; and (2) the launch of the European Ground Motion Service (EGMS) as part of the Copernicus Programme.

On the climatic extremes, and according to the IPCC (2019), 'the scale of recent changes across the climate system as a whole—and the present state of many aspects of the climate system—are unprecedented over many centuries to many thousands of years'. The climate models and projections made by the IPCC (2018) point out to global surface temperature increases until at least mid-century under all greenhouse gas emissions scenarios considered (including CO_2 emissions declining to net zero around or after 2050). In this context, increases in the frequency and intensity of hot extremes, marine heatwaves, heavy precipitation, agricultural and ecological droughts and even cyclones should be taken as a new reference across the Atlantic region of Europe. The social dimension also needs to be considered, in particular the impact of climate change awareness on behavioural changes. Greta Thunberg and the intense media coverage of the *Fridays For Future* movement she created in 2018 boosted the debate around climate change and sustainability both among the population and in the political arena (Rucht & Sommer, 2019).

Until recently, the monitoring of ground-instability was site-specific and administered by local authorities using conventional means available (e.g., line levelling, electronic distance measuring, GPS) in connection with a particular known local geohazard. This reactive approach changed with the setup of the European Ground Motion Service (EGMS) as part of the Copernicus Programme (source). This service, designed to provide for free consistent, regular, standardised, harmonised and reliable information regarding natural and anthropogenic ground motion phenomena over Europe and across national borders, with millimetre accuracy (Copernicus, 2020), was launched in 2020.

3.3.3 Community Support

The uptake and capitalisation of AGEO's results is fundamentally linked to the earliest development of connections and solid partnerships with relevant stakeholders or networks that could build upon project outputs after the funding period. The citizens in the area where a disaster happens and their local governments and voluntary agencies are the first to have to cope with the damage of a disaster; local governments/municipalities normally act as the primary "first provider" of emergency response services, and they often coordinate assistance with public and private organisations and agencies. In this framework, the role that local governments/municipalities and disaster response authorities play in response to a natural disaster in the Atlantic region, including the assessment of risks, the response and the recovery stages, is crucial to minimise impacts and save lives.

An assessment made by the pilot leaders, concluded that the agents vital to sustaining the five Citizen Observatories created, willing to receive training (for capacity building) and lead or support the uptake of the Observatories were[2]:

1. Landowners and citizens with a detailed and all-inclusive knowledge of their territory, who need to feel empowered to participate actively in the prevention and response to natural hazards;
2. Local community leaders who have a strong influence on landowners and citizens' behaviour;
3. Civil protection/disaster response agencies, who need trustable local information from people that have a detailed and all-inclusive knowledge of their territory/area;
4. Regional and local authorities (e.g., regional councils, county councils, municipalities, cities) who have direct responsibility for land regulations, civil protection and response to crisis requirements.

After identifying the four main groups of stakeholders interested in the uptake of the Observatories, the main benefits (for each stakeholder) associated to maintaining and enhancing (i.e., managing) the Citizen Observatories were discussed, and generic performance indicators were outlined (Table 3.4).

Relevant civil protection/disaster response authorities for the capitalisation of AGEO in the European Atlantic region include National Disaster Management Authorities, Environment Agencies, public infrastructure (e.g. highways, railways, water dams) private owners, managers and utility providers (gas, electricity, water). The support of these entities played a crucial role in the project implementation and boosted public participation. This support came in the form of partnerships with community-based organisations, and promotion of the AGEO app, to support decision-making processes. By endorsing AGEO, civil protection and disaster response authorities in the area of the five pilots increased public trust in the project and encouraged greater public participation.

3.4 Conclusions

Whatever the range of influences (temperature, rainfall events, snowmelt, frosting, cyclones, etc.), it is certain that climate change effects will

[2] The listing of these groups, identified in the five pilots, does not exclude the possible interest of other stakeholders (such as insurance companies, professional networks and science-based organisations) in the uptake of a specific Citizen Observatory.

Table 3.4 Learning outcomes tailored to AGEO's agents of change

Stakeholder groups interested in the uptake of the Observatories	Main benefit for the stakeholder	Generic performance indicator to measure the benefit
Landowners and citizens that have a detailed and all-inclusive knowledge of their territory	Greater awareness about disaster's risk and vulnerability and its relevance to prevention and response	Community support for the Observatory Incentives for compliance
Local community leaders	Increased understanding of disaster risk assessment and management	Stakeholders' engagement Transparency of information to stakeholders
Civil protection/disaster response agencies	A better understanding of the needs and concerns of citizens	Stakeholders engagement Transparency of information to stakeholders
Regional and local authorities	Enhanced land use planning and formulation/implementation of regulations	Improved compliance. Better civil protection and response to crisis requirements

impact the type, extent, magnitude and possibility of changes in rock and ground stability conditions, triggering geohazards.

With the advent of new and advanced communication and Earth Observation technologies, the potential for proactive citizen involvement in disaster preparedness, mitigation, and warning has increased, particularly in the context of the ongoing effects of climate change. Citizen science can make a significant contribution by providing a large volume of data that can enhance our understanding of natural hazards and the associated risks. This information can inform decision-making processes related to disaster preparedness and response, ultimately reducing the risk of harm to both communities and ecosystems. Furthermore, citizen science can serve to increase awareness and encourage community engagement in disaster risk reduction efforts, promoting a more active role in protecting both themselves and their communities.

In this vision, technological advances and capacity-building actions would enable integrating risk authorities and citizen communities' viewpoints, combining theoretical and empirical perspectives with real-world, on-the-ground applications, thus creating resilient societies with a shared understanding of risks, roles and responsibilities between risk agencies and risk publics. This change would probably be gradual, starting with information-sharing mechanisms (such as the already adopted emergency warnings sent to mobile phones within one region) and moving on to collaborative environments where experts and specialists and the lay publics that they serve work together to jointly tackle complex compound-risk scenarios and multi-agent threats.

The identification and engagement of relevant stakeholders is paramount in establishing a successful foundation for citizen participation in disaster preparedness, mitigation, and warning. Overcoming linguistic and cultural barriers and establishing trust and credibility are crucial, yet challenging, aspects of the process, which require a collaborative, participatory approach throughout the design and implementation of the project. Strengthening stakeholder engagement can be achieved through building partnerships and fostering relationships with the most relevant groups of stakeholders, resulting in increased participation, impact, and sustainability.

References

Copernicus. (2020). European Ground Motion Service. https://land.copernicus.eu/pan-european/european-ground-motion-service.

IPCC (2018) Global warming of 1.5 °C: An IPCC Special Report on the impacts of global warming of 1.5 °C above pre-industrial levels and related global greenhouse gas emission pathways, in the context of strengthening the global response to the threat of climate change, sustainable development, and efforts to eradicate poverty [V. Masson-Delmotte, P. Zhai, H. O. Pörtner, D. Roberts, J. Skea, P.R. Shukla, A. Pirani, W. Moufouma-Okia, C. Péan, R. Pidcock, S. Connors, J. B. R. Matthews, Y. Chen, X. Zhou, M. I. Gomis, E. Lonnoy, T. Maycock, M. Tignor, T. Waterfield (eds.)]. https://www.ipcc.ch/sr15/.

IPCC (2019) Climate change and land: an IPCC special report on climate change, desertification, land degradation, sustainable land management, food security, and greenhouse gas fluxes in terrestrial ecosystems [P.R. Shukla, J. Skea, E. Calvo Buendia, V. Masson-Delmotte, H.-O. Pörtner, D. C. Roberts, P. Zhai, R. Slade, S. Connors, R. van Diemen, M. Ferrat, E. Haughey, S. Luz, S. Neogi, M. Pathak, J. Petzold, J. Portugal Pereira, P. Vyas, E. Huntley, K. Kissick, M. Belkacemi, J. Malley, (eds.)]. https://www.ipcc.ch/srccl/.

Rucht, D., & Sommer, M. (2019). Fridays for Future. Vom Phänomen Greta Thunberg, medialer Verkürzung und geschickter Mobilisierung: Zwischenbilanz eines Höhenflugs. *Internationale Politik, 4*, 121–125.

Liu, H., Grossberndt, S., & Kobernus, M. (2017). Citizen science and citizens' observatories: trends, roles, challenges and development needs for science and environmental governance. In: Foody, G., See, L., Fritz, S., Mooney, P., Olteanu-Raimond, A., Fonte, C., & Antoniou, V. (Eds.) Mapping and the citizen sensor, pp. 351–376. London: Ubiquity Press. https://doi.org/10.5334/bbf.o.

Mendelow, A. L. (1991). Environmental scanning: The impact of the Stakeholder concept. Proceedings from the Second International Conference on Information Systems, pp. 407–418. Cambridge, MA. https://aisel.aisnet.org/icis1981/20/.

National Geographic Society. (2020). Citizen science. Resource Library, Encyclopedic Entry. https://education.nationalgeographic.org/resource/citizen-science.

Osterwalder, A. Pigneur, Y., Bernarda, G., Smith, A., & Papadakos, T. (2014). Value proposition design. Strategyzer series. Wiley, New Jersey. ISBN: 978-1-118-96805-5. https://media.wiley.com/product_data/excerpt/50/11189680/1118968050-15.pdf.

Silvertown, J. (2009). A new dawn for citizen science. *Trends in Ecology & Evolution, 24*(9). https://doi.org/10.1016/j.tree.2009.03.017.

Copernicus Services and Geohazards Management. Lessons Learnt from Citizens' Observatories

4

Isabel Montoya-Montes, Lídia Quental, Inés Galindo, Eoghan P. Holohan, Marion Jaud, Kieran Parker, Domingos Rodrigues, Pedro Gonçalves, Nieves Sánchez, Inmaculada García, Nicolas Le Dantec, Kirstin Lemon, Elsa Ramalho, Susana Machado, Iwan Le Berre, Fatima P. Gouveia, Claudia Pinto, Marta Béjar-Pizarro, Gerardo Herrera and Rui Carrilho Gomes

4.1 Introduction

Copernicus is the Earth Observation programme,[1] of the European Union (EU). It is aimed at developing European information services based on satellite Earth Observation and in situ data collection (Jutz & Milagro-Pérez, 2018). The objectives of this programme are: (i) to monitor and forecast the state of the environment on land, sea and in the atmosphere; (ii) to support climate change mitigation and adaptation strategies, (iii) to increase the efficient management of emergencies and (iv) to improve the security of every citizen. By providing information on natural disasters, such as forest fires or floods, Copernicus helps to

[1] https://www.copernicus.eu/en/about-copernicus/copernicus-detail (accessed 06/02/2023).

I. Montoya-Montes (✉) · I. Galindo · N. Sánchez
Department of GeoHazards and Climate Change, IGME-CSIC, 35003 Las Palmas, Spain
e-mail: isabel.montoya@ulpgc.es

I. Galindo
e-mail: ines.galindo@csic.es

N. Sánchez
e-mail: n.sanchez@igme.es

I. Montoya-Montes
Instituto de Oceanografía y Cambio Global (IOCAG), Universidad de Las Palmas de Gran Canaria, Las Palmas de Gran Canaria, Spain

L. Quental · P. Gonçalves · E. Ramalho · S. Machado
Laboratório Nacional de Energia e Geologia, Estrada da Portela, Bairro do Zambujal, Apartado 7586, Alfragide, 2610-999 Amadora, Portugal
e-mail: lidia.quental@lneg.pt

P. Gonçalves
e-mail: pedro.goncalves@lneg.pt

E. Ramalho
e-mail: elsa.ramalho@lneg.pt

S. Machado
e-mail: susana.machado@lneg.pt

E. P. Holohan
UCD School of Earth Sciences, University College Dublin, Dublin 4, Ireland
e-mail: eoghan.holohan@ucd.ie

M. Jaud · N. Le Dantec
UMR 6538 GeoOcean, Institut Universitaire Européen de La Mer, University of Brest-CNRS, Rue Dumont d'Urville, 29280 Plouzané, France
e-mail: marion.jaud@univ-brest.fr

R. C. Gomes et al. (eds.), *Citizens' Observatories on Geohazards*, Geoenvironmental Disaster Reduction,
https://doi.org/10.1007/978-3-031-53371-6_4

prevent the loss of lives, property, and damage to the environment.[2]

The Copernicus program is led by the European Commission (EC), acting on behalf of the EU, and is in charge also of the users' requirements and of the implementation of the services component. The European Space Agency (ESA) is responsible for the technical coordination of the space-based component, which comprises two types of satellite missions: the dedicated Sentinel missions, developed by ESA specifically to meet the Earth observation needs of Copernicus users, and the Contributing Missions, several existing and planned Earth observation satellites that are operated by other space agencies or organizations which provide some data to the program. Both the Sentinel and the Contributing Missions ensure the long-term availability of critical observations for operational and scientific applications. The Sentinel satellites carry a range of technologies, such as radar and multispectral imaging instruments, to cover all environmental domains (Jutz & Milagro-Pérez, 2018).

Vast amounts of global data from satellites and from ground-based, airborne and seaborne measurement systems are being used to provide information to help service providers, public authorities and other international organisations to improve the quality of life for the citizens of Europe and beyond. Most of the information delivered by Copernicus is made available and accessible to any citizen and to any organisation around the world on a free, full, and open basis.

Copernicus provides essential information for six thematic domains in the form of the following services:

-**Copernicus Marine Environment Monitoring Service (CMEMS)**—provides regular and systematic reference information on the physical and biogeochemical state, variability and dynamics of the ocean and marine ecosystems. The observations and forecasts support all marine applications, including marine safety (oil-spill detection and transport), coastal and marine environment applications, water quality, weather forecasting and climate.

-**Copernicus Land Monitoring Service (CLMS)**—relevant to the land environment, includes monitoring for water management, agriculture and food security, land-use change, forest monitoring, soil quality, urban planning and natural protection services. A European Ground Motion Service product portfolio was released in 2022. This new service measures ground

[2] Copernicus Emergency Management Service. https://emergency.copernicus.eu/ (accessed 06/02/2023).

N. Le Dantec
e-mail: nicolas.ledantec@univ-brest.fr

K. Parker · K. Lemon
Geological Survey of Northern Ireland, Dundonald House, Upper Newtownards Road, Ballymiscaw, Belfast BT4 3SB, UK
e-mail: kiepar@bgs.ac.uk

K. Lemon
e-mail: klem@bgs.ac.uk

D. Rodrigues · F. P. Gouveia
University of Madeira, Funchal, Portugal
e-mail: domingos@uma.pt

F. P. Gouveia
e-mail: fatima.gouveia@staff.uma.pt

I. García · M. Béjar-Pizarro · G. Herrera
Department of GeoHazards and Climate Change, IGME-CSIC, 28003 Madrid, Spain
e-mail: i.garcia@igme.es

M. Béjar-Pizarro
e-mail: claudia.pinto@cm-lisboa.pt

G. Herrera
e-mail: g.herrera@igme.es

I. Le Berre
UMR 6554 Littoral-Environnement-Télédétection-Géomatique, University of Brest, CNRS, Institut Universitaire Européen de La Mer, Rue Dumont d'Urville, 29280 Plouzané, France
e-mail: iwan.leberre@univ-brest.fr

C. Pinto
Lisbon Municipality, Campo Grande 25, Lisbon, Portugal
e-mail: claudia.pinto@cm-lisboa.pt

R. C. Gomes
CERIS, Instituto Superior Técnico, Universidade de Lisboa, Av. Rovisco Pais 1, 1049-001 Lisbon, Portugal
e-mail: rui.carrilho.gomes@tecnico.ulisboa.pt

displacements, including landslides and subsidence, as well as deformation of infrastructure.

-**Copernicus Atmosphere Monitoring Service (CAMS)**—provides continuous data and information on atmospheric composition, including monitoring for air quality and ultraviolet radiation forecasts, greenhouse gases and climate forcing.

-**Copernicus Emergency Management Service (EMS)**—provides all actors involved in the management of natural disasters (floods, forests fires, volcanoes, earthquakes, etc.), of human made emergency situations, or of humanitarian crisis with timely and accurate geo-spatial information derived from satellite remote sensing. The remotely sensed data are complemented by available in situ or open data sources.

-**Copernicus Service for Security (CSS)**—provides information in response to Europe's security challenges. It improves crisis prevention, preparedness and response in three key areas: border surveillance; maritime surveillance and support to EU External Actions.

-**Copernicus Climate Change Service (C3S)**—supports society by providing authoritative information about the past, present and future climate in Europe and the rest of the World. C3C cross-cuts all of the above domains.

With the AGEO Project focused on the assessment and management of geohazards, particular attention was paid to Copernicus EMS, as it supports all phases of the emergency management cycle: preparedness, prevention, disaster risk reduction, emergency response and recovery. The Copernicus EMS has been operational since 2012. It is composed of two main components: (i) on-demand mapping and (ii) early warning and monitoring. The on-demand mapping component is the EMS Mapping Service.[3] This provides a "Rapid Mapping" option for emergency response and a "Risk and Recovery Mapping" option for disaster prevention and planning. The early warning and monitoring component comprises three elements currently: the European Flood Awareness System (EFAS), the European Forest Fire Information System (EFFIS) and European Drought Observatory (EDO).[4]

The EMS Rapid Mapping (EMSR) option provides geo-spatial information within hours or days from its activation in support of emergency management activities immediately following a disaster. Local authorities can demand activations of the service through their Copernicus national contact point, i.e. the authorised service user (QSG, 2015) while emergencies are occurring. This activation will allow access to several products such as: (i) pre-event maps (reference maps), (ii) affected areas' identification and assessment (first estimate product), (iii) event impact and affected geographical extent (delineation products), and (iv) damage intensity assessment (grading products). The workflow, guidelines and products portfolio are available in QSG (2015), Dorati et al. (2018), MOP (2018) and Joubert-Boitat et al. (2020).

The EMS Risk and Recovery Mapping (EMSN) option consists of the on-demand provision of geospatial information in support of disaster management activities not related to immediate response, i.e. in a non-rush mode. This applies to activities dealing with prevention, preparedness, disaster risk reduction and post-event recovery phases. The EMSN can also be demanded by local authorities in a similar procedure to that for the Rapid Mapping. Tailored studies can be undertaken, compatible with a different timescale, as well as predefined set of standardised products.[5] Thus, Copernicus EMS enables monitoring of geological hazards and provides tools to prepare for emergency response. The mapping services intended to homogenize and centralize the collection of

[3] https://www.copernicus.eu/sites/default/files/documents/Copernicus_EmergencyMonitoring_Feb2017_0.pdf (accessed on 20 February 2023).

[4] https://edo.jrc.ec.europa.eu/edov2/php/index.php?id=1000https://edo.jrc.ec.europa.eu/edov2/php/index.php?id=1000 (accessed on 20 February 2023).

[5] https://emergency.copernicus.eu/mapping/ems/risk-and-recovery-mapping-portfolio (accessed 02 March 2023).

data with a set of indicators such as date, location, type and damage, providing factsheets as part of the service. In this sense, the analysis of Copernicus products and services highlights the importance of the necessary cooperation among civil protection, public authorities and private companies when emergency plans are developed, as key players giving suggestions and recommendations to spread awareness of these natural disasters.

The timeframe of the AGEO project has allowed us to review the Copernicus products and to promote their use by the various citizen observatories that have been established in the project. Not only has the use of the Copernicus program been disseminated, but it has also been actively used in the AGEO pilots through activations of the Copernicus EMS service and through the use of satellite images for hazard assessment or for estimating ground deformations.

Ground movements such as subsidence, uplift or landslides can be detected through radar image processing techniques, such as Interferometry of Synthetic Aperture Radar (InSAR) (Hansen, 2001; Hooper, 2008; Barra et al., 2017). Since Sentinel-1 provides free data access and achieves short revisit times (6–12 days), its images have been widely used for hazard monitoring and assessment related to landslides (Auflič et al., 2023; Reyes-Carmona et al., 2020), subsidence impacts (Hussain et al., 2022; Nduji et al., 2023) or geotechnical issues (Scoular et al., 2019). In addition, optical and multispectral data available from Sentinel-2 also highly contribute to hazard monitoring (Caballero et al., 2019).

Thus, within the AGEO project, both radar (Sentinel-1) and optical images (Sentinel-2) were processed. The former was used to estimate the deformation of some pilot areas with mass movements hazard, subsidence, uplift, or geotechnical problems (see Chap. 6). In the case of mass movements, radar images not only were used for the detection of slow movements, but also to validate the method with faster movements. Furthermore, Sentinel-1 has also been processed to attempt to map the extent of flooding for two different events.

Several Sentinel-2 images were processed to gain better knowledge of the pilot areas. The data were used to create several false-colour band combination maps that highlighted different aspects of the pilot areas, such as vegetation, built-up areas, water features, etc. Moreover, multivariate data analysis with algorithms such as Principal Component Analysis (PCA) and minimum noise fraction (MNF) were used for visualisation.

In addition, a step forward in risk assessment was taken with the characterisation of the elements at risk obtained for each AGEO pilot, the results of which may have important implications for geohazard management and emergency planning (Papathoma-Köhle et al., 2007).

In this chapter the uptake of the Copernicus EMS is reviewed for the AGEO partner countries in the EU Atlantic Region from 2012–2022. Surveys undertaken during the AGEO project highlighting the perception and awareness of stakeholders, as well as their suggestions for improvements or new requirements, are summarised. Uptake of Copernicus and Earth Observation (EO) products in the frame of the citizen observatories on geohazards is also addressed. Finally, strategic recommendations for strengthening the use of Copernicus products and services, and for new Copernicus products for the Atlantic Region, are given.

4.2 Status Review on the Use of Copernicus Products and Services in the Atlantic Region

The Atlantic region is exposed to a range of low-probability/high-impact events and to various hazard/risk scenarios. Due to the low likelihood of occurrence and/or the high cost of mitigating action, many parts of the Atlantic region may lack the level of preparedness for effective monitoring and response. Copernicus aims to make useful products, services and tools available to the public and to managers dealing with natural hazards, with the aim of improving the management of natural hazard emergencies.

Table 4.1 Summary of the overall responses from the stakeholders' survey

Question to the stakeholders	Possible answers	France	Ireland	Portugal	Spain	UK
User interface interpretation	Easy		x	x	x	
	Difficult	x				x
Preferences regarding Copernicus EMS information products	Download of Copernicus data and further processing of the data to produce other value-added products			x	x	x
	Display/share/redistribute the Copernicus service information products without any modification	x	x			
	Further process of the Copernicus service information to produce other value-added products		x			
Preferred way to get Copernicus EMS information from the Web Portal	Automated mode			x	x	x
	Manual mode	x	x			

One aim of the AGEO project was to assess the knowledge that end-users have about Copernicus products and services and about the extent of the use of those products and services. Additionally, the project aimed to identify potential user needs and to elicit suggestions for improvement of Copernicus services.

Thus, questionnaires were designed within the AGEO project to check the level of engagement and satisfaction with service providers, training, data access and the user interface. A total of 128 stakeholders responses, gathered in 2020, revealed that although most of the stakeholders used satellite data, they did not use Copernicus services. An exception is the case of Portugal, where half of the participants used Copernicus services and demonstrated a medium to high knowledge about Copernicus and EO systems. However, a common point among the whole group of survey participants, except in Portugal, is that most of them did not acquire any training on the use of data from the Copernicus Programme. In Portugal, half of the stakeholders surveyed did not acquire specific training.

Table 4.1 summarizes the overall responses from the stakeholders' survey. Globally, stakeholders' feedback let us know about data they demand, interaction quality with Copernicus services, involvement degree, tools usefulness, users' interests, needs and difficulties, so it contributes relevant information about the efficiency of the service. In France, Ireland and Spain, just a small number of participants had requested data from Copernicus services; however, in UK and Portugal most of the stakeholders' organizations did already request Copernicus data. Those who have done, requested data mainly related to risk prevention, emergency responses and geological risk research. Research projects (in UK and Portugal) and unspecified reasons (in France and Spain) are the main motivations that led the users to request Copernicus data. Most of the stakeholders are not involved in the development or implementation of any tool in the Copernicus Emergency Management Service. The Web Portal was the preferred means of display for Copernicus EMS information. All the participants agree with the purpose to use Copernicus EMS products: risk management. Table 4.1 summarizes the overall responses from the stakeholders' survey.

Stakeholders' organizations in Portugal and UK had already downloaded data from Copernicus EMS, but in any of the participating regions it is not clear the reason that gave rise to

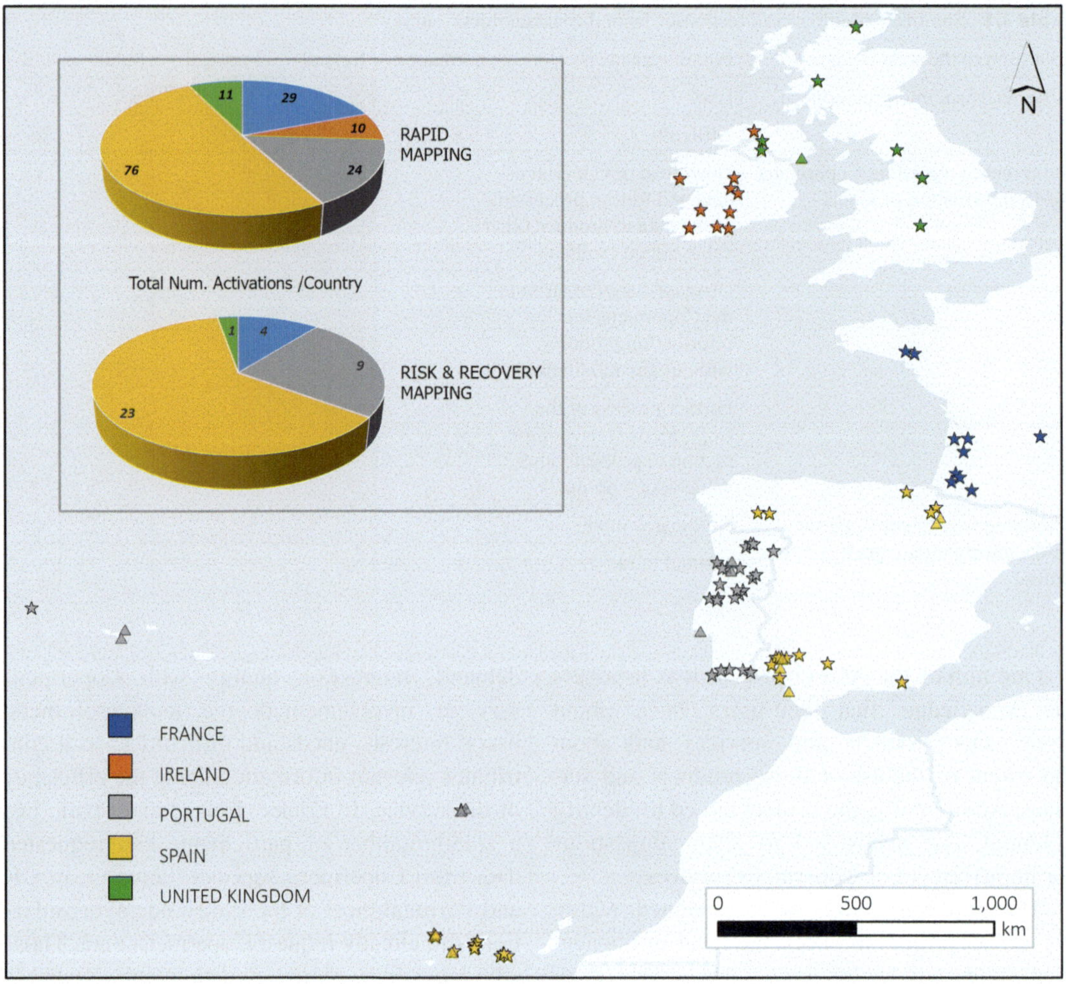

Fig. 4.1 Map of Copernicus EMS activations in the Atlantic Area from 2012 to 2022. Stars correspond to EMS Rapid Mapping activations; Triangles correspond to EMS Risk and Recovery activations. Pie charts present the total number of activations in the participating countries. Notice that not all the activations in France, Portugal, Spain and United Kingdom, are in the EU Atlantic area. (Data compilation from https://emergency.copernicus.eu/mapping/list-of-activations-rapid# accessed 02 March 2023 and from https://emergency.copernicus.eu/mapping/list-of-activations-risk-and-recovery accessed 02 March 2023)

the impediments to downloading information. It is in general thought that the information provided by Copernicus EMS is useful for an emergency planning and response and that the experiences with Copernicus are considered useful.

Complementarily, to assess the use of Copernicus services and products in the Atlantic Area, we reviewed the Copernicus EMS activations for the last 10 years (2012–2022) for the EU Atlantic region, while paying attention either to the EMS Rapid Mapping or to the Risk and Recovery Mapping Activations.[6] Results are shown in Figs. 4.1 and 4.2 respectively.

The review of the Copernicus EMS activations in the participating EU Atlantic Area countries, revealed that 150 EMSR activations and

[6] https://emergency.copernicus.eu/mapping/list-of-activations-rapid; https://emergency.copernicus.eu/mapping/list-of-activations-risk-and-recovery accesed on 20 February 2023.

Fig. 4.2 Copernicus Emergency Management Service activations in the Atlantic area from 2012 to 2022: **a**, **c** Rapid Mapping (EMSR). Data compilation from https://emergency.copernicus.eu/mapping/list-of-activations-rapid# accessed 02 March 2023). **b**, **d** Risk and Recovery (EMSN). Data compilation from https://emergency.copernicus.eu/mapping/list-of-activations-risk-and-recovery accessed 02 March 2023). **a** and **b** represent number of activations along time differentiated by type of hazard. **c** and **d** represent number of activations differentiated by type of hazard and by country

37 EMSN activations, were implemented from 2012 to 2022, being that 78 (EMSR) and 19 (EMSN) correspond strictly to EU Atlantic Area (Figs. 4.1 and 4.2). Rapid Mapping activations have thus far greatly outnumbered Risk and Recovery Mapping activations (ratio of 4:1). Floods or fires accounted for 91% of the Rapid Mapping activations in the EU Atlantic Area. The rest of the activation types were storms, windstorms, and other anthropogenic or biogenic hazards. Portugal and Spain were the countries with highest number of rapid

activations (75%). The total number of activations has increased through the years, revealing the growing participation and engagement of the countries (Fig. 2a, b). This fact gives Copernicus EMS Programme significance and demonstrates its usefulness as a risk management tool. For the activations of Copernicus EMS in the Atlantic Area in the timeframe 2012–2022 (Fig. 4.1 and 4.2), a few details per country are given below. Details and in-depth analysis can be found in the technical reports and geodatabases of all these activations available on the EMS website.[7]

4.2.1 France

The EMSR was activated 29 times in the country and 10 times in the EU Atlantic area between 2012 and 2022 (Figs. 4.1 and 4.2). Most of these Atlantic area activations concerned floods (70%) and wildfires (30%). The Risk and Recovery Mapping (EMSN) mode was only activated 4 times in the country, but none of them were in the Atlantic area. The activations of the Copernicus EMS in the Atlantic area resulted in 89 products: 4 'First Estimate', 7 'Reference', 71 'Delineation' and 7 'Grading' products. The number of activations has increased in the last years, namely 2021 and 2022. For France, the authority empowered to activate the Copernicus EMS is the *Centre Opérationnel de Gestion Interministérielle des Crises* (COGIC).

4.2.2 Ireland

There have been 10 activations of the Copernicus EMS for Ireland to end-2022. All activations were for the EMSR mode of Copernicus EMS. It appears that the EMSN mode of Copernicus EMS, has yet to be activated for Ireland. All of the past activations of Copernicus EMS for Ireland are associated to meteorological hazards, principally storms and floods. Activations have been made at a rate of one or two per year in the years 2014–2022, except for 2018–2019, when no activations were made. Geographically, the activations relate predominantly to emergencies or impacts in the Northern, Western and Southern parts of Ireland (Fig. 4.1). This reflects the facts that the average precipitation and wind speeds in Ireland are stronger along the island's western seaboard and diminish eastward, and that low-pressure weather systems and storms from the Atlantic Ocean typically migrate eastward and across Ireland. Overall, the Copernicus EMS in rapid mapping mode has so far delivered a total of 184 mapping products related to activation in Ireland. Of the four rapid mapping products that can be supplied by the Copernicus EMS Rapid mapping mode, by far the most delivered for Ireland has been the 'Delineation' product, which comprises 172 of the total. In addition, 8 'Reference' and 4 'Grading' products have been supplied for three of the events in Ireland. No 'First Estimate' products have been supplied. The National Directorate for Fire and Emergency Management (NDFEM) acts as the national focal point for the Copernicus EMS in Ireland.

4.2.3 Portugal

The EMSR was activated 24 times, with most (79%) of the activations concerning wildfire/forest fire, followed by floods (13%), with one humanitarian and one storm events. The EMSN was activated nine times and in this type of activation fires are more balanced with respect to geological hazards. The geological hazards result from internal or external geodynamic processes, such as volcanic hazard, seismic, mass movement, coastal flood risk or multiple natural risk assessment, i.e. seismic, lava flow, tsunami and storm surge, landslide, coastal erosion, flood, etc., as demanded by the Autonomous Regions of Azores and Madeira for risk assessment

[7] https://emergency.copernicus.eu/mapping/ems/risk-and-recovery-mapping-portfolio (accessed 02 March 2023).

products for planning and recovery. The latter led to the production of maps of exposure, vulnerability and risk for the various hazards under analysis. For the multiple natural hazard and risk assessments in both archipelagos of Madeira and Azores, 2285 maps were produced.

Hazards of *s.s.* geological origin only correspond to 5 EMSN activations (018, 020, 034, partially 031, and 129), and exceptionally floods, as mixed natural hazard, in EMSR activation 072. The *Autoridade Nacional de Emergência e Proteção Civil* is the national focal point for Copernicus EMS in Portugal.

4.2.4 Spain

There were 33 EMS activations in the Spanish Atlantic area (27 EMSR and 9 EMSN). Collectively, 86% of these activations related to wildfire hazards. Activations were mainly requested in the Andalusia region, with a few in the Canary archipelago the Basque Country and the Navarre region. 85% of EMSR activations were concerning fire hazard. The rest are related to floods, volcanic activity and a collapse of a dumping site. The climatic characteristics of this country, with dry and arid regions, create the ideal conditions for wildfire occurrence. Seasonal heavy rains are also frequent in some regions, especially in Mediterranean areas, where flood events normally occur. For the EMSN service, on its own, activations are related to fire hazard (45%), volcanic activity (33%) and rockfall hazard (11%). The vast majority of products generated are delineation and grading products. Both are derived from images acquired as soon as possible after the emergency event. These products are very interesting and useful for emergency managers, because updates to the maps can be supplied, so that the evolution of the event can be monitored. Even with a high number of activations if compared with other countries, Copernicus services are not used by the majority of stakeholders as it would be desirable. However, Copernicus EMS has been activated every year from 2012 to 2022, and an increase in the number of activations is observed through time, which means that the service is progressively being integrated in the national hazard management systems. The Ministry of Interior—*Centro de Coordinación Operativa* (CECOP)—*Dirección General de Protección Civil y Emergencias* is the national focal point for Copernicus EMS in Spain.

4.2.5 United Kingdom

United Kingdom authorities activated Copernicus EMSR 11 times from 2012 to 2022, and 7 of them were in the Atlantic area. 57% were related to floods and 43% related to fires. From 2012 to 2018, EMSR was activated nearly every year in the UK Atlantic area (except for 2014). In 2021, the EMSN was activated due to wildfire emergency for the first and only time to date. The activations generated a total of 81 products. 25 of them were 'Reference' products, 51 were 'delineation' products, and 5 were 'grading' products. The United Kingdom Cabinet Office—Civil Contingencies Secretariat and the EU Intelligence Analysis Centre were the authorised users in United Kingdom.

4.3 Copernicus Services in the Frame of Geohazards Citizens' Observatories

Citizens' observatories on geohazards in the AGEO project encouraged the local use of innovative EO products and services provided by European data infrastructures, in particular Copernicus. Valuable information can be obtained for different geohazards (Table 4.2), to complement data from citizens and stakeholders and to contribute to the hazard assessment in each observatory. In addition, its use serves to promote and disseminate the possibilities it offers for any organisation involved in geohazard management.

Thus, considering the risk assessment process (Table 4.2): (i) during the pre-event phase, Copernicus can contribute with a general characterization and, in some cases, with early warning systems through the weather

Table 4.2 Attributes derived from Copernicus products and services*, for each stage of emergency management and risk analysis stage for each specific geohazard

Geohazard	Pre-event stage	During-event stage	Post-event stage
Floods	Precipitation, vegetation, weather forecast, EFAS, digital elevation model (slope), elements throughout the sensitivity area, sea level, waves	Affected areas, elements for crisis management	Affected areas map with different levels of characterization, maps
Rockfalls	Precipitation, vegetation, new constructions (update landcover), fires burning (loss vegetation, temperature), land use, geomorphology, digital elevation model, deformation, settlements, landscape dynamics monitoring in coastal zones, soil moisture index, exposed elements	Precipitation	Affected area map
Coastal erosion	Exposed elements, settlements, landscape dynamics monitoring in coastal zones, Digital Elevation Model, water and wetness, land cover, riparian zones	Wave height, wind, sea level	Affected area map
Earthquakes	Area prone to deformation, tectonic plate movement	Tectonic plate movement, wave height, sea-level, affected areas, elements for crisis management, deformation	Affected area map (access for rescue teams)
Landslides	Area prone to movement or instability, precipitation, vegetation cover, weather forecast, new construction, land use, geomorphology, digital elevation model, settlements, soil moisture index, exposed elements	Precipitation, deformation, affected areas, revised digital elevation models elements for crisis management	Affected area map
Volcanic eruptions	Area prone to deformation, satellite images, digital elevation model, temperature, gas emissions, exposed elements	Deformation, wind, temperature, emissions, affected areas, elements for crisis management	Affected area map
Geotechnical risks	Area prone to deformation, tectonic plate movement, satellite images, precipitation, vegetation cover, weather forecast, new construction	Precipitation, deformation, affected areas, elements for crisis management	Affected area map

* Copernicus (Marine Environment Monitoring Service-CMEMS, Land Monitoring Service-CLMS, Atmosphere Monitoring Service-CAMS, Emergency Management Service-EMS, service for Security, Climate Change Service-C3S)

forecast or ocean condition; (ii) in the during-event phase, affected areas might be identified with Copernicus data and also the elements for crisis management, and variables to consider to manage the emergency; (iii) during the post-event phase, Copernicus again may provide the users with the affected areas map with different levels of characterization.

4.3.1 Copernicus Emergency Management Service Activations in AGEO

Activations of Copernicus EMS were requested by some of the pilots in the frame of AGEO project, thus involving the citizens' observatories directly in the use of Copernicus products. The Canary Islands and Madeira citizens' observatories requested the activation of the service, each for their specific purposes.

Canary Islands Observatory (Spain)

Two activations of Copernicus EMS were requested in the frame of AGEO project in the Citizens' Observatory on Rockfalls in the Canary Islands. The Spanish General Directorate of Civil Protection and Emergencies (CENEM), on behalf of the Geological and Mining Institute of Spain (IGME-CSIC), activated the service.

- *Activation EMSN093 La Gomera*. An important rockfall occur on 14th November 2020 in a cliff at SW coast of La Gomera. The objective of this EMS activation within the Canary Islands Pilot was to contribute to the Copernicus products usage at local level and it was intended to demonstrate that Copernicus data could be used to study small-medium sized rockfalls.

Thus, the Risk & Recovery Flex Mapping Service was activated, and the following products were requested[8]:

- PR1. Reference data;
- PR2. Pre-event digital elevation model (DEM);
- PR3. Post-event DEM;
- PR4. Delineation of source and deposit area extent;
- PR5. Estimation of the volume of mass moved;
- PR6. Assessment of rockfall/avalanche hazard in area of interest (Fig. 4.3);
- PR7. Exposure of services and population to rockfall/avalanche hazards;
- PR8. Rockfall/avalanche risk assessment;
- PR9. Ground deformation analysis.

Every product includes topographical elements (place names and administrative boundaries, hydrology, physiography), population (based on recent and/or alternative census data), settlements (urban, suburban, rural, etc.) with buildings, road network and other related transport networks, location of hospitals, schools, local services, food and distribution infrastructure, location and type of industry or potential industrial accident sites, and finally, land use coverages.

Since the result constitutes a tailor-made product, numerous exchanges of information were needed to approach the question in the most effective way. The time length taken to produce the whole new dataset lasted from Feb 21 to Oct 21. Volume estimation of the mass movement was not satisfactory due to difficulties with the pre-event DEM. However, from a combination of Sentinel-1 images interesting results were achieved from deformation analysis by applying Differential Interferometry and Multi-Temporal interferometry. Several local displacements were identified. In particular, at the cliff where the November 2020 rockfall event occurred, the time series revealed that acceleration had started several months before the fall occurred (CEMS, 2021). Therefore, it highlights the enormous potential of this tool for monitoring potentially hazardous slopes.

- *Activations EMSN112/124, EMSR546 and EMSN119*. Due to the volcanic emergency on La Palma on 19 September 2021, Copernicus EMS was activated in its different modalities:

[8] https://emergency.copernicus.eu/mapping/ems-product-component/EMSN093_01GOMERA_ROCKFHAZ_OVERVIEWA1/2 (accessed 25th February 2023).

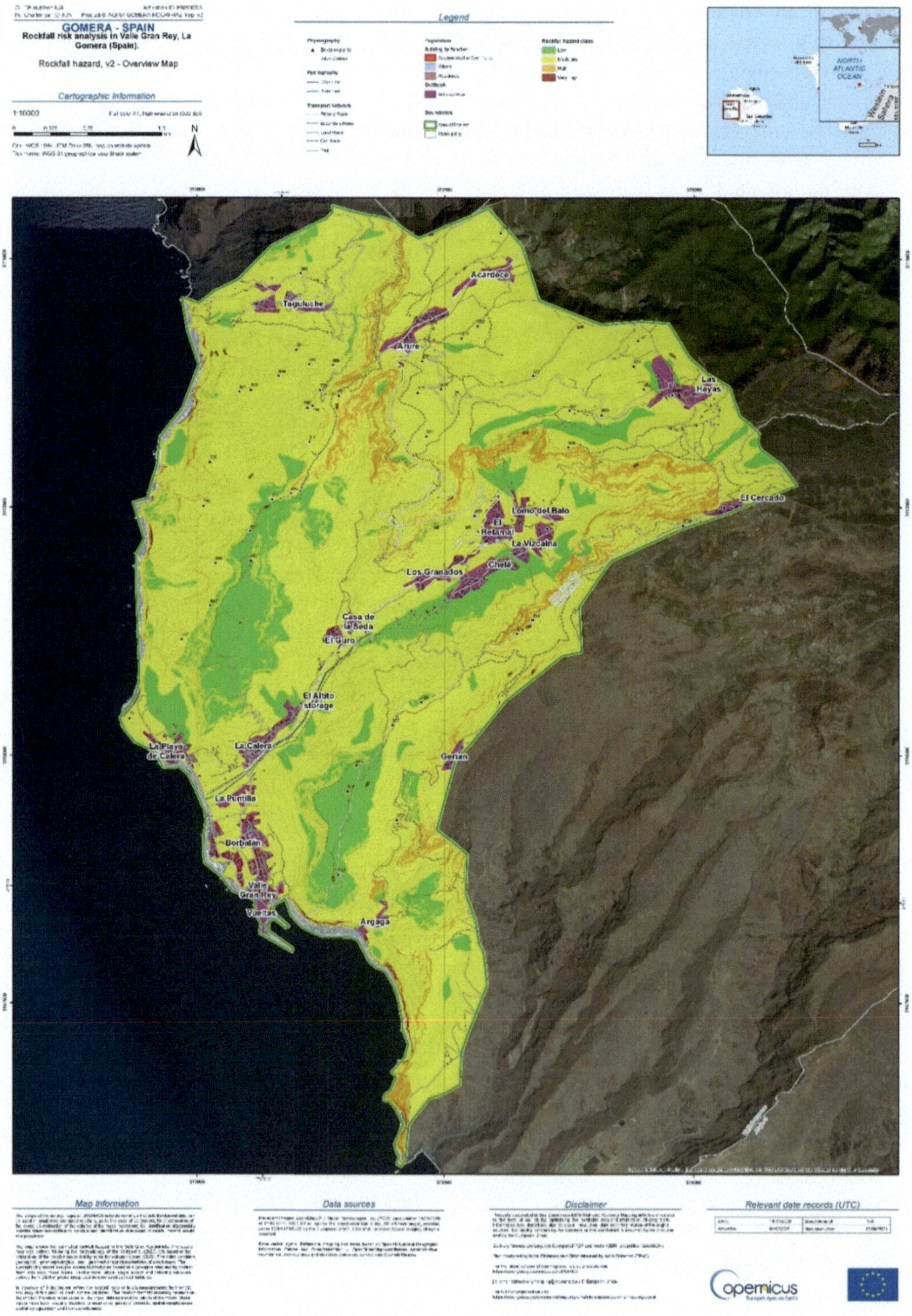

Fig. 4.3 Rockfall hazard map in Valle Gran Rey region (La Gomera Island), a product from Copernicus EMSN093 Activation. [https://emergency.copernicus.eu/mapping/ems-product-component/EMSN093_01GOMERA_ROCKFHAZ_OVERVIEWA1/2]. Copernicus Emergency Management Service (© 2021 European Union), EMSN093

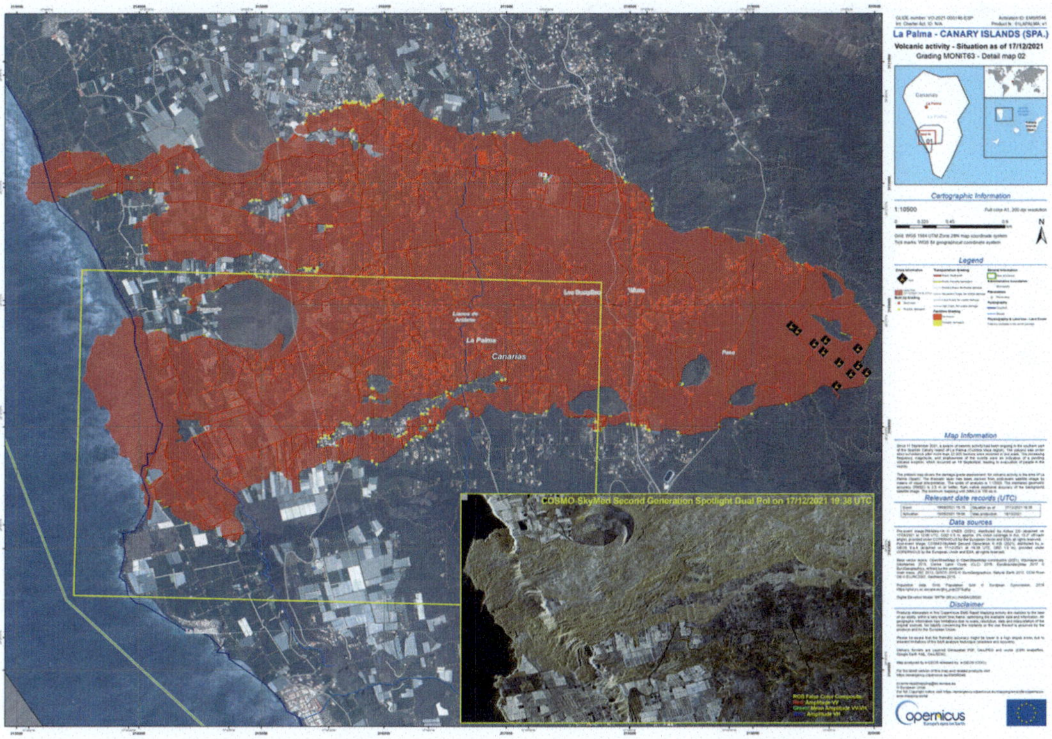

Fig. 4.4 Volcanic activity in La Palma Island. Map of situation as of 17/12/2021; grading product from Copernicus EMSR546 Activation. [https://emergency.copernicus.eu/mapping/list-of-components/EMSR546]. Copernicus Emergency Management Service (© 2021 European Union), EMSR546

(a) Rapid Mapping, obtaining the following products: delimitation of the extent of the lava flows, estimation of affected buildings and services (Fig. 4.4).
(b) Risk and Recovery Standard Mapping, obtaining the following products: ground deformation monitoring, mass movement risk map (pre- and post-eruption) and soil erosion risk map (pre- and post-eruption);
(c) Risk and Recovery Flex Mapping, obtaining the following products: DEM of the cone and lava flows, changes in ground elevation, volume of lava flows, discharge rate: DEM of the cone and lavas, changes in ground elevation, volume of lavas, discharge rate.

The products received provided valuable information during the emergency and were very useful for many national and international institutions. Specifically, the IGME-CSIC working group that collaborated during the emergency used the Copernicus data to: (i) complement the topographic data obtained by the Island Government in La Palma, Emergency and Rescue Group of Canary Government (GES) and IGME-CSIC´s Aerial Works Service (STA) during the emergency and obtain a more detailed mapping of the advance of the lava flows; (ii) analyse ground deformation; (iii) analyse and understand morphological changes in the volcanic cone; (iv) detect changes in the lava flows; (v) analyse mass movements associated with seismic activity.

Madeira Observatory (Portugal)

- *Activation EMSN089.* On 25th December 2020 and 7th January 2021, a storm brought strong winds and heavy rain to Madeira Island. Flooding and landslides occurred in the municipality of São Vicente, 20 people were evacuated in Ponta Delgada and 7 in Boaventura.

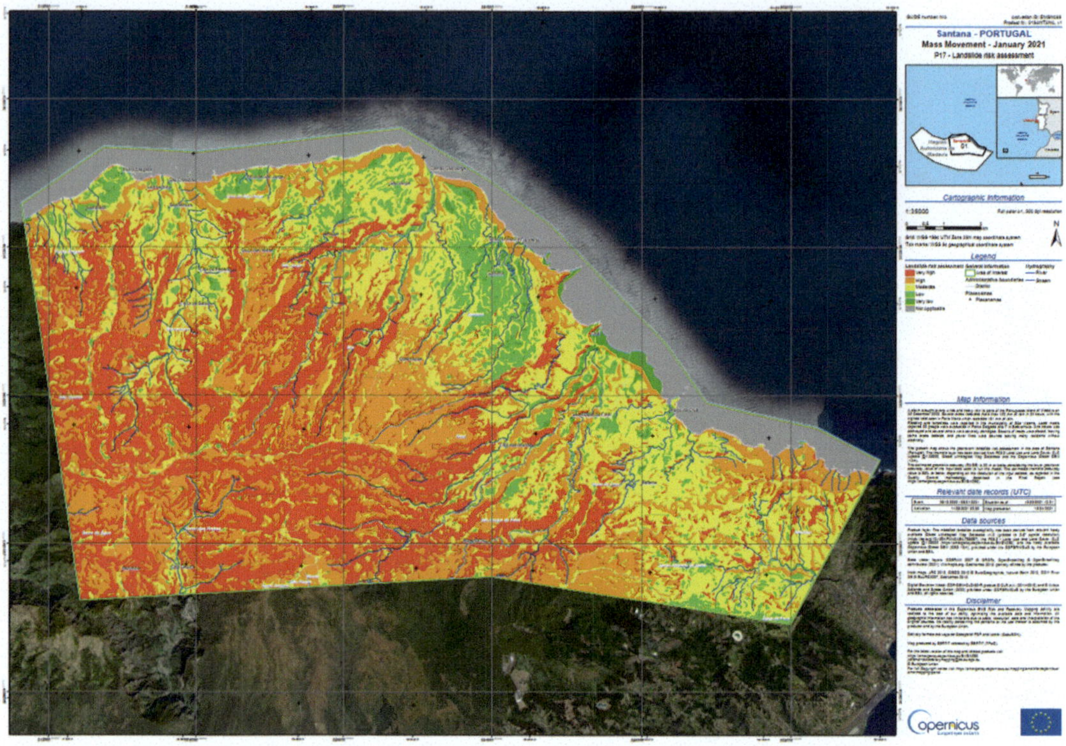

Fig. 4.5 Post-disaster landslide susceptibility assessment map in São Vicente region (Madeira, Portugal). Map of situation as of 17/12/2021; product from Copernicus EMSN089 Activation. [https://emergency.copernicus.eu/mapping/list-of-components/EMSN089] accessed 02 March 2023. Copernicus Emergency Management Service (© 2021 European Union), EMSN089

One house was destroyed, and several others were severely damaged. Dozens of roads were closed, leaving some areas isolated, and power lines were downed, leaving many residents without electricity. Several areas recorded more than 100 mm of rain in 24 h on 25 December 2020, with the highest total seen in Porto Moniz, which recorded 161 mm of rain.

A request for activation of the RRM (Risk and Recovery Mapping) service was made.[9] The goal of the activation was to get a more accurate picture of the mass movements induced by the heavy rainfall that occurred over the period ranging from 25/12/2020 to 08/01/2021. The products generated within this activation were:

- P01.1 Digital Surface Model (DSM) that required a new VHR tri-stereo acquisition, and hence was part of a separate delivery.
- P02.1a Reference dataset was provided over AOI01 Santana.
- P03.2 Land Use and Land Cover dataset, which was required as an input for landslide risk assessment.
- P08.1 Detailed damage assessment analyses over affected areas based on VHR imagery acquired on the 10/02/2021.
- P09 Reconstruction monitoring was provided for AOI01 Santana based on new VHR imagery acquired.
- P17 Post-disaster landslide risk assessment was provided (Fig. 4.5).
- 2 × P14 Impact assessment/exposure analyses on assets and population is provided based on both P08.1 (impact assessment) and P17 (exposure).

[9] https://emergency.copernicus.eu/mapping/list-of-components/EMSN089https://emergency.copernicus.eu/mapping/list-of-components/EMSN089 (accessed 02 March 2023).

4.3.2 Other Uses of Copernicus Services and Products for Geohazards

Undoubtedly, Copernicus EMS represents an extremely valuable service in the management of geohazards, as depicted in previous sections. However, interesting data are also provided from this platform through other services such as the European Ground Motion service (EGMS)[10] (EGMS, 2017), launched in 2022, as part of the Copernicus Land Monitoring Service (CLMS).

The EGMS provides products oriented to analyze hazards related to ground deformation such as landslides, subsidence, volcanic unrest, on local, regional and national scales, through use of InSAR processing techniques on Sentinel-1 images. Three types of products are offered: basic, calibrated and ortho. The Basic products provide points with the average displacement velocity referred to a local reference point. Calibrated products give same information, but they are absolute, referenced to the European GNSS positioning framework. Ortho products provide purely vertical and purely horizontal motions through the combination of different satellite view angles. Basic and Calibrated products are provided in the satellite line of sight. Moreover, 50×50 km gridded velocity model is also available (European Union, 2022).

Complementing these products, in the framework of the AGEO project, one of the consortium partners, University College of Dublin (UCD), developed an InSAR processing toolkit – EZ-InSAR – to cover the specific needs of some pilot areas of the citizen' observatories on geohazards created in the framework of the project (Hrysiewicz et al., 2023). Details about the EZ-InSAR tool can be found in Chap. 6.

Moreover, activities and actions were organized by the partners of AGEO project in order to promote knowledge and use of Copernicus, to contribute to a better use of Copernicus products, to increase the use of the available services, and subsequently to help also to identify possible bottlenecks in the use of the programme. Each of the pilot observatories implemented the use of EO products as explained below.

Canary Islands Observatory (Spain)

The characterisation of the pilot areas, in terms of stability, was essential for further studies in the framework of the Citizen's Observatory on Rockfalls in the Canary Islands. In addition to the Copernicus EMS activations mentioned in section before (La Gomera and La Palma), the identification of instabilities from Sentinel images and LIDAR data was carried out for the regions of San Bartolomé de Tirajana (Gran Canaria), La Laguna (Tenerife) and Valle Gran Rey (La Gomera) (see Chap. 9). Source areas mapping, susceptibility maps (considering slope, lithology, land use, precipitation and proximity to public works), and deformation analyses were obtained (Fig. 4.6).

Results were satisfactory for the identification of ground movements with LIDAR, because it detects variations in ground height quite well. However, results for identification of rockfalls with radar data were not very promising due to several factors involved in the final results: inadequate scale, velocity of the process, displacement orientation. Some details are presented in Chap. 9.

Madeira Observatory (Portugal)

The Citizens' Observatory in Madeira, focused in landslides and rockfalls, selected some areas of interest to evaluate the feasibility of using InSAR for monitoring slow mass movements. The main findings from the InSAR observations on the creep movements in talus deposits that take place in the municipality of Machico (Madeira), are explained below.

The identified point targets are mainly concentrated in the Machico valley and Ponta de S. Lourenço (eastern part of the Madeira island). Few PS (Persistent Scatterers) targets were identified at the mountainous region, where the topography is steep and the vegetation coverage

[10] https://land.copernicus.eu/pan-european/european-ground-motion-service (accessed 1st March 2023).

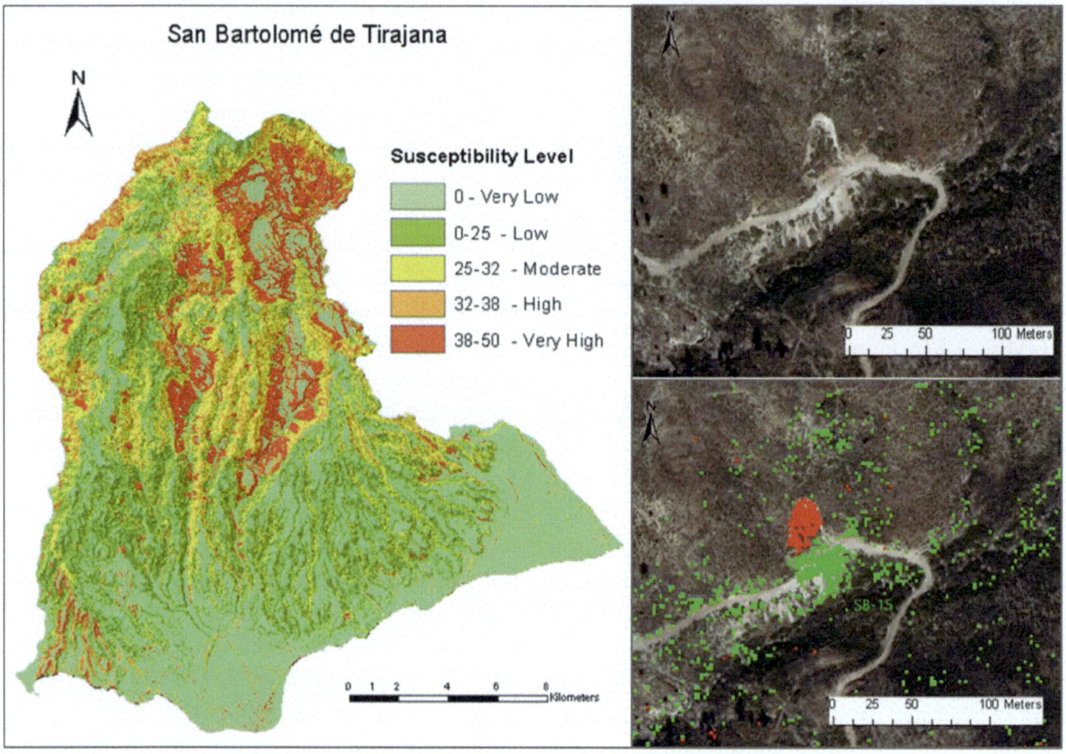

Fig. 4.6 (Left) Rockfalls susceptibility map for San Bartolomé de Tirajana municipality in Gran Canaria. (Right) Instabilities identification with LIDAR data in San Bartolomé de Tirajana municipality

is dense. At the Machico valley, the movements on the east side of the valley are overall more active than those on the west side. The maximum slope-parallel movement velocity reached is about 7 mm/yr over the three identified slope units. Compared to the landslide susceptibility map released by Rodrigues and Ayala (1994), our InSAR measurements confirmed some of the unstable slopes (creep movements) at the Machico valley. However, some small ones were not identified, likely due to the moderate resolution of Sentinel-1 SAR images and the low displacement of these movements.

At Ponta de S. Lourenço with a different geological setting than in the Machico valley the observed displacements are well correlated with active erosion process in some areas or the presence of loose sediments (eolian sands).

Moreover, interesting use of Sentinel Playground was carried out for educational purposes in the frame of the Observatory.

Lisbon Observatory (Portugal)

The Citizens' Observatory in the city of Lisbon is a pilot project focusing on seismic risks, urban flooding, landslides and geotechnical risks.

A spatial database of relevant information from Copernicus products was created to improve the characterisation and information about the pilot site. Various information themes from the Copernicus Land Portfolio were used, such as urban atlas, land cover and land cover changes, impervious surfaces, building heights, etc., ranging from very high (Hot Spot) to medium/low resolution, resulting in the creation of several maps.

Several Sentinel-2 images were acquired and processed to gain better knowledge of the pilot area. The data was used to create several false colour combination maps highlighting different aspects of the pilot area, such as vegetation, built-up area, water features, etc. Multivariate

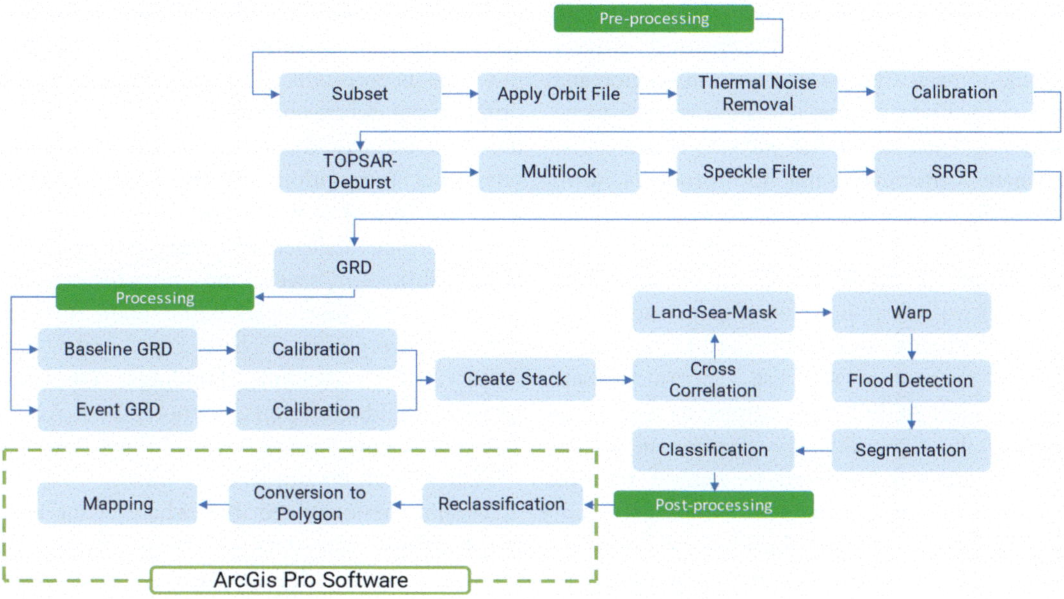

Fig. 4.7 Methodology used to process Sentinel-1 data in order to map flood extent

data analysis such as Principal Component Analysis (PCA) and Minimum Noise Fraction (MNF), were used for visualisation and to reduce the dimensionality of data. PCA aims to reduce the dimension of correlated variables (such as bands) to a new set of uncorrelated variables (PC's) that can be used for a variety of purposes such as visualisation, compression, feature extraction, segmentation and classification. MNF aims to reduce noise in the dataset while preserving important information that can be used for image classification, target recognition and data visualisation.

The Lisbon pilot site is prone to flooding, especially urban flooding. To attempt to map the extent of flooding for two different events, Sentinel-1 data were used. The events of 18/02/2017 and 13/12/2022 were chosen as the availability of data close to the event dates increases the chance of producing accurate post-event data.

The process of converting the data from Sentinel-1 to the extent of inundation is shown in Fig. 4.7. In this case, the data from the pre-event image must be compared with the post-event image to determine the areas potentially affected by flooding. The data was processed in SNAP[11] and then visualised in desktop mapping software.

The resulting data shows the areas potentially affected by flooding, with the most accurate representation of the extent of flooding being found on large flat areas, such as airport runways or football field, where water accumulated for the longest period. Smaller areas, on the other hand, are not mapped well, possibly due to two reasons: the spatial resolution of the sensor and the image acquisition by the satellite did not take place during the event, but 10/18 h afterwards.

Brittany Observatory (France)

Since the Brittany pilot (OSIRISC observatory) is focused on the analysis of coastal flooding and erosion hazards, the use of Copernicus EMS was not envisaged because coastal hazards need regular and long-term monitoring rather than a rapid activation for an emergency, as these phenomena occur progressively. In addition, although

[11] Https://earth.esa.int/eogateway/tools/snap Accessed 02 March 2023.

the quality of the Pleiades images would enable to identify the hazards of submersion or erosion and the impacted issues, the erosion phenomena are too localised for the activation request to be validated. Therefore, other ways to use and disseminate Copernicus and EO products were explored.

Firstly, the method for developing indicators proposed in OSIRISC observatory (see Chap. 9) was re-applied, using Copernicus data to export them to other pilots. This application allowed to explore and use Copernicus services to meet AGEO expectations, as well as to promote the Copernicus programme in a new context: elements exposed to natural hazards (housing buildings, economic buildings, cropland, port, road, airport). In addition, considering an exploratory focus, an attempt at enriching the products offered by Copernicus services was pursued through the use of Pleiades satellite imagery in the same context: the characterisation of elements at risk. Both databases, Copernicus and Pleiades, were explored to answer the following questions: what is the potential of Copernicus services and infrastructures in the evolution and characterisation of coastal issues exposed to natural hazards? and, what is the potential contribution of Pleiades imagery in this process?

Thus, an inventory of available databases, specifying their availability for each AGEO pilot, has been produced. A table to classify the different datasets according to the theme studied has been created, focusing mainly on land use typologies (vegetation, urbanization, coastline, hydrography….) as well as the scale covered by the Copernicus datasets and their spatial resolution. Additionally, several indicators for exposed elements were developed based upon relevant Copernicus products (Coastal zones and Urban Atlas 2018): economic buildings, cropland, housing buildings, ports, airports and roads. These indicators have been tested for each of the AGEO pilot observatories (Fig. 4.8).

Results demonstrated that, whatever the hazard, the Copernicus database has great potential for characterising elements at risk when working at the right scale. Local scale Copernicus datasets show limited interest when regional and national reference data are available. On the other hand, on a larger scale, national or European, the potential of these indicators becomes interesting and useful to produce a harmonised diagnosis of different territories and thus to envisage an initial identification of risk areas. Pleiades imagery, for its part, has great potential on a local scale. Its temporal and spatial resolution, which is more efficient than that

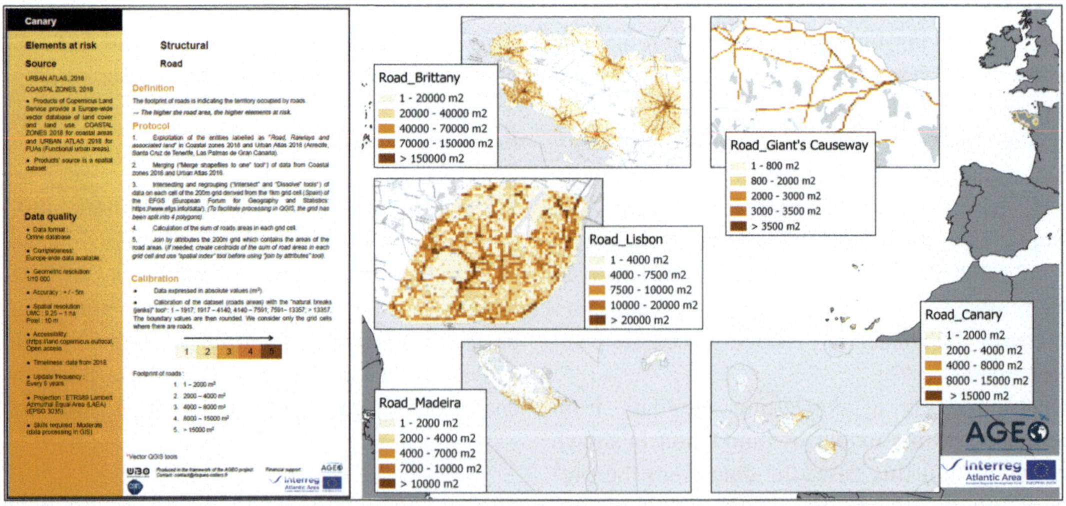

Fig. 4.8 Left: Technical factsheet for the indicator "Road". Right: Maps for the indicator "Road" in the different AGEO project pilots (From Bonnier, 2021)

of the Copernicus services, gives the opportunity to better identify numerous elements at risk and to monitor coastline evolution.

Finally, the Pleiades imagery and the Copernicus database complement each other, one being of interest on a local scale while the other is of interest on a larger scale, whether regional, national or European. It could therefore be interesting for observatories to have annual access to Pleiades imagery to complete and update their databases of elements at risk and coastline evolution at regional scale.

Causeway Coast Observatory (Northern Ireland—UK)

The Causeway Coast pilot is coordinated by Geological Survey of Northern Ireland (GSNI), which has not used Copernicus data and services but has used other Earth Observation techniques including satellite radar interferometry (InSAR) techniques to remotely assess small ground movements in geotechnical or structural infrastructure. The satellite radar data has been used to assess and monitor landslide instability and subsidence.

A Copernicus database of several thematic maps has been collected to improve the characterisation and information about the pilot site, similarly to the Lisbon Observatory, focusing on the regional framework and resulting also in the creation of several maps.

Since GSNI is staffed by scientists of the British Geological Survey (BGS), the Citizens'Observatory in NI also benefits from BGS experience in using Copernicus data and services. Thus, the BGS team played a major role in delivering the European Space Agency's (ESA) Terrafirma project. Terrafirma provides a pan-European ground motion hazard information service based on InSAR techniques in EU member states, and supports civil protection agencies, disaster management organisations and coastal, rail and motorway authorities in the process of risk assessment and mitigation. The BGS was also part of the Coastal Erosion Project within the Science for Society slice of the 5th Earth Observation Envelope Programme (EOEP-5) run by the ESA, exploring coastal change from space to prepare the ground for long-term exploitation of remote sensing from satellites, including Copernicus data, by large user communities.

Progressions have been achieved in the framework of the AGEO project related to Earth observation. The Northern Ireland Earth Observation Steering Group (NI-EOSG) and Northern Ireland

Earth Observation User Group (NI-EOUG) were established in 2021. The groups are adressed to increase the awareness of EO data, services and applications (GSNI, 2021), being that one element of focus is Copernicus Rapid Mapping Service for emergency planning & response. Moreover, NI Space Agency is more and more promoting EO data and on the other hand, increasing EO use to deliver tasks by Government departments i.e. agriculture, land mapping etc. In addition, NI Executive Office is Disaster Charter Authorised User, then, Copernicus Rapid Mapping Service is now included within Civil Contingency disaster response. Positively, there is a growing awareness of EO capability, but it is limited to specific users and at the research stage.

The experience gained from both GSNI and the wider network from BGS was used to develop the uptake of Copernicus data and services in Northern Ireland through the AGEO project framework. This assists in the culture change towards remote sensing and Earth observation for the assessment and monitoring of geological hazards in Northern Ireland.

4.4 Strategic Recommendations for Strengthening the Use of Copernicus Products and Services, and New Copernicus Products Suggestions for the Atlantic Region

Strengthening the use of Copernicus products and services is one of the pillars of this European program since there were established mechanisms to monitor user needs and ensure their adoption through mechanisms for programming, monitoring, reporting and evaluating Copernicus.

AGEO and its citizen observatories implemented and encouraged the use of Copernicus services to help managers and stakeholders to improve their efficiency in emergency management. At the same time, these users provided feedback to AGEO with valuable information about their needs and suggestions for improving Copernicus EMS. Thus, this two-way communication enriched both the risk management plans and the Copernicus program itself by allowing assessment of its limitations, and suggesting recommendations for its strengthening.

Assessment of the two types of satellite data available on Copernicus Programme.

Impressions and requirements of stakeholders in participating regions in the timeframe of AGEO project, helped the experts to analyse limitations on the two main types of satellite data available in the Copernicus Programme, captured with passive or active sensors: optical and radar products. Concerning the optical sensors, the 2.5 m pixel resolution provided within the European image mosaic is the better resolution achieved. This value can be adequate for several tasks carried out in risk management, such as site characterization, triggering factors analysis, vulnerability assessment and geohazards monitoring through Copernicus Early Warning & Monitoring systems that are already in place (e.g. EFAS). Services such as the EMS offer rapid risk and recovery mapping with very high resolution by using the data coming from unmanned aerial platforms or contributing missions. Moreover, it is interesting that in case of an alert of a potential flood hazard, the EFAS warning system has an alert that works as a trigger for satellite activation to capture images, expecting, in this way, to reduce the time required for obtaining post-event imagery.

Thus, regarding optical data, Sentinel-2 is useful for a general characterization of the regions but quite limited for monitoring tasks in some cases such as very active and small-sized processes (rockfalls). Some of the products offered could be very interesting and valuable for the analysis of triggering conditions, such as moisture index. However, its temporal resolution is limited if compared with the high temporal variability of this parameter in some areas so it would not be useful for such regions. Likewise, changes detection exploiting Sentinel-1 and Sentinel-2 could be a useful tool for coastal erosion analysis in those shorelines with high rates of retreat. However, spatial resolution is limited in those cases with low variability. Nonetheless, even with its limitations, Copernicus optical products, as shown in the AGEO project, can be used not only in a scientific way but also as part of dissemination activities with schools and general public.

Radar products through the measurement of ground deformation are useful in geohazards management to detect where the active geohazard is located, to monitor the evolution of deformation, to characterize the geohazard parameters, to model it and even to predict its impact in the infrastructures. There are opportunities to exploit Copernicus satellite images products through European Space Agency online processing tools of satellite images. Processing satellite SAR images with Radar Interferometry tools allows for monitoring ground surface deformation related to landslides and subsidence phenomena. However, for this purpose it is necessary to have some knowledge and expertise both for INSAR processing of satellite images and for the analysis of INSAR displacement data to study those geohazards. In order to attend users' necessities regarding ground motion issues, the European Ground Motion monitoring service was implemented in 2022.

The use of radar data (Sentinel-1) in the specific cases of the AGEO pilots is fully recommended in those cases in which deformations are slow and large enough to be detected by InSAR techniques. Landslides, subsidence or earthquakes are some examples of the geohazards present in the Atlantic regions that might be assessed. As part of the protocol of radar data usage on geohazard analysis, some important points should not be neglected:

(i) Validation of results. The use of radar data and its analysis should be carefully done and detailed validation protocol is needed in order to get results with guaranteed quality.

Sometimes results do not completely agree with what is seen in the ground, therefore, reliability analysis of the data is needed through the use of complementary data (GPS, piezometric information, etc.) to validate the results.

(ii)Resolution might be limited for some cases and loss of information could occur. For example, it would be possible to find regions with deformation data and without statistical significance, but if resolution is improved then those points could become statistically significant. Likewise, very active processes but with small dimensions could not be analysed with resolutions available up to now.

(iii) Costs and timing of processing procedures must be considered. Depending on the resolution required, processing costs could be very expensive.

(iv) Integrability on the management systems should be evaluated. Requirements on software and hardware needed to processing data, should be considered. GIS software, even if open source, is necessary in order to do spatial and temporal analysis once the displacement maps are obtained.

Assessment of the Copernicus products and procedures for the management and prevention of geohazards.

Limitations of Copernicus products and procedures for the management and prevention of geohazards through the experience of AGEO partners on its use on pilot areas were compiled. Suggestions for the improvement of the existing services and also for the introduction of new Copernicus products and services for geohazard risk reduction and monitoring in the Atlantic region are presented below. They are related to activation procedures of Emergency Management Service, data access, monitoring, training, spatial resolution and new products.

- *Activation procedures*. Clarification and simplification of activation protocols is highly demanded by users to ease the process, because, up to now, it is considered complex and bureaucratic. In addition, some users proposed that it would be interesting to enable a direct activation of Copernicus services by public research institutions or local authorities, with the knowledge of the focal point. However, it is worth noting that the focal point validates the data according to reliable rules, so its presence on the protocol is justified.

Improvements in timing and communications would be also desirable. Some delays on delivery time of mapping products have been detected. This could potentially compromise subsequent stages in the emergency management plans. However, those delivery time delays are expected to improve by reducing the time required for obtaining post-event imagery and by optimizing the acquisition time with respect to an event's evolution (MOP, 2018), as it happens with the procedure used in floods since 2017, using EFAS floods warning to pre-task image acquisitions without an official service activation (Wania et al., 2021).

European Flood Awareness System (EFAS) activation could include flash floods associated with pluvial or urban flooding besides flash flooding related only to river networks. Moreover, some stakeholders suggested a better connection with the national EFAS focal point in order to better understand the early warning system procedure and to know, for example, which is the threshold to issue an alert.

Easy access to high resolution products (for example data from contributing missions) would help for local event analysis, which requires detailed resolutions. However, those events are too localised for the activation request to be validated.

Ancillary data are needed to get a higher quality in EMS activations and resulting products. Those cases in which the data cannot be provided by the user, this will not be considered at the final product causing a lack of quality. Nevertheless, for the cases in which data are provided, the results are significantly improved (Quental and Machado, 2020).

- *Data access*. Global database with all the products derived from activations, instead of searching manually within each activation, is demanded. In addition, each Copernicus service is a different subsystem so they need different registration processes and passwords. It would be interesting to have a common user-register for all the services. Furthermore, to homogenize the web sites of each Copernicus service with commonalities on the interfaces, will contribute to the user friendliness and therefore to facilitate the information access.

- *Monitoring*. Unfortunately, for continuous monitoring, up to now, it is necessary to rely in user own capabilities. It is not possible rely on Copernicus EMS Mapping (EMSR and EMSN) because it does not have this continuous capability. Undoubtedly, its implementation would be very valuable. A higher temporal resolution of data offered would allow analysing active geological processes with high accuracy. However, fortunately, this capability is present in Copernicus EMS Early Warning and Monitoring component in EFFIS and EFAS and EDO systems, as well as in most of other Services.
- *Training*. Open training (webinars) in several languages, about Copernicus services and potentialities is the most demanded improvement among the users. Thus, interesting issues would be: promotion of EO capabilities and limitations to assist emergency response, promotion of how to access portals and integration to user systems and improved training for access, use, integration and interpretation of datasets. Furthermore, Massive Open Online Courses (MOOC's) about InSAR are also requested. A remarkable suggestion to spread EO knowledge would be to embed EO education in school/university. In this respect, the alignment with established networks such as Copernicus Academy would be very profitable given the synergies. Part of the goals of this network are the development of lectures, training sessions, traineeships as well as educational and training material to empower the next generation of researchers, scientists, and entrepreneurs with suitable skill sets to use Copernicus data and information services to their full potential. Moreover, it seems necessary to increase dissemination of platforms and sites such as Sentinel Playground, powered by Sinergise Lab., Education site in Sentinel Hub, EO Browser in Sentinel Hub, Copernicus Mooc and Research and User Support (RUS), that offer tools, educational resources and courses to improve the knowledge in Copernicus programme. Thus, investment in training and dissemination will lead to an increasing uptake of Copernicus products.
- *Spatial Resolution*. Improvement of the spatial resolution in the obtained products is one of the most requested issues. Local studies and small-sized geohazards require high resolution data. Available information and thematic mapping for local areas is also a highly demanded improvement. Up to date, the existing data allow general characterization of these local areas. Therefore, the better the resolution, the higher the number of users accessing Copernicus data, and the better the quality of the results.
- *New products*. Based on the experience of the stakeholders analysing Copernicus services, a few new products were suggested to be created or developed by Copernicus authorities:
  - InSAR processing and post-processing tools (interferogram atmospheric corrections and unwrapping).
  - InSAR processing tools integrated in fire detection capabilities
  - Real time fire detection.
  - Wider information on Coastal zones. The specific satellite from which the used product was derived is indicated, along with some elements on the method to compute the data product that is provided. However, there is very limited information on the precise methodology used to derive other products, so more details will be desirable.
  - Improvements in the tools offered by the viewers in order to enable wider possibilities to access and select the data available.
  - Database including historical geohazards events detected by Copernicus tools or from other ancillary sources.

Requirements and limitations on Copernicus EMS presented before could be partially solved with the products offered by Copernicus on its existing services. Thus, open access to data (like contributing missions) could be available when required by some specialised users not related to research or public administration. Likewise, the European Ground Motion Service could cover part of the requirements of the users regarding InSAR. Early Warning Systems linked to the Copernicus EMS can facilitate data acquisition by linking pre-tasking of satellite images to the Early Warning System without an official service activation, and thus minimising time gaps to improve the EMS response if activated (Wania, et al., 2021).

Moreover, synergies among the academic, institutional and economic spheres would ensure that Copernicus services are appropriated for risk managers and would support the development of new products derived from satellite image processing (Copernicus and others).

Therefore, it would be desirable to have the promotion of a policy of each National Authority/Response Agency developing inhouse EO skills, as well as the consideration of EO at development stage of national/local strategic action plans, emergency plans etc.

4.5 Conclusions

Copernicus is the European Union Earth Observation and Monitoring Programme, which helps the users to understand and manage the environment through satellite and in situ observations. Free and open data are provided to the users (citizens, public authorities and international organizations). The programme covers several information services such as atmosphere, marine, land, climate change, security and emergency.

The role of Copernicus EMS in monitoring geological hazards is to provide adequate tools for emergency response, through the rapid mapping (EMSR) and for prevention and planning, through the risk and recovery mapping (EMSN). The analysis of Copernicus products and services highlights the importance of the necessary cooperation with civil protection, public authorities and private companies when emergency plans are developed, as these key players give suggestions and recommendations to spread awareness of these natural disasters.

Copernicus EMS constitutes a powerful tool for stakeholders in their emergency management. The Copernicus program is continuously improving and developing new products following the stakeholders requirements. Users feedback, monitored by Copernicus EMS, provides useful information to enable the improvement of the service and, in addition, to strengthen the risk managements systems of the participating Atlantic regions. Thus, the stakeholder feedback questionnaires revealed that satellite data are normally used. However, Copernicus use is less extensive and largely restricted to academic or research institutions.

97 Copernicus EMS activations have been requested in the Atlantic Area by the participating countries from 2012 until 2022. Wildfires and floods are the most common hazards causing the activations, and Portugal and Spain are the countries with a higher number of activations. Timeline of activations reveals the growing participation and engagement of the countries in the use of the Copernicus Emergency Service.

Copernicus products offer a variety of possibilities in geohazards analysis, contributing to risk assessment in each of its stages (pre-event, during-event, post-event). Efforts to integrate Copernicus and EO products in the citizens observatories on geohazards implemented in the frame of AGEO project have been made. Thus, Copernicus EMS activations proposed by some of the AGEO partners contributed to enhance and reinforce the knowledge about the geological processes acting as well as to give support to the emergency in case of Rapid Mapping activations.

Projects and activities using Copernicus products and services presented in previous sections are some examples that highlight the multidisciplinary uses, different working scales, and varied purposes where Sentinel products have turned out to be useful. Thus, rockfalls source areas mapping, rockfalls susceptibility maps, slow mass movements detection, mapping extension of flooding, inventory and cartography of elements at risk and/or relevant for geohazard analysis, and landslide and subsidence monitoring, are examples of use cases that have been explored in the timeframe of AGEO project.

Optical and radar data provided by the Copernicus programme have been demonstrated to be very useful, as long as the scale of work is appropriate. Risk monitoring, as well as localization, characterization and modelling are some of the utilities that this type of data allows for in the assessment and management of geohazards. On one hand, Sentinel-2 provides useful optical data to obtain a general characterization of the acting geological process and environmental conditions. Monitoring will be possible whenever the spatial and temporal scale and variability of the process allow it. On the other hand, radar data from Sentinel-1 are useful to detect ground deformations which are slow and large enough to be detected by InSAR techniques. Special attention must be paid to validation of results, resolution limitations, costs and timing of procedures and integrability on management systems.

Regarding the assessment of Copernicus products and procedures for the management and prevention of geohazards, suggestions were collected from partners in Atlantic area. They were related to activation procedures of Emergency Management Service, data access, monitoring, training, spatial resolution, and new products. Easier activation procedures, open training and improvements in data resolution are the most demanded aspects by users in Atlantic region.

Knowledge of Copernicus products and services should be widely disseminated and popularised. This should lead to an increasing number of users and thus to the enrichment and growth of the Copernicus programme.

Acknowledgements The authors acknowledge stakeholders and citizens involved in the actions carried out in each of the Citizens' Observatories oriented to disseminate the use and uptake of Copernicus products and services and other EO data. We thank Copernicus Programme for their willingness to collaborate with the activation of the RR service in experimental mode, contributing positively to rockfall research. This work has been undertaken in the framework of the INTERREG Atlantic Area AGEO Project (EAPA_884/2018—AGEO).

References

Auflič, M. J., Herrera, G., Mateos, R. M., Poyiadji, E., Quental, L., Severine, B., Peternel, T., Podolszki, L., Calcaterra, S., Kociu, A., et al. (2023). Landslide monitoring techniques in the geological surveys of Europe. *Landslides, 2023*, 1–15. https://doi.org/10.1007/s10346-022-02007-1

Barra, A., Solari, L., Béjar-Pizarro, M., Monserrat, O., Bianchini, S., Herrera, G., Crosetto, M., Sarro, R., González-Alonso, E., Mateos, R., Ligüerzana, S., López, C., & Moretti, S. (2017). A methodology to detect and update active deformation areas based on sentinel-1 SAR images. *Remote Sensing, 9*(10), 1002. https://doi.org/10.3390/rs9101002

Bonnier, M. (2021). Le potentiel du service Europeen Copernicus pour evaluer les enjeux exposes aux risques naturels. Universite de Bretagne Occidentale, p. 68

Caballero, I., Ruiz, J., & Navarro, G. (2019). Sentinel-2 satellites provide near-real time evaluation of catastrophic floods in the west mediterranean. *Water, 11*(12), 2499. https://doi.org/10.3390/w11122499

CEMS (Copernicus Emergency Management Service). (2021). Final report EMSN-093: Rockfall risk analysis in Valle Gran Rey, La Gomera (Spain). Copernicus Emergency Management Service Risk & Recovery Mapping - Non-standardized products (FLEX) Lot 1, p. 129

Dorati, C., Kucera, J., Marí i Rivero, I., & Wania, A. (2018). Product user manual of Copernicus EMS rapid mapping, JRC Technical Report JRC111889. European Union, p. 41

EGMS. (2017). European Ground Motion Service (EU-GMS). A proposed Copernicus service element. White paper, p. 24. Retrieved 20 February, 2021, from https://land.copernicus.eu/user-corner/technical-library/egms-white-paper.

Hanssen, R. F. (2001). Radar Interferometry: Data Interpretation and Error Analysis. Kluwer Academic Publishers.

Joubert-Boitat, I., Wania, A., & Dalmasso, S. (2020). Manual for CEMS-Rapid Mapping Products, EUR 30370 EN, Publications Office of the European Union, Ispra, 2020, ISBN 978–92–76–21683–4, https://doi.org/10.2760/29876, JRC121741.

Jutz, S., & Milagro-Pérez, M. P. (2018). 1.06 - Copernicus Program. In: Liang, S (ed.). Comprehensive Remote Sensing, Elsevier, pp. 150–191. https://doi.org/10.1016/B978-0-12-409548-9.10317-3.

European Union. (2022). Copernicus Land Monitoring Service, European Environment Agency European Ground Motion Service - Copernicus Land Monitoring Service. Factsheet. Retrieved 2 March, 2023, from https://land.copernicus.eu/pan-european/european-ground-motion-service/EGMSFactsheetMay2022Final.pdf.

Geological Survey of Northern Ireland—GSNI. (2021). Geological Survey of Northern Ireland Annual Report 2020–2021. GSNI - Department for Economy, p. 37

Hooper, A. (2008). A multi-temporal InSAR method incorporating both persistent scatterer and small baseline approaches. *Geophysical Research Letters, 35*,. https://doi.org/10.1029/2008GL034654

Hussain, M. A., Chen, Z., Shoaib, M., Shah, S. U., Khan, J., & Ying, Z. (2022). Sentinel-1A for monitoring land subsidence of coastal city of Pakistan using Persistent Scatterers In-SAR technique. *Scientific Reports, 12*(1), 5294. https://doi.org/10.1038/s41598-022-09359-7

Hrysiewicz, A., Wang, X., & Holohan, E.P. (2023). EZ-InSAR: An easy-to-use open-source toolbox for mapping ground surface deformation using satellite interferometric synthetic aperture radar. Earth Sci Inform (2023). https://doi.org/10.1007/s12145-023-00973-1.

Wania, A., Joubert-Boitat, I., Dottori, F., Kalas, M., & Salamon, P. (2021). Increasing timeliness of satellite-based flood mapping using early warning systems in the Copernicus emergency management service. *Remote Sensing, 13*(11), 2114.

MOP. (2018). Copernicus Emergency Management Service—Mapping. Manual of Operational Procedures Guidelines for EC Services, Service Providers and Authorised Users. European Commission DG JRC, DG GROW, DG ECHO Version 1.3, p. 90. Retrieved 2 March, 2023, from https://emergency.copernicus.eu/mapping/sites/default/files/files/EMS_Mapping_Manual_of_Procedures_v1_3_final.pdf.

Nduji, N. N., Madu, C. N., Okafor, C. C., & Ezeoha, M. U. A. (2023). Geo-hazard risk assessment technique for analyzing impacts of surface subsidence within Onyeama mine. *South East Nigeria. Land, 12*, 575. https://doi.org/10.3390/land12030575

Papathoma-Köhle, M., Neuhäuser, B., Ratzinger, K., Wenzel, H., & Dominey-Howes, D. (2007). Elements at risk as a framework for assessing the vulnerability of communities to landslides. *Natural Hazards and Earth System Sciences, 7*(6), 765–779. https://doi.org/10.5194/nhess-7-765-2007

QSG (2015) Copernicus emergency management service quick start guide, p. 2. Retrieved 17 February, 2022, from ttps://emergency.copernicus.eu/mapping/sites/default/files/files/CopernicusEMS-Quick_Start_Guide.pdf.

Quental, L., & Machado, S. (2020). AGEO Deliverable D4.1: Status report on the use of Copernicus EMS in the participating regions, 1,4401.

Reyes-Carmona, C., Barra, A., Galve, J., Monserrat, O., Pérez-Peña, J., Mateos, R., Notti, D., Ruano, P., Millares, A., López-Vinielles, J., & Azañón, J. (2020). Sentinel-1 DInSAR for monitoring active landslides in critical infrastructures: The case of the rules reservoir (Southern Spain). *Remote Sensing, 12*(5), 809. https://doi.org/10.3390/rs12050809

Rodrigues, D., & Ayala, F. (1994). Landslides in the Machico area on Madeira Island. 7th Congress of the International Association of Engineering Geology, 1495–1500. Lisbon.

Scoular, J., Ghail, R., Mason, P., & Lawrence, J. (2019). The use of InSAR for identifying potential geotechnical hazards and establishing a baseline for urban tunnelling in London. Geophysical Research Abstracts, 21, EGU2019–9406–3. EGU General Assembly.

Geohazards Revealed by Multi-temporal InSAR Datasets at AGEO Citizen Science Observatory Pilots in Lisbon, Madeira and Northern Ireland

5

Alexis Hrysiewicz, Xiaowen Wang, Eoghan P. Holohan, Ana Paula Falcão, Domingos Rodrigues, Kieran Parker, Eleni Mangina and Rui Carrilho Gomes

Abstract

The European Union's Atlantic region is susceptible to a wide range of geohazards. Measuring ground surface motion is critical for detecting and monitoring many of these geohazards. We present observations of potential ground motion as defined by Interferometry of Synthetic Aperture Radar (InSAR) over areas encompassing three Citizen Observatory (CO) pilot sites of the AGEO project—the Lisbon Region, eastern Madeira Island, and Northern Ireland. These data are based on multi-temporal processing of C-Band Synthetic Aperture Radar (SAR) images that were acquired between 2014 and 2022 by Copernicus Sentinel-1 satellites. In the Lisbon region, our measurements reveal several subsidence hotspots with a peak velocity of about –5 mm/yr. An area of apparent ground uplift occurs at the northern part of Setúbal Peninsula. The observed motions are likely induced by groundwater over-pumping, urbanization on poorly consolidated sediments, or local tectonic fault motions. On Madeira Island, ground motions of up to 7 mm/yr are inferred on urbanized slopes in the Machico valley and the Canical areas; these are linked with slow landslides. In Northern Ireland, the InSAR data reveal potential subsidence of peatlands around the Causeway coast; any link between InSAR-detected motions and landslides there is ambiguous. Elsewhere in Northern Ireland,

A. Hrysiewicz · X. Wang · E. P. Holohan (✉)
UCD School of Earth Sciences, University College Dublin, Dublin, Ireland
e-mail: eoghan.holohan@ucd.ie

A. Hrysiewicz · E. P. Holohan
Science Foundation Ireland Centre for Research in Applied Geosciences (iCRAG), Dublin, Ireland

X. Wang
Faculty of Geosciences and Engineering, Southwest Jiaotong University, Chengdu, China

A. P. Falcão · R. C. Gomes
Instituto Superior Técnico, University of Lisbon, Lisbon, Portugal

D. Rodrigues
Faculty of Exact Sciences and Engineering, University of Madeira, Madeira, Portugal

CERIS, Institute of Earth Sciences, Lisbon, Portugal

K. Parker
Geological Survey of Northern Ireland, British Geological Survey, Belfast, UK

E. Mangina
UCD School of Computer Science, University College Dublin, Dublin, Ireland

R. C. Gomes et al. (eds.), *Citizens' Observatories on Geohazards*, Geoenvironmental Disaster Reduction,
https://doi.org/10.1007/978-3-031-53371-6_5

the data indicate subsidence of peatland west of Belfast city and of reclaimed land within the Belfast harbour area. These observations are attributed to peatland drainage dynamics and to urbanization-induced consolidation of estuarine sediments ('sleech'), respectively. Good agreement is shown between our InSAR results and those of the Copernicus program's new European Ground Motion Service (EGMS) for all three study areas. Both the EGMS and bespoke local InSAR datasets can thus provide a basis for focusing the ground-based efforts of Citizen Observatories for geohazards on areas of detected motion. InSAR-derived ground motion data can also contextualize Citizen Science-derived datasets.

5.1 Introduction

A main goal of the AGEO project (2019–2023) was to test and improve the uptake of Copernicus program products and services for monitoring of geohazards in the European Union (EU) Atlantic area. Using Sentinel-1 synthetic aperture radar (SAR) data to detect ground motion in Citizens Observatories was considered important to: (i) detect geohazards, principally in the form of active landslides (e.g. slope creep or slumps) and land subsidence; (ii) provide guidance on where to focus citizen science initiatives and (iii) to provide context for data collected by citizen scientists. Therefore, in addition to developing a more user-friendly SAR interferometry (InSAR) toolbox termed EZ-InSAR (Hrysiewicz et al., 2023a), as is outlined in Chap. 6 of this book, InSAR ground motion datasets were generated for several of AGEO citizens observatories.

Three Citizen Observatory pilot areas were selected (Fig. 5.1): the Lisbon region (Portugal), Madeira Island (Portugal), and Northern Ireland (UK). Here we present maps and selected time-series of the ground surface motions derived from Sentinel-1 SAR images acquired over these three pilot regions in the years 2014 and 2021 inclusive.

5.2 Method

Synthetic Aperture Radar (SAR) is an active remote sensing technique that uses a moving radar transmitter to emit microwaves, and then create images of the Earth's surface by using the returned (back-scattered) signal received at the moving antenna (Rosen et al., 2000). The back-scattered radar signal contains two components: (i) amplitude information representing the strength of the radar response; and (ii) phase information representing the waveform position of the radar response as a fraction of one complete sine wave cycle (a single SAR wavelength). The phase of the SAR image is determined primarily by the distance between the satellite antenna and the ground targets.

The principle of using repeat-pass SAR interferometry (InSAR) to detect ground motion is illustrated schematically in Fig. 5.2. Here two radar antennae are assumed to fly perpendicularly to the plane of view at different points in time and with a separation (baseline) of their positions in space. At each point in time, a SAR image of the ground area is acquired. The ground surface has moved in the time between the two satellite passes, however, such that the sensor-to-ground range is different from one SAR image acquisition time to the next. The range difference can be represented by the phase difference at each resolution cell (pixel) in the SAR images. A map (ground image) of these phase differences, termed an interferogram, can thus be generated to record the apparent range difference for each corresponding pixel of the SAR images. This information can then be used to define potential ground motions across the area of interest.

In addition to information on ground displacement, the phase difference measured by InSAR contains contributions from the Earth's topography (including a so-called 'flat Earth' contribution), from the satellites' spatial separation (baseline), from radar wave propagation delay due to atmospheric moisture, and from random noise. The contributions from the geometry of the Earth's surface can be removed by estimating them based on a reference ellipsoid for the Earth and a suitable digital elevation model (DEM) (Li

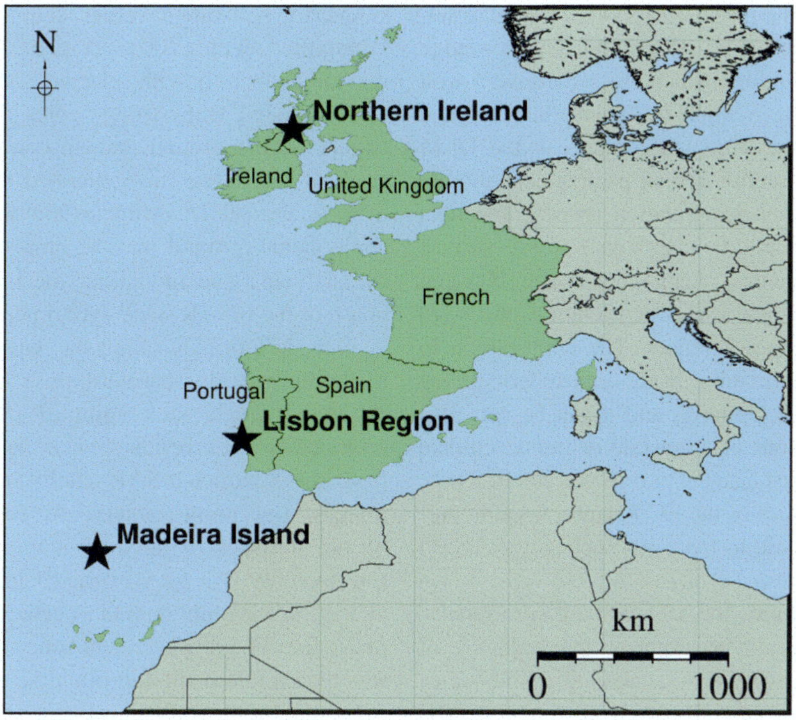

Fig. 5.1 Locations of three observatory pilots of the AGEO project. The black stars also roughly show the areas for which InSAR data were generated. Highlighted in bright green are the countries of the European Union Atlantic area. Coastline data are from Global Self-consistent Hierarchical High-resolution Geography (Wessel & Smith, 1996)

Fig. 5.2 Illustration of the concept of InSAR for measuring ground motion. Movement on a geological fault creates a displacement of the ground surface between the acquisition of two SAR images. D is the total ground displacement; V is its vertical component and H is its horizontal component. L is the component of ground motion as seen in the satellite's line of sight (LOS). L is measured as the difference in backscatter phase between corresponding pixels in each SAR image (see main text for further explanation)

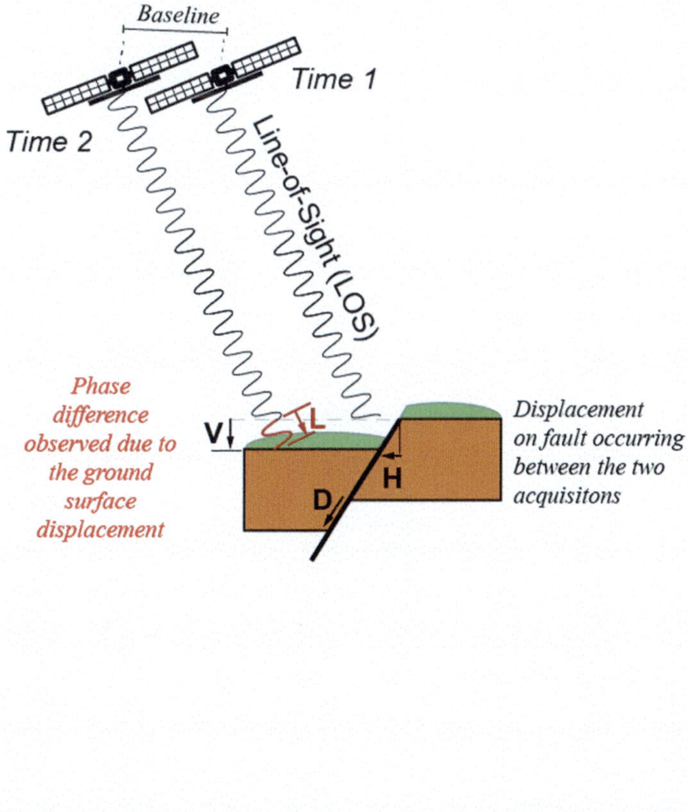

et al., 2022). The phase contribution from the satellite separation can be calculated from knowledge of the satellite orbits, which are usually very well constrained for Sentinel-1. Thereafter, residual errors may arise from inaccuracies in the DEM data or in the satellite orbital parameters. Another potential issue is decorrelation (poor coherence) of the backscattered radar signal if the surface reflective properties change too much between the SAR acquisitions. Decorrelation leads to unreliable phase differences. The contribution to the phase information from atmospheric delay is problematic to remove, and it can be very significant in regions of temperate or humid climate with ground displacements that are small. These are characteristics of the EU Atlantic Area, in particular areas close to the coast such as the AGEO project pilot sites.

Multi-temporal InSAR (MT-InSAR) analysis techniques aim to improve the accuracy of ground displacements by analyzing a time-series of many SAR images (Ferretti et al., 2001; Berardino et al., 2002; Crosetto et al., 2016; Even & Schulz, 2018). A major advantage of MT-InSAR is that atmospheric phase contributions, which are unstable in time, can be filtered out by considering them as noise in time. Here we used the Interferometric Point Target Analysis (IPTA) method provided by the commercial GAMMA software (Werner et al., 2000, 2003; Wegmüller et al., 2004) (Fig. 5.3). The IPTA method deals only with selected SAR image pixels associated with strong 'point-like' reflective targets on the ground that appear throughout a stack of interferograms generated from multiple SAR images. IPTA estimates the different phase components (ground motion, atmosphere, noise) by analyzing and modelling the interferometric phase of the InSAR stack over time.

Our MT-InSAR analysis with the IPTA method begins by assembling a stack of co-registered single look complex (SLC) images provided by European Space Agency (ESA) from the Sentinel-1 SAR satellites. From these images, the point targets are selected based on the statistics of intensity stability and spectral diversity. We then extracted interferometric phases on the point targets to generate a stack of phase list. The displacement time series analysis was then implemented through the steps of point triangulation, model regression, residual phase calculation, and spatiotemporal filtering for isolating the different phase components (Sadeghi et al., 2021; Werner et al., 2003; Wegmüller et al., 2004). An important aspect of the IPTA is the possibility of estimating the parameters iteratively for improving the accuracy of estimation of the different phase contributions.

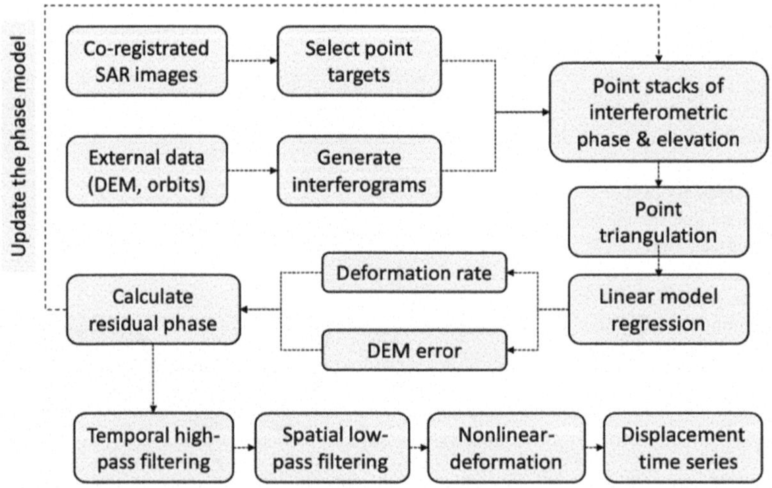

Fig. 5.3 IPTA SAR data processing flow chart of the Persistent-Scatterer IPTA processing. For a multi-reference network with multi-looked phases, deformation-rate estimation and nonlinear deformation estimation were not performed

In this study, we used two approaches to MT-InSAR—the Persistent-Scatterer (PSI) IPTA and the multi-reference IPTA. These differ in the use of reference images to which other SAR images are compared to calculate phase differences and hence ground displacements. The PSI-IPTA includes a single reference image at the center of the time series. The multi-reference IPTA includes several reference images at different positions within the time series and it also includes multi-look phases (i.e. phase difference information calculated for each pixel by averaging across several surrounding pixels). The multi-reference stack has a certain redundancy in the observations in that each time interval is covered by several InSAR observations. This strategy has an advantage of maintaining phase coherence at the areas with light vegetation, for instance at the AGEO pilots of Madeira Island and Northern Ireland (i.e., outside of Belfast city). The interferometric phase components were determined by using linear regression as defined by the IPTA method. A linear displacement rate is assumed in the Persistent-Scatterer IPTA approach; there is no such assumption in the multi-reference IPTA approach.

An important caveat of InSAR is that the phase-derived measurement of ground displacement is along the Line of Sight (LOS) direction of the SAR sensor (Fig. 5.2). This means that a measurable range change may be caused by either vertical or horizontal motion or a combination of these. One can, however, combine InSAR displacement measurements from different lines of sight to estimate the vertical and horizontal displacement components of ground motion (Wright et al., 2004).

We calculated the ground motion in the pilot areas of the AGEO project by using Sentinel-1 SAR images taken from both ascending and descending satellite viewing directions. We combined the differing LOS measurements to find horizontal (East–West) and vertical (Up-Down) components of ground motion (neglecting the N-S component). We processed the SAR images acquired between 2014 and 2022, with imaging mode of IW and polarization of VV. A total of 617, 515, and 672 Sentinel-1 SAR images were used in the MT-InSAR analysis for the Lisbon, Madeira Island, and Northern Ireland areas, respectively.

5.3 Results

5.3.1 Lisbon

Figure 5.4 shows the up-down (i.e. vertical) and east–west (i.e. horizontal) components of the ground motion velocity in the Lisbon region. The uncertainties for the vertical and east–west velocity components were estimated to be about 1 mm/yr. Broad subsidence at a rate of about 3 mm/yr can be observed over the Lisbon city and the east part of the Setúbal Peninsula. Subsidence is also evident at the margins along the Tagus River, on the eastern Setubal Peninsula, and in the vicinity of the Arrábida Mountain. A zone showing a small uplift at a rate of about 1 mm/yr—i.e. close to uncertainty level—is shown for the north part of Setúbal Península (Caparica). The magnitude of east–west velocity is faint compared to the vertical velocity and is close to the uncertainty level.

Figure 5.5 shows the vertical displacement rates of four selected zones where ground subsidence signals are well-defined and well above the uncertainty levels (see the black rectangles Z1 ~ Z4 in Fig. 5.4a). Land-use in these four regions includes residential, transport infrastructure, farmland, and construction. The maximum subsidence rate is ~ 5 mm/yr.

Figure 5.6 shows vertical displacement time series averaged within a 100 m radius of each of four selected points P1- P4 (for locations, see Fig. 5.5). Each displacement time series shows a subsidence trend between 2014 and 2021. At point P2, which represents an area of logistical buildings constructed incrementally between 2002–2011 in low ground directly adjacent the River Tagus, the ground motion also exhibits periodic seasonal oscillations superimposed on the overall steady subsidence trend. At P3, which represents a set of industrial buildings constructed around 2009–2010, the ground

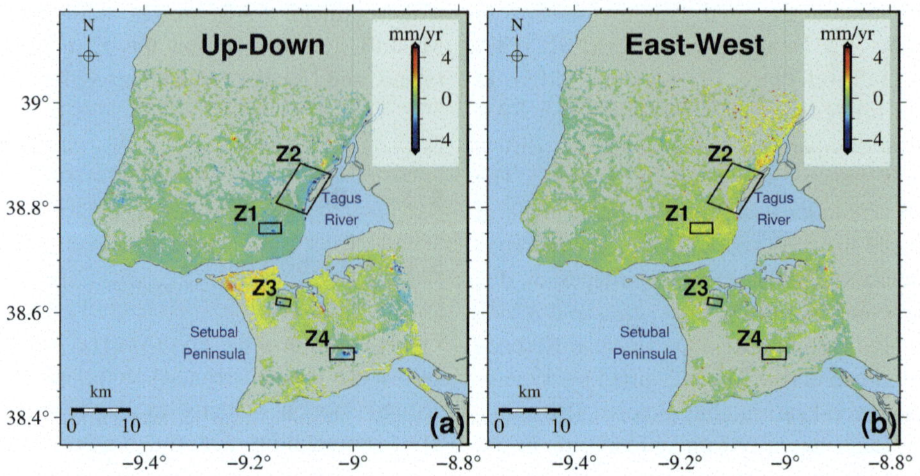

Fig. 5.4 InSAR-derived ground motions in the Lisbon region. **a** InSAR vertical velocities, positive for uplift, negative for subsidence. **b** East–West velocities, positive for eastward, negative for westward. These results were generated by using the PSI IPTA approach and span the years 2015–2020 inclusive. Coastline data is from Global Self-consistent Hierarchical High-resolution Geography (Wessel & Smith, 1996)

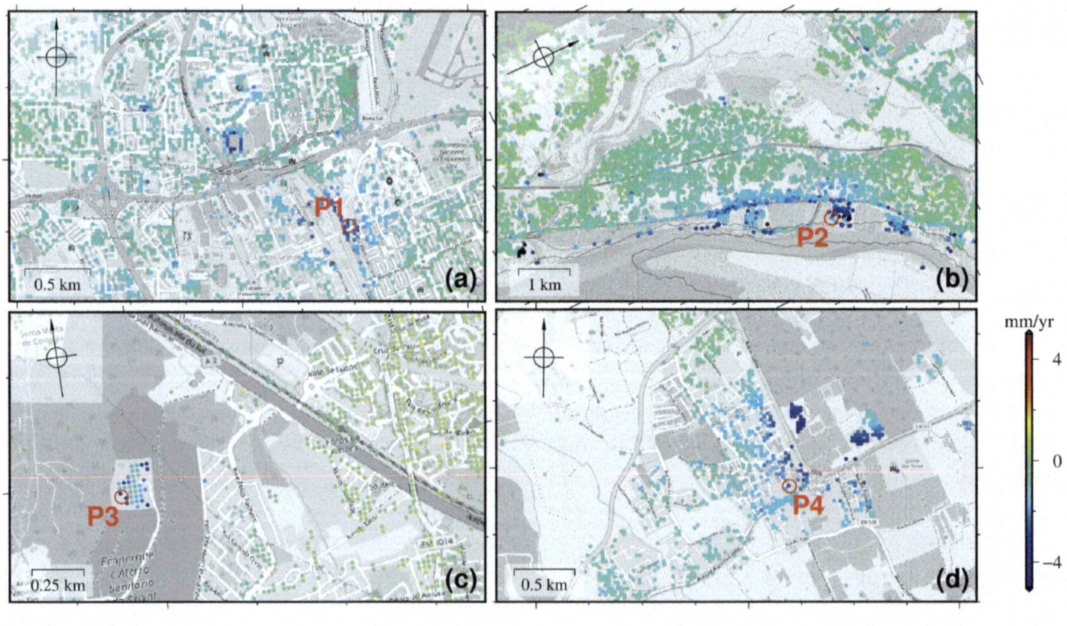

Fig. 5.5 Close-up maps of InSAR-derived vertical ground motion rates in parts of the Lisbon region. For locations, see the zones Z1-Z4 in Fig. 5.4. **a** Zone 1 corresponds to Laranjeiras, a neighborhood in the center of Lisbon. **b** Zone 2 corresponds to a factory area in Santa Iria da Azoia (Loures municipality), near the Tagus River. **c** Zone 3 corresponds to an urbanized area in Amora parish (Seixal municipality). **d** Zone 4 corresponds to an urbanized area in Azeitão parish (Setúbal municipality). Base map: ©OpenStreetMap contributors

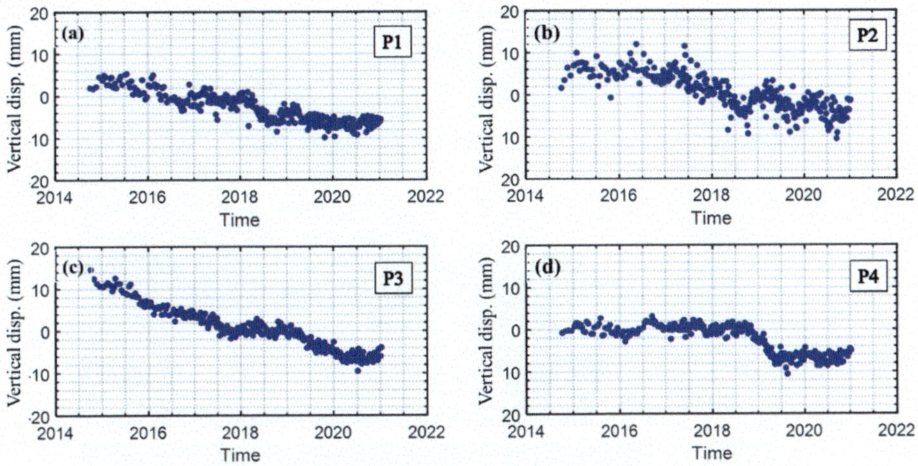

Fig. 5.6 Time series of ground motion derived from InSAR at selected points in the Lisbon region. For the location of each point, see Figs. 5.4a and 5.5.

underwent near-linear subsidence that apparently paused in late 2017 and resumed early 2019. At P4, which represents an area of mainly residential and commercial buildings built prior to 2002, the ground has in contrast been apparently quite stable for most of the observation period but underwent a subsidence of 7–8 mm between July 2018 and May 2019. The different temporal deformation patterns at the four points indicate that ground motions in the Lisbon region are probably influenced by site-specific factors.

5.3.2 Madeira

Figure 5.7 shows the LOS displacement rates obtained for the eastern edge of Madeira Island. The InSAR data points are mainly concentrated in the Machico valley and the areas of Caniçal and Ponta de S. Lourenço in the east part of the island. Few points occur in the mountainous and densely vegetated regions to the north and southeast of the Machico valley. LOS subsidence occurs in several areas of the Ponta de S.Lourenço peninsula, especially near the "Prainha do Caniçal" (see the dashed ellipse in Fig. 5.7). The peninsula is delimited to the north by steep cliffs of between 70–130 m in height and the areas of apparent motion are underlain by aeolian deposits and easily eroded volcaniclastic deposits that produce local 'badland' topography.

Assuming that the ascending and descending InSAR measurements show LOS motion at local slopes, we calculated the down-slope horizontal displacement based on relationships between the SAR look direction and local topographic parameters (i.e., slope angle and slope aspect). Figure 5.8 shows the inferred horizontal surface displacement component (arrows) and the vertical component (colored points) along the slope. In the Machico valley, the slopes on the northeast valley side appear to show greater and more consistent horizontal motion than those on the west side. The maximum slope-parallel movement velocity reached about 7 mm/yr.

Figure 5.9 shows the ascending and descending LOS displacement time series at a selected point (P1) on a slope in the Machico valley that is moving toward the southwest. The surface slope angle around point P1 is 13°. The positive displacement trend given in the ascending pass data and negative in displacement trend given in the ascending pass data indicates a large westward component of motion at this point. Both the ascending and descending LOS displacement time series at P1 show strong seasonal

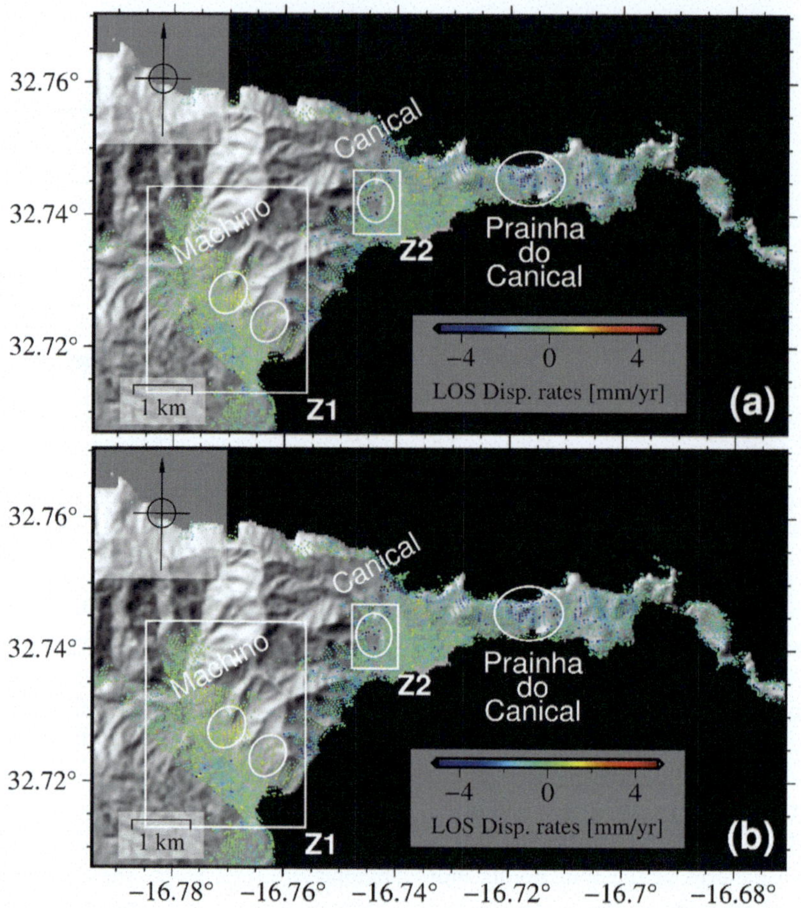

Fig. 5.7 InSAR-derived rates of ground motion in the northeast of Madeira Island. **a** Line of sight (LOS) motion rates measured from the ascending satellite path and **b** LOS motion rates measured from the descending satellite path. Negative means movement away from satellite. These data were generated by multi-reference IPTA processing and span the years 2015–2021 inclusive. Basemap: Copernicus DEM

oscillations. These seasonal signals are plausibly related to periods of high rainfall and thus elevated soil moisture and/or groundwater levels—a major factor in landslides.

5.3.3 Belfast

The AGEO Citizens Observatory site in Northern Ireland is the Causeway Coast. Our InSAR data covered this area, as well as Garron Peninsula and the Belfast area. The InSAR data around the Causeway coast revealed mainly subsidence related to peatland lying inland of the coast. The main hazard in the area is landslides, especially rockfalls. However, the InSAR data could not be clearly related to slope instability, which we suggest is a consequence of the steep cliffs that make up the Causeway coast being predominantly north facing. Northward slope motions are difficult to detect from the east–west looking Sentinel-1 satellites.

Figure 5.10 shows the inferred ground motion velocity derived for the Belfast area. Ground motion is dominantly vertical. The rural Divis and Black Mountain area west of Belfast exhibits widespread subsidence, with maximum subsidence rate up to –8 mm/yr. The area is characterised by superficial deposits of glacial till and peat. Near the city centre, we can

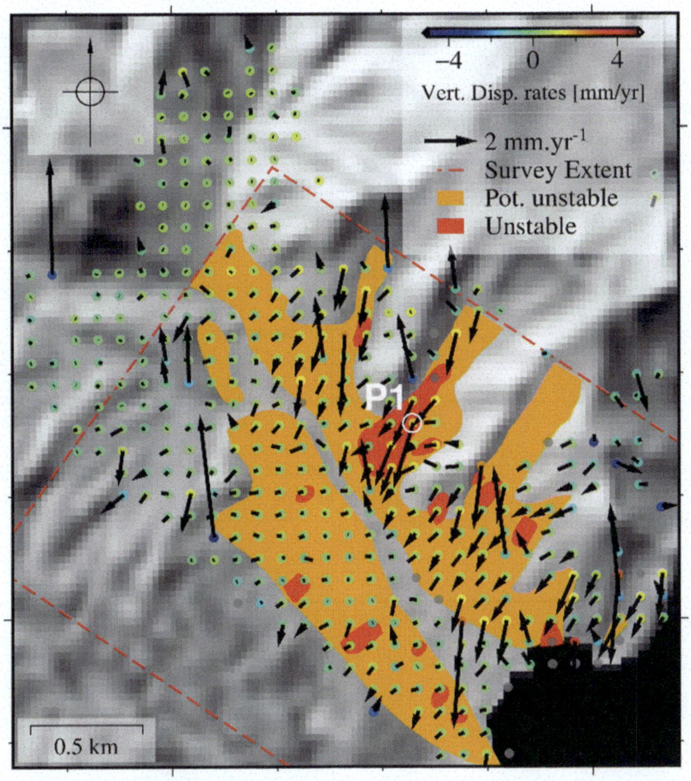

Fig. 5.8 The InSAR-derived horizontal and vertical displacement rates in the lower Machico valley (Z1 on Fig. 5.7). Right is a map of landslide susceptibility derived from past field observation (adapted from Rodrigues & Ayala-Carcedo, 1994). Several slopes on the valley sides were identified as unstable (red colour) with field evidence of slope creep on weak soils, mainly seen as damage to buildings and infrastructure. Based on mapping of underlying deposits, other areas in the valley were identified as potentially unstable (orange color). Base map: Copernicus DEM

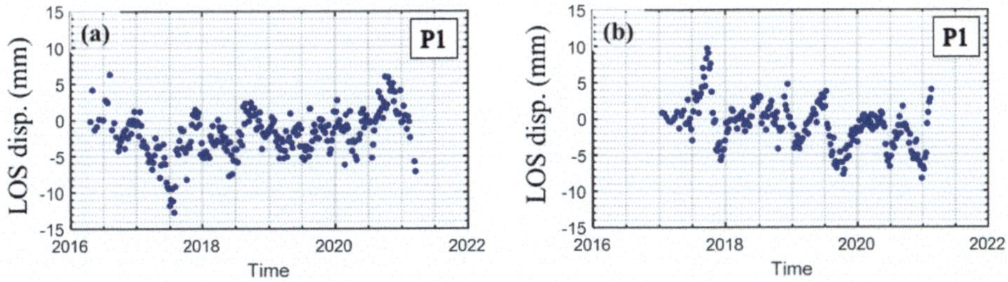

Fig. 5.9 Time series of ground motion for an active landslide in the Machico valley, Maderia. Components of ground motion are in the satellite line of sight (LOS) from **a** ascending path and **b** descending path. See Fig. 5.8 for location of point P1

observe ground uplift surrounding the Royal Victoria Hospital peaking at a rate of 4 mm/yr, while in contrast the adjacent Westlink road shows subtle ground subsidence. In the Belfast harbour area, we mainly observe subsidence with a peak velocity of about –6 mm/yr. Displacement occurs the commercial/industrial urban areas near the Dargan crescent road and the several areas of reclaimed land near the estuary, many of which were underwent construction or other modification between 2001 and 2016.

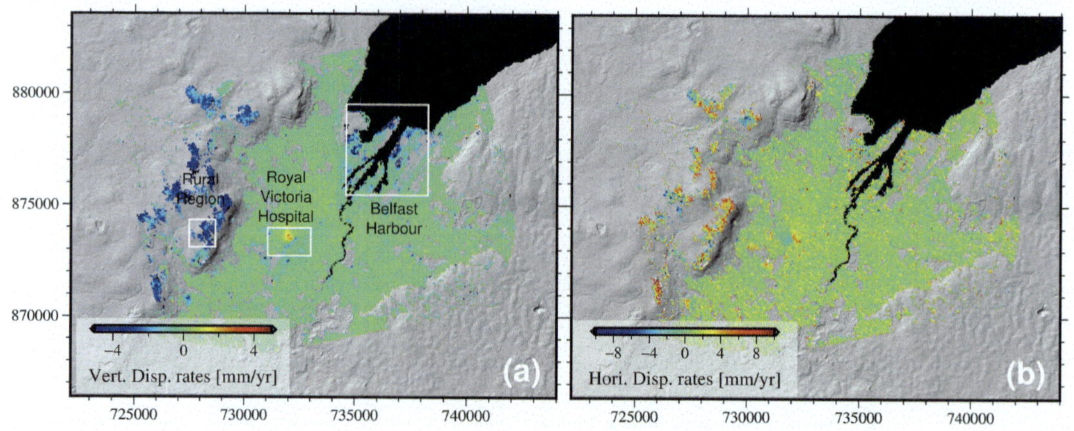

Fig. 5.10 InSAR-derived maps of ground motion rate inferred for the Belfast area. **a** vertical and **b** east–west components. These data were generated by multi-reference IPTA processing and span the years 2015 to mid-2021. Base map: Copernicus DEM

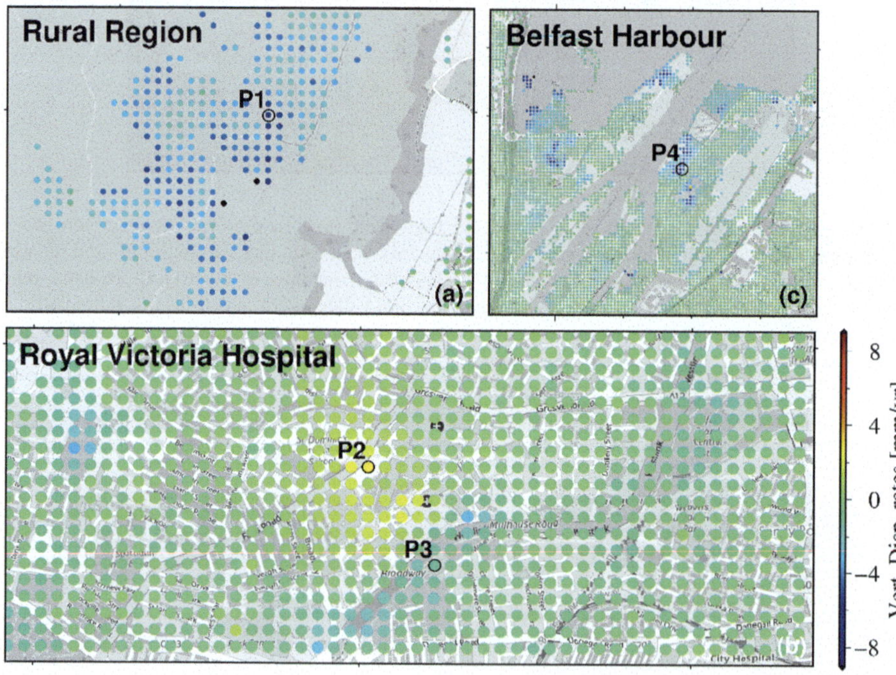

Fig. 5.11 Close-up maps of vertical ground motion rate in three parts of Belfast. See Fig. 5.10 for locations of each area. Base map: ©OpenStreetMap contributors

Figure 5.11 shows close-ups of each of the areas of ground motion, and Fig. 5.12 shows representative time-series at selected points. Point P1, P3 and P4 show near-linear subsidence with time, with some superimposed seasonal oscillations. P1 shows a linear displacement–time relationship consistent with peatland drainage and/or peat compaction. Point P2 at the Royal Victoria Hospital shows an uplift trend overall, although it was relatively stable between 2015 and 2017 and appear to have begun to rise near-linearly after 2018, coinciding

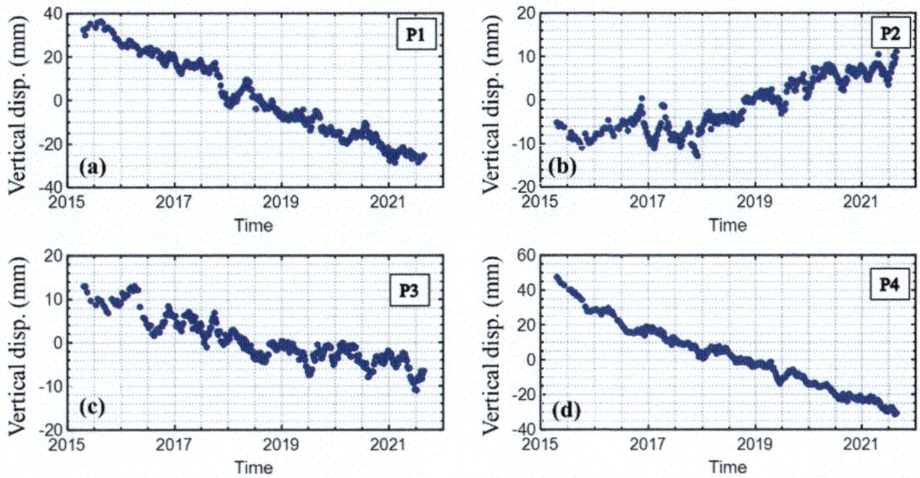

Fig. 5.12 Time-series of InSAR-derived vertical displacement in Belfast. The locations of the selected points P1-P4 are shown in Fig. 5.11

with a phase of construction nearby on the hospital grounds between 2018–2020. Point P3 at the Westlink A12 is where the Forth River and Backstaff River converge—both rivers are concealed below ground and have, in the past, resulted in major flooding on the A12 (in 2010). It is thus possible that the seasonal fluctuations in the time series of ground motion are influenced by water levels. P4 displays near-linear vertical motion in an area of industrial use on reclaimed ground that was formerly occupied by estuarine sediments.

5.4 Discussion

5.4.1 Validation of Results

Ground motion in the Lisbon region has been investigated intensively before, so we selected this region as the main validation site for our MT-InSAR results. We compared our results with: (i) the ground deformation products provided by the first release of the European Ground Motion Service (EGMS), which were also produced from Sentinel-1 SAR images (Costantini et al., 2021); (ii) previous InSAR measurements produced from ERS and ENVISAT SAR images acquired between 1995 and 2010 (Catalão et al., 2015; Heleno et al., 2011); and (iii) contemporaneous in-situ ground motion data from GNSS observations (Fonseca et al., 2021).

The distribution and rates of subsidence in Lisbon city derived from our IPTA processing and from the EGMS are very similar (Fig. 5.13), although our data cover an additional 18-month time frame. The EGMS results show that the subsidence area at Laranjeiras has a maximum velocity of –5.4 mm/year, which differs from our MT-InSAR measurements by only 0.3 mm/yr. The displacement time series of EGMS there also exhibits a linear subsidence pattern (EGMS, 2022) like our results (see Figs. 5.6 and 5.13).

Compared to older (1995–2010) InSAR data for Lisbon, the ground motion measured from Sentinel-1 data shows a similar subsidence rate at the Laranjeiras area (Heleno et al. 2011; Catalão et al. 2015), but the extent of the affected area has slightly expanded. Earlier InSAR investigations (Heleno et al., 2011) reported ground motion along the west edge of the Tagus River, as we see also. Finaly, Fonseca et al., (2021) compared the ground displacement velocities at 22 continuous GNSS stations in the Lisbon region to our IPTA-derived InSAR data. Cross-plots between the InSAR and GNSS measurements, available in the supplementary

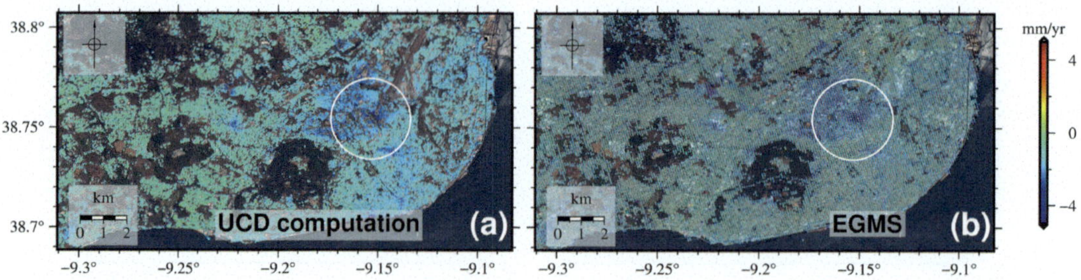

Fig. 5.13 Comparison of IPTA InSAR results with those of the Copernicus EGMS. Vertical ground motion rates in Lisbon city from **a** our IPTA processing and **b** the EGMS. The Laranjeiras subsidence area is indicated by the white circle. The EGMS motion rates span January 11th, 2016 to June 06th, 2020 (EGMS, 2022). Base map: Sentinel-2 imagery (May 2024)

material to that paper, yield correlation coefficients of 0.11 and 0.82 for the vertical and east–west velocities, respectively.

It can be seen in the EGMS online viewer (https://egms.land.copernicus.eu/) that the results of the EGMS are also strikingly like ours in both Madeira and Belfast. The distributions of ground subsidence over the Belfast harbour region from Sentinel-1 InSAR data shown here are consistent also with those derived by Cigna et al., (2017), who found land subsidence of about –3 mm/yr in the harbour region based on ERS-1/2 SAR data acquired between 1992 and 2000.

5.4.2 Geohazard Implications for the Lisbon Region

Tectonic activity may contribute to ground motion in the Lisbon region (Fonseca et al., 2021). The southern sector of the Lusitanian basin, SW Portugal, has been the locus of relevant seismicity in historical time (Vilanova & Fonseca, 2004). The list of known destructive earthquakes affecting the region and the adjacent continental shelf ranges in time from 1344 to 1909, with catastrophic occurrences in 1356, 1531, and 1755. Based on displacement observations at 22 GNSS sites and our InSAR data, Fonseca et al. (2021) argued that a likely explanation for the complex pattern of deformation near Lisbon concerns the connection of the Lower Tagus Valley fault with the offshore fault system.

Groundwater pumping taking place in a coarse-grained soil layer can cause ground subsidence, which has been reported extensively in cites that use groundwater for industrial or agriculture use (Bell et al., 2008). In-situ piezometer measurements performed in several boreholes throughout the Lisbon region have revealed a groundwater table lowering since the 1990s (Heleno et al., 2011). The lowering of the water table has been described as the main reason for the ground deformation observed at the Laranjeiras area between 1995 and 2008 (Catalão et al., 2015). Also, the Laranjeiras area is close to the two subway lines, excavations of which, between 1988 and 2002, may contribute to the ground subsidence as well.

We therefore suggest that the observed local ground subsidence signals in Lisbon are likely associated with anthropogenic activities, such as groundwater pumping, subway construction, and urbanization, while the tectonic activities may also partially contribute to the observed deformation.

5.4.3 Geohazard Implications for Northeast Madeira Island

Madeira Island is the emergent part of the much larger Miocene-Holocene age Madeira-Desertas shield volcano, which rises from the ocean floor at about –3500 m to 1861 m above sea level. The island is composed of subaerial sequences of volcanic rocks, mainly lava flows of basanites and alkali basalts, pyroclastic deposits, and volcaniclastic gravity sediments (landslides, debris flows, and lahars).

Urban development on Madeira Island is highly concentrated in the southern coastal areas, namely the capital, Funchal, and other municipalities like Machico. In the Machico Valley, the urban center has increasingly occupied former agricultural areas. The consequences of this change in land use are, among others, the disruption of the natural drainage network and the disappearance of the vegetation cover. On talus deposits in the slopes of Machico valley, highly plastic clays that facilitate slope movements are present. These creep movements, which have rates of up to several cm per year, have lead to damage of the houses and infrastructure (Rodrigues & Ayala-Carcedo, 1994). The InSAR-derived displacement signals attained here on the known areas of slope motion in Machico (and Caniçal) demonstrates that InSAR measurements can be successfully used to remotely identify and monitor unstable slopes on Madeira Island (Dai et al., 2020).

5.4.4 Geohazard Implications for the Belfast Area

Belfast city stands on the banks of the River Lagan and around its estuary (Belfast Lough). To the west, the city is bordered by the Divis and Black Mountain area. The subsidence observed on the uplands west of the city likely relate to peatland dynamics, in which the organic soils swell and contract in tandem with the rise and fall of groundwater levels (Fiaschi et al., 2019; Hrysiewicz et al., 2023b, 2024). While peatlands on high ground can be sources of landslides, the horizontal motion components indicated in the InSAR data are small and so cannot be unambiguously interpreted as indicating creep.

Our observations reveal localised ground subsidence signals in the urbanized areas around Belfast Lough, mainly located in the area adjacent the estuary. A mixture of soft clay, silt, peat and mud that lies beneath the city and the estuary has the local name of 'Belfast sleech'. It is well-known locally as a weak substrate that causes geotechnical problems (Gregory & Bell, 1991). We thus suggest that both urbanization and soil consolidation—particularly related to the occurrence of the Belfast 'sleech'—is responsible for the observed ground subsidence in the vicinity of the estuary and on reclaimed land. Our results tally with earlier ERS-1/2-derived InSAR data for the period 1992–2000 (Cigna et al., 2017), which also show subsidence rates in Belfast harbour of 3–4 mm/yr. The Belfast harbour area has thus experienced long-term ground subsidence, which is ongoing, and should therefore be routinely monitored for future risk mitigation in light of forecasts of rising sea levels.

5.5 Summary and Conclusions

We investigated ground motions related to geohazards at three AGEO Citizen Observatory pilot sites in the Lisbon region, on Madeira Island, and in Northern Ireland by using the Sentinel-1 SAR images acquired between 2014 and 2021. The GAMMA IPTA method was used to implement a multitemporal-InSAR analysis.

Ground motions inferred through InSAR in the three study regions have velocities smaller than 10 mm/yr. Our InSAR measurements agree well with past InSAR data from ERS and Envisat satellites, with available GNSS data and with the recently released first results of the Copernicus European Ground Motion Service (EGMS). Our results thus represent a corroboration of the EGMS data in the studied areas.

The ground motions at each pilot area have their own spatiotemporal features and probably have diverse driving factors. These driving factors may link to the human activities, soil or sediment compaction, environmental changes (i.e., variations in precipitation, groundwater water levels, etc.), and, in the case of Lisbon, tectonic motions.

Parts of Lisbon city and the east of the Setúbal Peninsula are undergoing subsidence at rate of about 3–5 mm/yr. We identified a subsidence bowl near Laranjeiras confirming the subsidence identified there since 1990s. A subsidence region was also identified at the east edge of Tagus estuary.

Only the northern part of Setúbal Peninsula shows a slight uplift signal. An active tectonic regime evidenced by GNSS data and characterised by ENE–WSW extension and NNW-SSE shortening may play a role there. Groundwater over pumping and ground settlement due to urbanization are also likely to be significant driving factors of ground motions in the Lisbon area.

In the northeast of Madeira, our InSAR data indicate slope-parallel movement at rates of up to 7 mm/yr on several slopes in the Machico valley that were previously identified from field observations as creeping. This demonstrates that InSAR can be successfully used to remotely identify and monitor unstable slopes on Madeira.

In Northern Ireland, the InSAR data reveal potential subsidence of peatlands around the Causeway coast; any link between InSAR-detected motions and ongoing landslides there is ambiguous. Elsewhere, the data indicate subsidence of peatland west of Belfast city and subsidence of reclaimed land around the Belfast harbour area. These observations are attributed to peatland drainage dynamics and to urbanization-induced consolidation of estuarine sediments (the Belfast 'sleech'), respectively.

This study therefore highlights the utility of Sentinel-1 data—from e.g. the European Ground Motion Service or other sources—to support Citizen Observatories for Geohazards. In particular, InSAR data can enable: (i) detection and monitoring of geohazards, principally active land subsidence and slow landslides (e.g. slope creep); (ii) spatial focusing of citizen science initiatives and (iii) clearer context for data collected by citizen scientists.

Acknowledgements This work was undertaken as part of the Platform for Atlantic Geohazard Risk Management (AGEO) project, funded by the Interreg Atlantic Area Programme through the European Regional Development Fund. (https://ageoatlantic.eu/, Grant No. EAPA_884/2018). AH and EPH also acknowledge support from iCRAG, the Science Foundation Ireland (SFI) Research Centre in Applied Geosciences (iCRAG-Phase 2—Grand Code: 13/RC/2092_P2), and the Higher Education Authority of Ireland's Covid-19 fund as administered by University College Dublin (UCD). XW also received some funding during the manuscript review phase from the National Natural Science Foundation of China (No. 42071410). We acknowledge the European Space Agency (ESA) for the Sentinel-1 and Sentinel-2 data, and NASA for SRTM DEM.

Data Availability All InSAR datasets are available on request from the authors, and those presented in the figures here are available on the Zenodo open data repository (Hrysiewicz, A. (2025). InSAR-derived datasets of ground surface motion for Lisbon (2015–2020), East Madeira (2016–2021) and Belfast (2015–2021) [Data set]. Zenodo. https://doi.org/10.5281/zenodo.14783149). Wessel et al. (2019) and Wessel and Smith (1996) developed and made freely available the GMT6 software and related shoreline database. Copernicus Sentinel 1 data for 2015–2023 were processed by ESA and were retrieved from Alaska Server Facility DAAC. Copernicus Sentinel 2 data processed by ESA were retrieved from Copernicus Data Space mid-2024). EGMS datasets are retrieved from EGMS Explorer web portal. The Open-Street-Map basemap is made available under the Open Database License: http://opendatacommons.org/licenses/odbl/1.0/. Any rights in individual contents of the database are licensed under the Database Contents License: http://opendatacommons.org/licenses/dbcl/1.0/. The Copernicus DEM GLO-30 Public is available on a free basis for the general public under the terms and conditions of the Licence found here, data were retrieved from Open Topography.

References

Bell, J. W., Amelung, F., Ferretti, A., Bianchi, M., & Novali, F. (2008). Permanent scatterer InSAR reveals seasonal and long-term aquifer-system response to groundwater pumping and artificial recharge. *Water Resources Research, 44*.

Berardino, P., Fornaro, G., Lanari, R., & Sansosti, E. (2002). A new algorithm for surface deformation monitoring based on small baseline differential SAR interferograms. *IEEE Transactions on Geoscience and Remote Sensing, 40*, 2375–2383.

Catalão, J., Nico, G., Lollino, P., Conde, V., Lorusso, G., & Silva, C. (2015). Integration of InSAR analysis and numerical modeling for the assessment of ground subsidence in the city of Lisbon, Portugal. *IEEE Journal of Selected Topics in Applied Earth Observations and Remote Sensing, 9*, 1663–1673.

Cigna, F., Banks, V., Donald, A., Donohue, S., Graham, C., Hughes, D., McKinley, J., & Parker, K. (2017). Mapping ground instability in areas of geotechnical infrastructure using satellite InSAR and Small UAV surveying: a case study in Northern Ireland. *Geosciences, 7*.

Costantini, M., Minati, F., Trillo, F., Ferretti, A., Novali, F., Passera, E., Dehls, J. F., Larsen, Y., Marinkovic, P., Eineder, M., Brcic, R., Siegmund, R., Kotzerke, P., Probeck, M., Kenyeres, A., Proietti, S., Solari, L., Andersen, H. S. (2021). European Ground Motion Service (EGMS). IEEE International Geoscience and Remote Sensing Symposium IGARSS, pp. 3293–3296.

Crosetto, M., Monserrat, O., Cuevas-González, M., Devanthéry, N., & Crippa, B. (2016). Persistent Scatterer Interferometry: A review. *ISPRS Journal of Photogrammetry and Remote Sensing, 115*, 78–89.

Dai, K., Li, Z., Xu, Q., Bürgmann, R., Milledge, D. G., Tomás, R., Fan, X., Zhao, C., Liu, X., Peng, J., Zhang, Q., Wang, Z., Qu, T., He, C., Li, D., & Liu, J. (2020). Entering the Era of earth observation-based landslide warning systems: A novel and exciting framework. *IEEE Geoscience and Remote Sensing Magazine, 8*, 136–153.

EGMS. (2022). European Ground Motion Service. Retrieved from https://egms.land.copernicus.eu/.

Even, M., & Schulz, K. (2018). InSAR deformation analysis with distributed scatterers: A review complemented by new advances. *Remote Sensing, 10*, 744.

Ferretti, A., Prati, C., & Rocca, F. (2001). Permanent scatterers in SAR interferometry. *IEEE Transactions on Geoscience and Remote Sensing, 39*, 8–20.

Fiaschi, S., Holohan, E. P., Sheehy, M., & Floris, M. (2019). PS-InSAR Analysis of Sentinel-1 Data for Detecting Ground Motion in Temperate Oceanic Climate Zones: A Case Study in the Republic of Ireland. *Remote Sensing, 11*(3).

Fonseca, J. F. B. D., Palano, M., Falcão, A. P., Hrysiewicz, A., & Fernández, J. (2021). Interseismic strain accumulation near Lisbon (Portugal) from space geodesy. *Geophysical Research Letters, 48*.

Gregory, B., & Bell, A. L. (1991). Geotechnical properties of quaternary deposits in the Belfast area. *Geological Society, London, Engineering Geology Special Publications, 7*, 219–228.

Heleno, S. I. N., Oliveira, L. G. S., Henriques, M. J., Falcão, A. P., Lima, J. N. P., Cooksley, G., Ferretti, A., Fonseca, A. M., Lobo-Ferreira, J. P., & Fonseca, J. F. B. D. (2011). Persistent Scatterers Interferometry detects and measures ground subsidence in Lisbon. *Remote Sensing of Environment, 115*, 2152–2167.

Hrysiewicz, A., Wang, X., & Holohan, E. P. (2023a). EZ-InSAR: An easy-to-use open-source toolbox for mapping ground surface deformation using satellite interferometric synthetic aperture radar. *Earth Science Informatics*. https://doi.org/10.1007/s12145-023-00973-1

Hrysiewicz, A., Holohan, E. P., Donohue, S., & Cushnan, H. (2023b). SAR and InSAR data linked to soil moisture changes on a temperate raised peatland subjected to a wildfire. *Remote Sensing of Environment, 291*, 113516.

Hrysiewicz, A., Williamson, J., Evans, C. D., Jovani-Sancho, A. J., Callaghan, N., Lyons, J., White, J., Kowalska, J., Menichino, N., & Holohan, E. P. (2024). Estimation and validation of InSAR-derived surface displacements at temperate raised peatlands. *Remote Sensing of Environment, 114232*.

Li, H., Zhao, J., Yan, B., Yue, L., & Wang, L. (2022). Global DEMs vary from one to another: an evaluation of newly released Copernicus, NASA and AW3D30 DEM on selected terrains of China using ICESat-2 altimetry data. *International Journal of Digital Earth, 1–20*.

Rodrigues, D., & Ayala-Carcedo, F. (1994). Landslides in the Machico area on Madeira Island. 7th Congress of the Internatinal Association of Engineering Geology". Lisboa, pp. 1495–1500.

Rosen, P. A., Hensley, S., Joughin, I. R., Li, F. K., Madsen, S. N., Rodriguez, E., & Goldstein, R. M. (2000). Synthetic aperture radar interferometry. *Proceedings of the IEEE, 88*, 333–382.

Sadeghi, Z., Wright, T. J., Hooper, A. J., Jordan, C., Novellino, A., Bateson, L., & Biggs, J. (2021). Benchmarking and inter-comparison of Sentinel-1 InSAR velocities and time series. *Remote Sensing of Environment, 256*.

Vilanova, S. P., & Fonseca, J. F. (2004). Seismic hazard impact of the Lower Tagus Valley Fault Zone (SW Iberia). *Journal of Seismology, 8*, 331–345.

Werner, C., Wegmüller, U., Strozzi, T., & Wiesmann, A. (2000). Gamma SAR processor and interferometry software. In Proceedings of the ERS-ENVISAT Symposium, Gothenburg, Sweden.

Werner, C., Wegmuller, U., Strozzi, T., & Wiesmann, A. (2003, July). Interferometric point target analysis for deformation mapping. In *IGARSS 2003. 2003 IEEE International Geoscience and Remote Sensing Symposium. Proceedings (IEEE Cat. No. 03CH37477)* (Vol. 7, pp. 4362–4364). IEEE.

Wegmüller, U., Werner, C., Strozzi, T., & Wiesmann, A. (2004). Multi-temporal interferometric point target analysis. In *Analysis of multi-temporal remote sensing images* (pp. 136–144).

Wessel, P., & Smith, W. H. F. (1996). A global, self-consistent, hierarchical, high-resolution shoreline database. *Journal of Geophysical Research, 101*(B4), 8741–8743.

Wessel, P., Luis, J. F., Uieda, L., Scharroo, R., Wobbe, F., Smith, W. H. F., & Tian, D. (2019). The generic mapping tools version 6. *Geochemistry, Geophysics, Geosystems, 20*(11), 5556–5564. https://doi.org/10.1029/2019gc008515

Wright, T. J., Parsons, B. E., & Lu, Z. (2004). Toward mapping surface deformation in three dimensions using InSAR. *Geophysical Research Letters, 31*.

Zhang, Y., Heresh, F., & Falk, A. (2019). Small baseline InSAR time series analysis: Unwrapping error correction and noise reduction. *Computers & Geosciences, 133*.

Zhang, Y., Zhang, J., Wu, H., Lu, Z., & Guangtong, S. (2011). Monitoring of urban subsidence with SAR interferometric point target analysis: A case study in Suzhou, China. *International Journal of Applied Earth Observation and Geoinformation, 13*, 812–818.

AGEO Citizens' Observatories App and Platform

6

Eleni Mangina, Levent Görgü, Alexis Hrysiewicz, Xiaowen Wang and Eoghan P. Holohan

Abstract

Citizens' observatories are web-based platforms designed to increase citizens' participation and knowledge of their surrounding environment by collecting different formats of data and information. This chapter presents the design and development of a standard set of customizable software artefacts for a citizen observatory for geohazards that support the functional needs of the citizen observatories and have been made available for use by other Atlantic Interreg initiatives via an online catalogue. Both AGEO observatory and mobile apps require only local action from the citizens that can have regional, national, and global impact. These tools are complemented by an easy-to-use open-source toolkit for mapping ground motion via Interferometry of Synthetic Aperture Radar (EZ-InSAR). These software tools developed within AGEO have prioritised ease of use and immediate accessibility to enable citizens to monitor geohazards and thereby potentially contribute to local or regional risk management.

Introduction

The main purpose behind AGEO Citizens' Observatory Platform's goals is to bring citizens and authorities together and hopefully encourage and also enable communication, interaction between each other. The main objective of the AGEO Citizens' Observatory Platform is considered from two perspectives, the first being a web platform for citizens to examine and monitor their observations of natural hazards. The second is that it targets being a platform for authorities to manage citizens' observations by enabling the platform managers to direct citizens' observations to the correct people who are responsible for the risk mitigation actions.

Information collected from project partners via surveys and recommendations led to a roadmap of exploitation for the AGEO platform or the AGEO mobile applications (Sects. 6.1, 6.2 and 6.3). This current implementation of the AGEO platform is built on the research described in Görgü et al. (2022) and is developed on the initial implementation also explained in Schubert et al. (2017). Several

E. Mangina (✉) · L. Görgü · E. P. Holohan
School of Computer Science, UCD, University College Dublin, Dublin, Ireland
e-mail: eleni.mangina@ucd.ie

A. Hrysiewicz · X. Wang
School of Earth Sciences, UCD, University College Dublin, Dublin, Ireland

A. Hrysiewicz · E. P. Holohan
Science Foundation Ireland Centre for Research in Applied Geosciences (iCRAG), Dublin, Ireland

R. C. Gomes et al. (eds.), *Citizens' Observatories on Geohazards*, Geoenvironmental Disaster Reduction,
https://doi.org/10.1007/978-3-031-53371-6_6

critical issues were addressed carefully in the software development; the resulting platform ensures data masking by adopting anonymization or pseudonymization in line with GDPR requirements. It also delivers a map-based interface using Google Maps as the vehicle which enables data to be viewed collectively. Provision for multimedia data collection is offered and this is stored in a cloud-based service. However, the software includes a rudimentary data quality filter that thwarts poor quality, incomplete or malicious/rogue data polluting the integrity of the data corpus.

The actions undertaken in the AGEO project and described within this chapter include:

Action 1: ***Design and development of a simple easy to use participatory sensing capability to facilitate the capture of citizen observations***. This is an app and associated web platform that supports characterization by citizens of observations of geohazards or of other natural hazards by choosing from several pull-down menu options. Annotation of the observation is supported through the upload of multimedia content (image), geo and temporal tagging of the observation. The Tool suite is based on harmonised procedures for interoperability ready to be deployed elsewhere in the Atlantic.

Action 2: ***Provision of a map-based interface for social engagement of a map-based interface through which citizens can observe the collective citizen observed geohazard data.*** Sustaining citizen engagement is challenging and as such a view of combined observed data is often seen as a mechanism for sustaining user engagement. The aim is to maintain the provision of voluntary human resources over an extended period for co-generation of new geohazard knowledge, making tacit geohazard knowledge explicit, so that it can be shared. The map-based interface of AGEO Platform is explained in Sect. 6.2. Geohazard knowledge will be shared, and risk awareness will be raised across the local communities. Members of the Observatory will be encouraged to act as multipliers, thus enhancing the learning capacity of a community and so its resilience.

Action 3: ***Provision of an online catalogue of citizen science resources.*** The aim of this task is to embed resources in the app and platform to stimulate the effective use of Citizens' Observatories for geohazards monitoring under operational conditions as an integral part of regional/national risk management systems. The observatory includes a comprehensive citizen science resource on Atlantic geohazards risk assessment, preparedness, mitigation and prevention.

Action 4: ***Provision of an easy-use, open-source, multi-temporal InSAR processing toolkit for future (post-AGEO) geohazard monitoring.*** This action aims to support project partners and Citizen Observatory Pilots with capacity for automated generation and updating of maps and time-series of ground motion. This action will also assist in the wider uptake of Copernicus products and services (specifically Sentinel-1A/B). The toolkit enables citizens to automatically access, process, and visualise multi-temporal InSAR data from Sentinel-1.

The structure of this Chapter is as follows:

Section 6.1: AGEO Citizens' Observatory Web Platform that brings together resources and tools developed by AGEO for public engagement and for citizens' observatories on geohazards.

Section 6.2: Map-based interface for sharing and collective management of citizen-observed geohazard data across all pilots.

Section 6.3: AGEO Citizens' Observatory Mobile App for streamlined collection of data by citizens on geohazards plus access to relevant educational materials.

Section 6.4: A toolkit called EZ-InSAR for automated accessing, processing and visualising multi-temporal InSAR data from Sentinel-1 within an easy-use graphical user interface (GUI).

Section 6.5: A guide to the set of MATLAB scripts that make up the EZ-InSAR toolkit.

Section 6.6: Conclusions and final remarks.

6.1 AGEO Citizens' Observatory Web Platform

On one hand, the AGEO Project aims to create a platform (AGEO Citizens' xxxx) that citizens can use to report, monitor, review and examine their observations of geohazards. On the other hand, the AGEO Platform should enable administrators or authorities to manage their citizens' observations. On AGEO Platform, regular users are expected to be the citizens of their locations. The visitors of AGEO Web Platform, whether they are registered/logged in or not, can use the platform either to list the citizens' observations by going into the Event Management tab which will list the events in a tabular form with their details or they can use it to view the events that have been reported to the AGEO by surfing over the map-based interface on the AGEO Web Platform. These users can also use the filters to search and check for specific events. Event Management details are explained in next Sect. 2.2. Map-based interface is explained in Sect. 6.3.

6.1.1 User Types

There are 3 types of users implemented within the AGEO Citizen Observatory Web Platform: Admins, Managers and Citizens (API Users). These user types should login to the platform with their credentials (see Fig. 6.1) to be able to carry out operations which are explained further in the next paragraphs. Any user can log in if s/he has credentials for the AGEO Web Platform or any person can select to sign up for the platform anytime and become a registered user of the AGEO Web Platform.

Registering for the platform will get all users a Citizen account which will have API access to the AGEO Platform's reported event list. AGEO Web Platform users gain access to an API key by default which is available to use to interact with AGEO Event reports if they need to and integrate in 3rd party systems to gather information about all the event reports that are reported on AGEO Web Platform. This is the reason that the API User and Citizen is used interchangeably in this document. Citizens (API Subscribers)

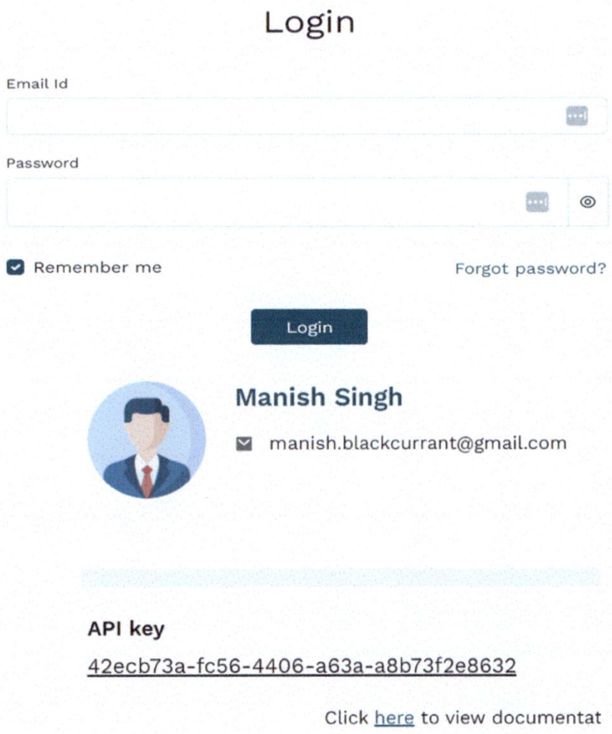

Fig. 6.1 Login Screen (top) and Citizen's Profile (middle) of the AGEO Citizens' Observatory Web Platform

can view their API keys in their profile at the platform and can generate a new API key. A link to the API documentation which contains the details about the API usage is supplied at users' profile as well (Fig. 6.1).

Moreover, there is also the possibility of being an unknown user for the platform. Unknown users are users without any login but still can utilise the web platform to investigate reported events. All visitors can view the reported events using the Map view (Sect. 6.3) or the tabular view at the Event Management part of the platform (Sect. 2.2).

On the other hand, the Admin type user account can be defined only during the initial setup phase of the web platform or can be added on the system level later on. Furthermore, any Citizen account can be upgraded to a Manager account by the Admin. Details about AGEO Managers are given in Sect. 3.1 and Admins in Sect. 3.2.

6.1.2 Event Management

All user types, including the unknown users, can filter and search for the events reported to the AGEO Web Platform by citizens using the AGEO Mobile Observatory Application. The built-in searching filters are based on the type of event, status of event, and the dates. These filters are implemented as dropdown menus on the Event Management (Fig. 6.2).

Users can utilise these built-in dropdown menus on top of the results table to filter the events based on their type, date, or status and narrow down on the results in the table. The AGEO Web Platform users are able to view all the events reported in a tabular view with a selection of parameters from their details consisting of the ID, type, location, date and status of the event. They can see the name of the Manager who verified the event if the event is

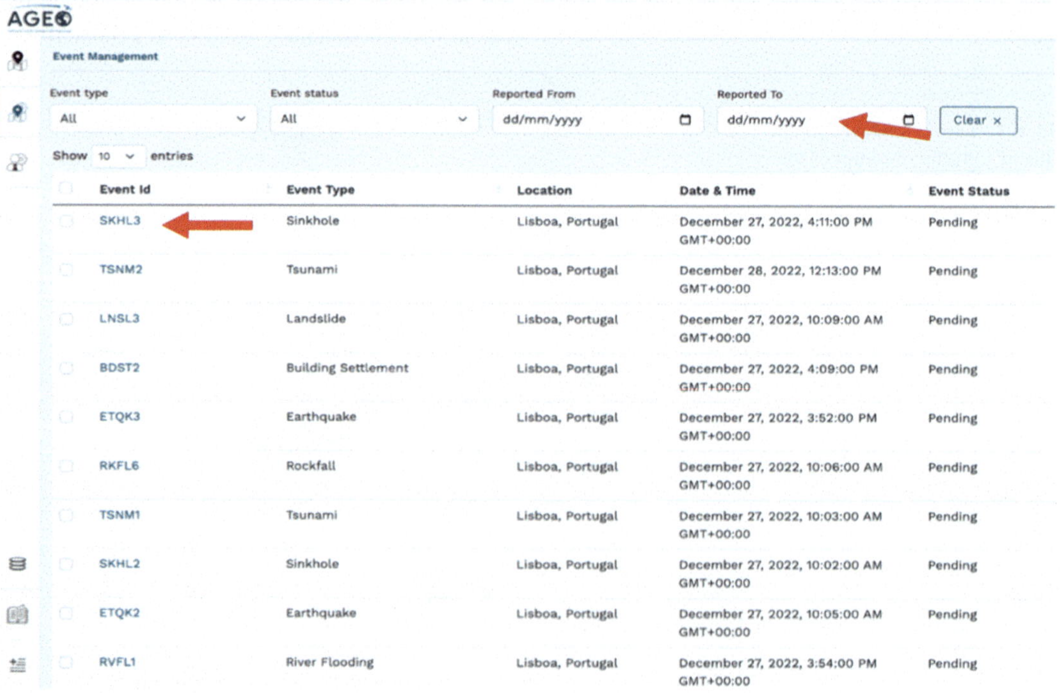

Fig. 6.2 The event management has dropdowns to filter based on type, status, and date. Clicking an ID opens the event details with the selected event's data

Approved within the AGEO Platform by AGEO Manager converting its status to "Accepted".

Citizens also have the possibility to sort the resulting list in the table based on table parameters. The ID, type, location, date, and status of the event parameters are presented as the column titles within the table. The users can pick a column and click on the title to sort the results table based on the parameter at their column title selection.

AGEO users can also search via a search box by giving keywords which simply can be the ID of the event or a location. Users can also select the Event by clicking on the Event ID to see the details about this event. Explanations about the Event Details page and details about different event types and statuses are presented below in Sect. 2.3.

Anonymous users and registered citizens have only the status options "Approved", "Rejected" and "Pending" to choose from in the event status-based filtering drop-down menu. Admins and Managers have additional options to view and filter the archived events with "Archived" status. Date based filtering is implemented by accepting a start date and end date from the user as a query and filter the results based on these dates.

AGEO Web Platform also supports other basic table functionalities. The users can choose to view paginated results for the list of events based on their filtering and search query. AGEO Web Platform can provide some related metadata to the user about their resulting events list, such as number of total results, filtered results and displayed results.

6.1.3 Event Details

The Event Details screen can be opened by the AGEO users by clicking on any event's ID on the Event Management (see Fig. 6.2). All the details of and the information about the user's selected event will be displayed on the screen and presented to the user on the Event Details page for their selected event (see Fig. 6.3).

These event details include the status and the type of the event, the photos captured (if any), all the sensor data collected during the photo capture, all the answers that the user gave to the common questions and all the answers gathered from the user for the event specific questions. These questions are asked to the users while they are reporting an event using their mobile device and the AGEO Mobile Observatory application. Details about the AGEO Mobile Observatory are given in Sect. 6.4. Furthermore, the user shall see a status tag on top of the event details which shall say "Rejected" in red, "Accepted" in green or "Pending" in yellow colour (see Fig. 6.4).

The Event Details page also presents the thumbnail formats of the photo(s) captured by the user during the reporting of the event and the AGEO users can select any of these photo thumbnails. These thumbnails are clickable and if the users prefer, they can see the full-size image of the photo by clicking on their selected thumbnail. The selected photo will be presented with its full-size image on the screen. An important detail is that AGEO Web Platform accepts reports without any photo capture in case of a "Quick Report" used while reporting by the citizen on the AGEO Mobile Citizen Observatory application. If a "Quick Report" is the case, the Event Details will have no photo thumbnails to show but the other details about the event will be presented. The "Quick Report" feature is further explained in Sect. 4.1.

While examining the information about a single event, users can also share the event in their preferred way which will simply send the direct URL for the selected event's details page to citizen's selected contacts via citizen's preferred social media tool or email. Moreover, users will be able to examine the details of the location for the event. The location details include the GPS coordinates and a view of that location on the map. AGEO Platform also checks for the name of the location using the reverse GPS lookup and if it finds the name for the location, the platform also displays this location information to the users (see Fig. 6.3).

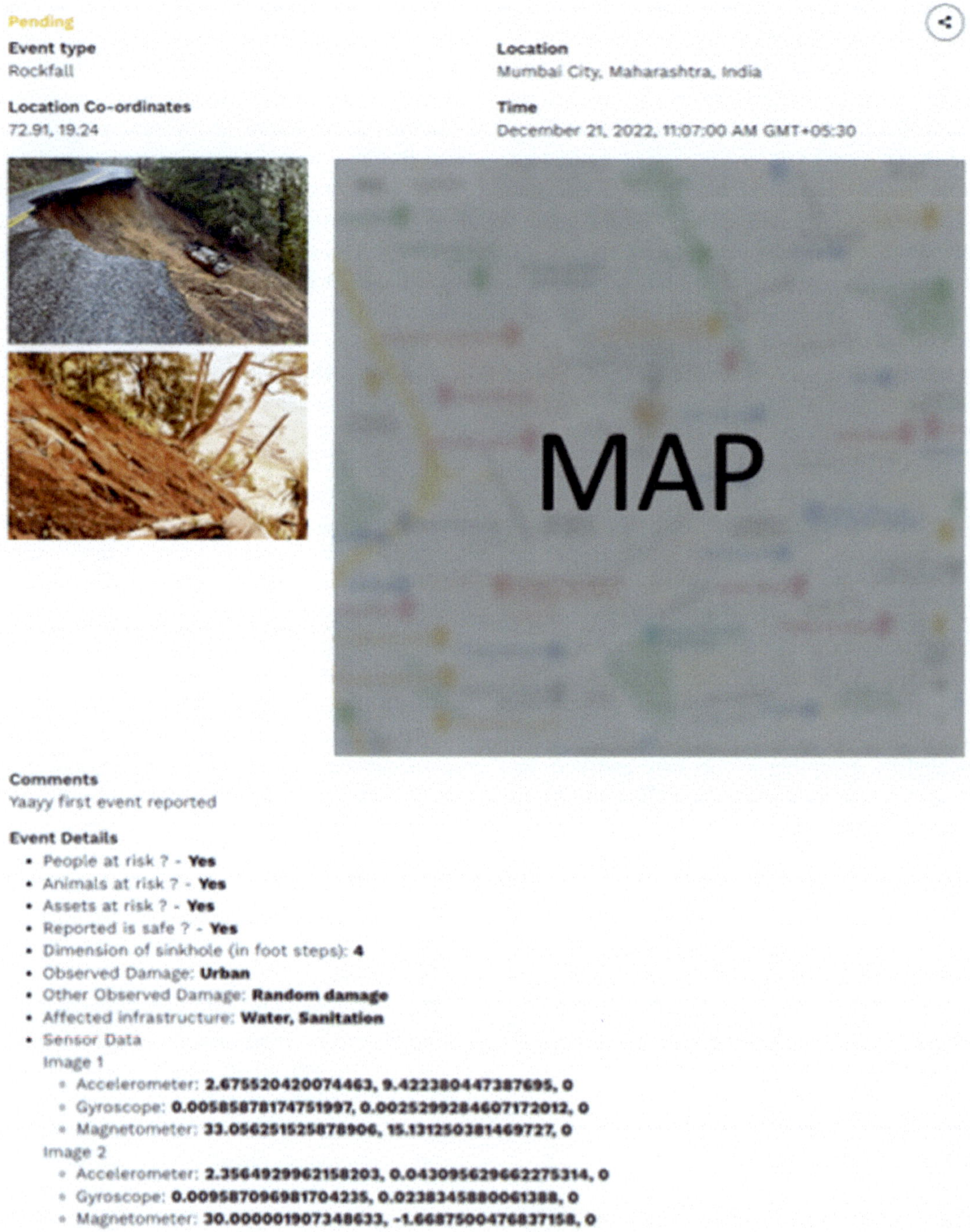

Fig. 6.3 Event details present the event status, type and location (top), thumbnails of the photos captured (top-left), location on the map (top-right), citizen's answers to common/event specific questions, all the sensor data (bottom)

Fig. 6.4 Event Status "Pending" is color coded with yellow, "Approved" with green and "Rejected" with red on the Event Details screen of AGEO Platform

Additionally, the Event Details page also shows every data collected by the user manually and automatically. Manual information is all the answers that the users give to the common questions and event specific questions while they are reporting the event. The automatically collected data includes the sensor data gathered from users' devices by the AGEO Mobile Citizens' Observatory Application. Details about AGEO Mobile Observatory are given in Sect. 6.4 and information about the "Quick Report" feature is included in Sect. 4.1.

6.1.3.1 Event Status

In its current implementation, AGEO presents the concept of "status" of the observation to its users. Events can have 4 different statuses in AGEO Platform being Pending, Approved, Rejected and Archived. When any AGEO Mobile Observatory user reports an event, it is created as an event with a Pending status by default within the AGEO Platform. Thus, an event's status is always Pending after the initial report from the citizen. These events in Pending state are not verified by any of the AGEO Managers.

The AGEO Managers can convert any Pending event's status to Approved or Rejected. All the Approved (Accepted), Rejected or Pending events shall be visible to all the users, irrespective of their user type and role within the AGEO Platform and whether they are being logged in or not. AGEO Managers can choose to approve an event directly or can reject an event after previously accepting it. AGEO Managers can also choose to reject an event directly.

On the other hand, an AGEO Manager can choose to archive any event, no matter if it is "Accepted" or "Rejected" before. An Archived event shall not be visible to the regular users at all. As the Rejected events are still viewable on the map and table view for the AGEO platform's registered or anonymous users, AGEO Managers can select to "Archive" any "Rejected" event to remove it from the regular users' view. This feature can be used as soft delete which will actually only hide the "Archived" events from citizens' view but not hard delete their information from the AGEO platform's database. Archived events will not be viewable to anyone except AGEO Managers. The AGEO Managers

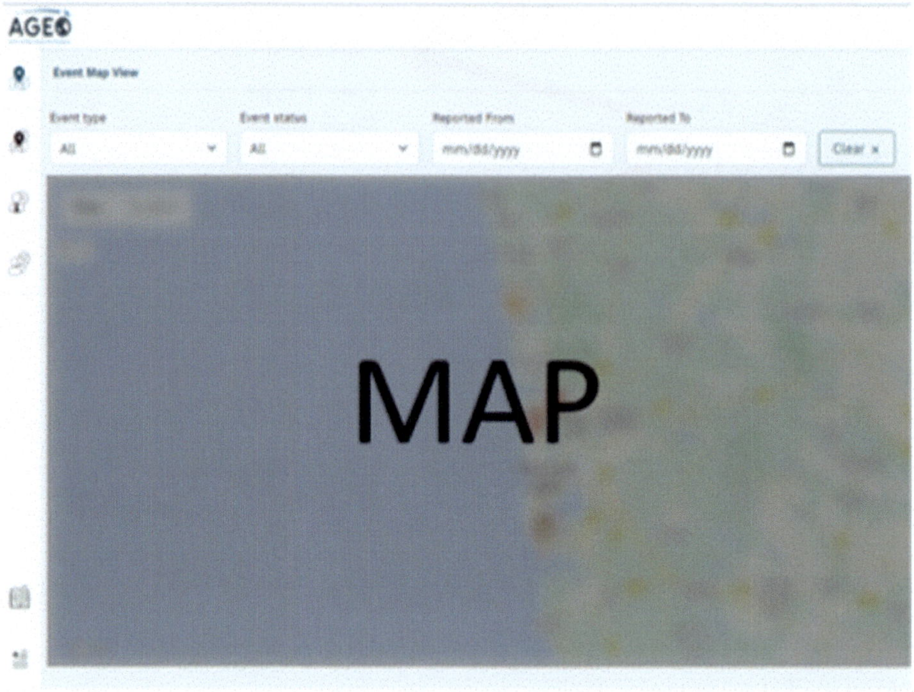

Fig. 6.5 Map-based interface of AGEO citizen observatory web platform

will still be able to see them in the tabular (Event Management) and map-view.

Archived events' status can be changed later and converted to "Unarchived" status by AGEO Managers. This will make the event visible to the users again and the event will appear to regular citizens with Approved status.

6.2 Map-Based Interface for Citizens' Observatory

Any visitor of AGEO Web Platform (AGEO Citizens' xxxx) can observe events on the map that have been reported by global users of AGEO Mobile Citizen Observatory application. The map-based interface of AGEO Web Platform paints a clearer picture of the events that were captured and makes it simple to locate the event over the map. The users can select to view the events either in Map or Satellite view they can also pan and zoom through the map for looking at different events. There are various filters present, like the Event Management tabular-view section, to fetch the event results for the desired set of events. These filters are based on event type, status and reported time frame. They can be used to narrow down on the resulting pins on the map.

Users can view all the events reported plotted on a map at AGEO Web Platform Map-based interface (see Fig. 6.5). Platform will list the most recent events around the world at its initial load. Initially, the platform will ask for permission to use the location of the user. If the user allows location access, the map will open zoomed and centred with the user's location. If the location access is denied by the user, the map will be centred around Europe by default.

Each event is represented by a separate pin. Multiple pins are clustered together and the total number of clustered pins is displayed to the user. The events are identified by different icons within the AGEO platform. Both AGEO Citizens' Observatory Web Platform and AGEO Mobile Citizens' Observatory Application support 11 event types in total which can be seen in Fig. 6.6 with their pins used in the platform.

Fig. 6.6 List of event types and their pins used in AGEO web platform and mobile observatory applications

Users can filter events based on the event status. Any user, logged in or not to the platform, has the option to choose to search between event statuses, Approved, Rejected or Pending. They can also filter events in a date frame. Users can choose the start date and end date for their preferred period.

6.2.1 Additional Functionalities for AGEO Managers

One significant feature that AGEO Platform presents to its Managers via its Map-based UI is to be able to draw polygons over the map for event filtering (Fig. 6.7). After their selection is finished, AGEO managers can download their selected event(s)' data using the download button (Fig. 6.8). AGEO supports data downloading in 3 formats.

Manager accounts can also view the event's details as in the citizen accounts. Additionally, they can accept ("Approve") or "Reject" any event report via the Event Details screen. They also can archive or unarchive any reported event when/if required (Fig. 6.9).

Currently any event reported to the platform via the mobile observatory app starts its journey as status "Pending". Only the managers of the platform can change this status to "Approved" or "Rejected". There is also an "Archived" status implemented which kind of soft deletes the event from the platform, meaning it can only be seen by managers of the platform and will not be presented to the regular citizens when they visit the platform, but the details of the event will be still held in the platform database. This way managers of the platform can then further change the status of this event if they need to, which brings flexibility on the managerial side

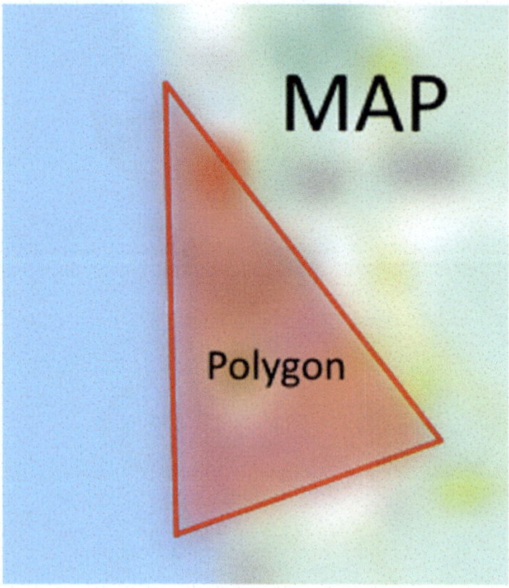

Fig. 6.7 Managers can draw polygons over the map to select events and download their data

of the platform and prevents any unwanted data loss of an event's detail in case of deletion by mistake.

Managers can change the status of an event on the Event Details screen which can be opened for the selected event by clicking on the Event ID. On the Event Details screen, in addition to the sensor data, images and other information which are also shown for the regular citizen accounts, at the very bottom of the page, Managers can see the options to "Accept" or "Reject" the event (see Fig. 6.9). After any Manager of AGEO platform has updated the status of an event as Accepted or Rejected, they can also archive and unarchive the event report and change the event's status between "Archived" to "Accepted" or "Rejected" again (see Fig. 6.9).

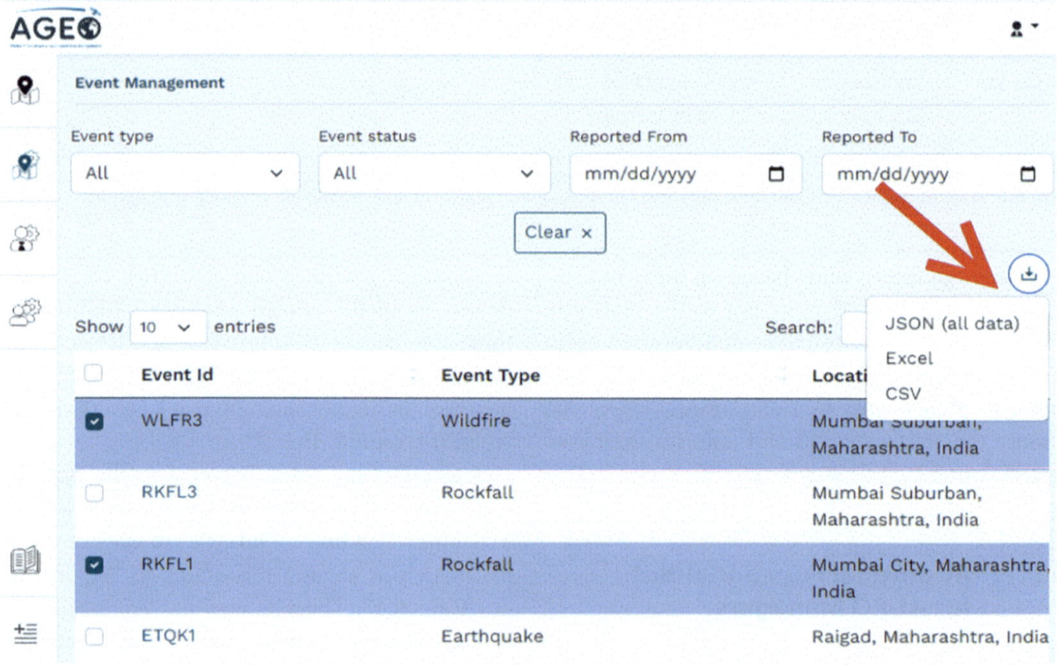

Fig. 6.8 Managers can download their selected events' data in their preferred format (JSON, Excel, or CSV) on event management screen of AGEO web platform

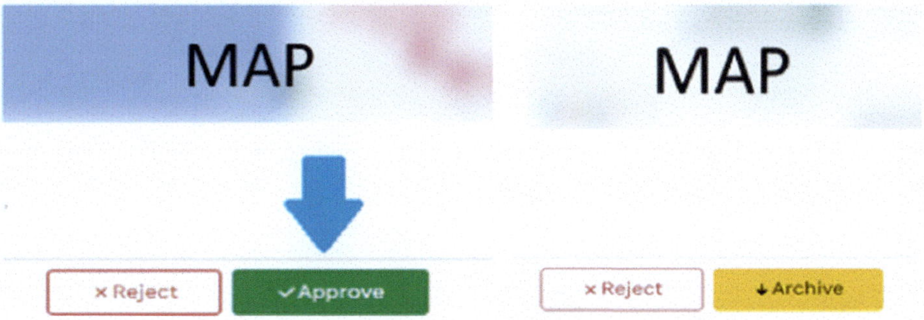

Fig. 6.9 Managers can "Approve" or "Reject" events on the event details page. Events can be Achieved or Un-archived by Managers anytime if needed

While using the Event Management section of the AGEO Web Platform, Managers can also search for events, same as the regular citizen account, based on the type of event, status of event, and according to the dates by using the filters implemented as dropdown menus. On the AGEO web platform only Admins and Managers have additional options to view and filter Archived Events. In addition to this search feature, Manager accounts can also download their selected list of events' data as a CSV, Excel, or JSON file. Downloading JSON data format supports complete event data dumping. Only structured data based on the common event questions is downloaded in Excel and CSV formatted downloads. Users' answers to the event specific questions are not included in Excel and CSV files. JSON format is currently used as the preferred format for data sharing for AGEO Platform as it uses MongoDB maintaining its unstructured data for different event types supported within the web platform and mobile application.

6.2.2 Additional Functionality for Admin

When the users sign up to the AGEO Web Platform, they receive Citizens accounts. Admins can see all the users as listed along with their Email IDs, the API keys and their membership Status at the "API Subscribers" section of their dashboard. The admins can change the status of the user between active and inactive states. Moreover, the Admins can also create new Managers or upgrade an existing Citizen (API User) account to AGEO Manager account by using the Manager Management section of their dashboard. An Admin can create, update and delete a manager account.

Admins and Managers can download the event data using the download button placed on the top right of the tabular view in Event Management part of the AGEO Web Platform (see Fig. 6.9). This is also the case in the map-based interface where the Admins and Managers can also select events and then download these events' data via the download button.

6.2.3 Educational Content

There are different sections available on the AGEO web platform such as the Privacy Policy section giving the details about AGEO's privacy policies both for the web platform and the mobile applications. Contact section allows the end user to send messages to AGEO Consortium.

Another service that AGEO Platform offers to citizens is its educational content which gives a short but detailed help guide for the user to prepare and steps to follow in case of an event type supported by the AGEO platform (see Fig. 6.10).

This educational content is accessible by both the mobile application and the web platform. On

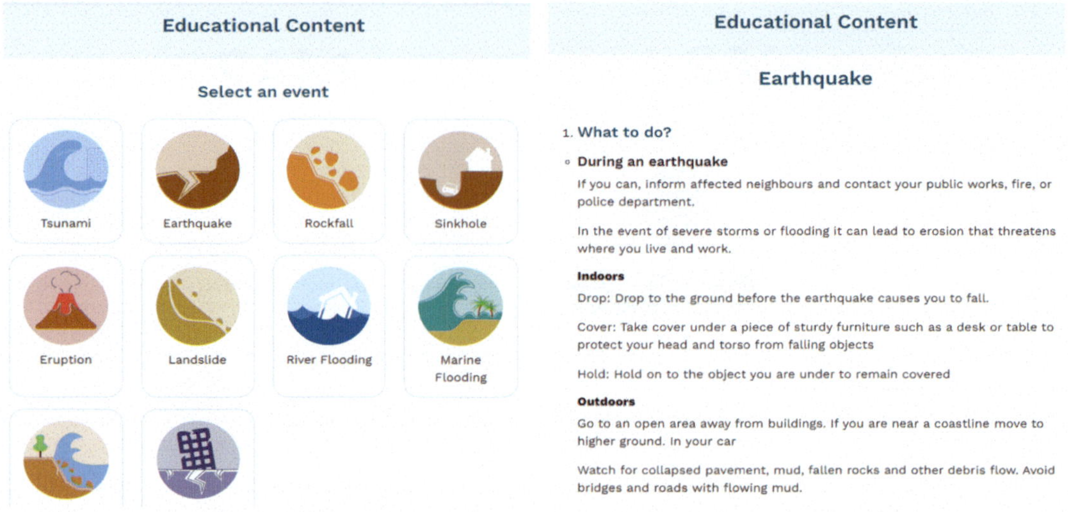

Fig. 6.10 Educational content on Web platform (left) Earthquake educational content (right)

the mobile platform the educational content is available for the user even if the user is offline without any internet connection in case of an emergency (see Fig. 6.11).

6.3 AGEO Citizens' Observatory Mobile App

Any anonymous user can report a new event occurrence in their area. The AGEO Mobile Citizens' Observatory application can be downloaded for both Android devices (AGEO xxxx) and iOS devices (AGEO xxxx). The application welcomes the users with a starting screen which has two options (Fig. 6.12). "Monitor an event" is the regular and preferred way to report about an event. There is also a "Quick Monitor" option which is added for emergency situations where the users quickly want to inform about an urgent situation. It allows the users to skip filling most of the event details and instantly report without a photo capture and answering all the event related questions if they want to. A lot of details remain untracked in this mode to be able to deal with emergency situations.

6.3.1 Monitor an Event

Citizens can use AGEO Mobile Observatory to monitor and report an event they are facing. The preferred way of receiving information to AGEO Web Platform via Mobile Observatory is to use the "Monitor an event" option. As the first step of reporting an event, the user will select the event type. AGEO Observatory supports various events which are briefly described in previous Sect. 6.3.

After the initial event type selection screen at the start of the journey, a location screen will automatically detect the user's location and show it on the screen to the user. The user can change the location of the occurrence of the event on the map displayed on the screen by clicking on the map to drop the pin where they want to (Fig. 6.13).

After the location detection and data gathering, the application will present the event related common questions to the user which are common for all the event types. In this section of reporting, the user is asked to capture one, two or three photos about the event or select them from the phone's gallery (see Fig. 6.14).

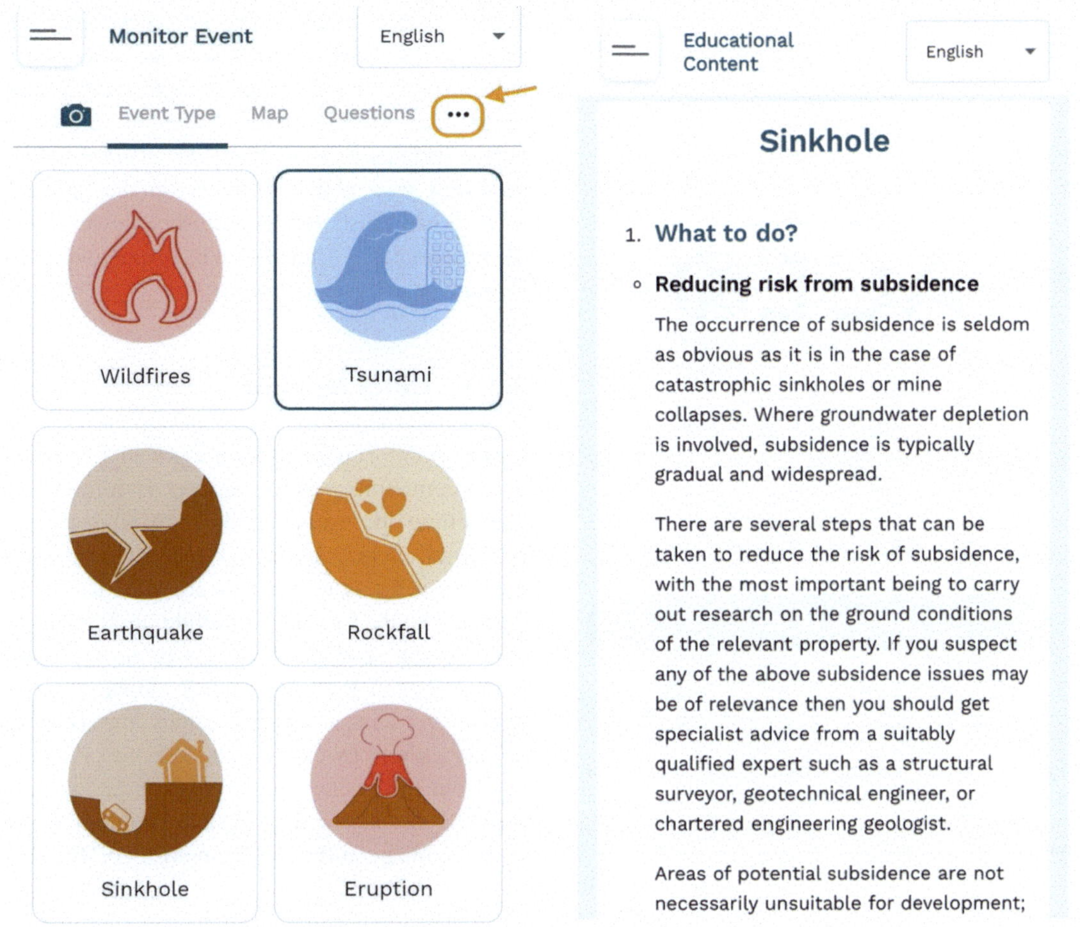

Fig. 6.11 Educational content on mobile application (left) Sinkhole educational content (right)

Capturing at least a photo is mandatory in this mode, but reporting without any photo capture is allowed for the "Quick Monitor" section which is also always accessible via the camera icon on the top left of the screen every time. This is aiming for urgent emergency situations.

AGEO Mobile Observatory Application also allows its users to pick previously captured photos from their device without any new capture which aims to allow interested citizens to wait for Wi-Fi connection as they may prefer not to send large photos over the 3/4/5G connections because of the data rates or other reasons.

During the photo capture, AGEO Mobile Observatory Application automatically gathers the accelerometer, gyroscope, and magnetometer sensors data from the mobile device of the user. After that screen, the mobile application presents the users with event specific questions which are different for each event type. Users can submit their report once they answer the event types related questions. Once the users finish filling their report, they can send it to AGEO Web Platform by clicking on the Submit button (Fig. 6.14).

Finally, after a successful submission, an URL for the just reported event's details on the AGEO Web Platform is shown to the users which is clickable and sharable on various social media platforms. Clicking the link will open a web browser with that URL and show the

Fig. 6.12 At the starting screen of the AGEO mobile citizens' observatory applications users are presented with two different ways to inform AGEO about events they face

event's details to the user via the AGEO Web Platform (see Fig. 6.15). Users are also easily able to monitor the status of their event report on the platform via this link. Clicking the share button enables users to share this event via their preferred way which depends on the applications already installed on the users' device, such as Facebook, Twitter, or WhatsApp and even through email.

6.3.2 Quick Monitor

Quick monitoring is a feature added for emergency situations where the user is allowed to skip filling the event details form while monitoring an event. Although some details remain untracked in this mode, it is accepted as a very critical feature for AGEO mobile application to allow urgent and quick access to its reporting feature. When the user selects Quick Monitor option. A screen which enables the user to upload a photo by capturing a new one or selecting from the phone's gallery. If a user captures a photo the sensor data is also captured along with the image at the same time (Fig. 6.16). This

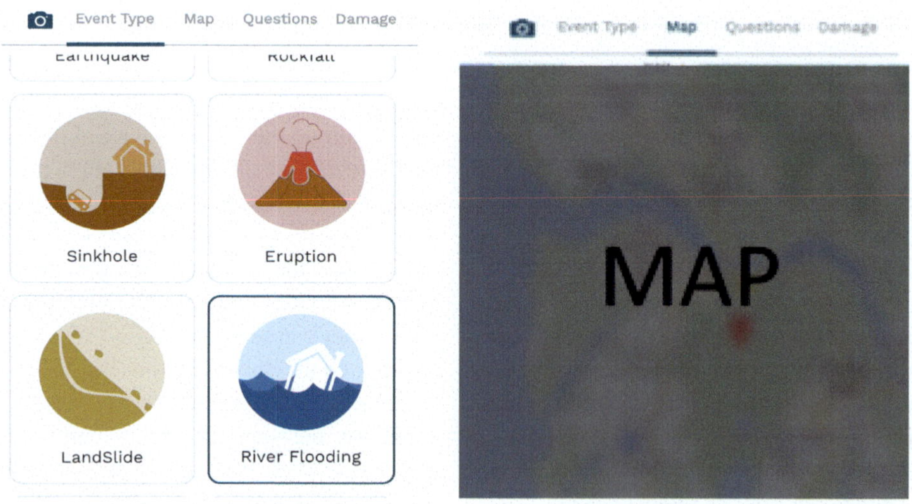

Fig. 6.13 At the starting screen of the AGEO mobile citizens' observatory applications users are presented with two different ways to inform AGEO about events they face

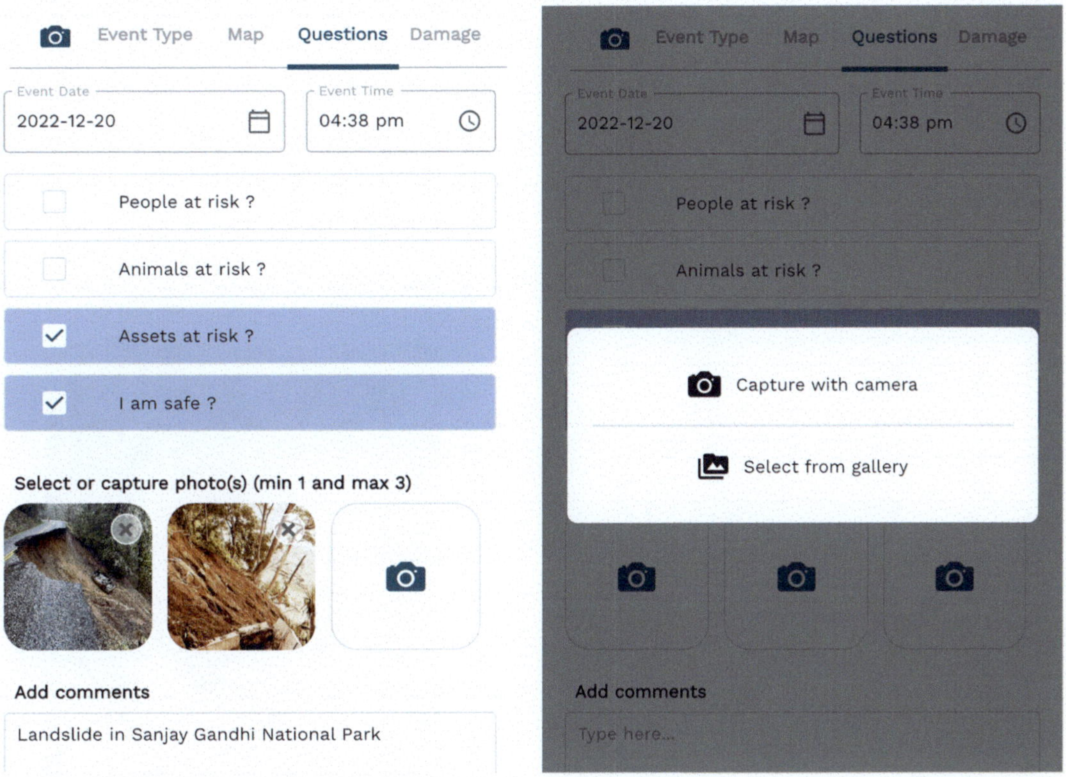

Fig. 6.14 At the starting screen of the AGEO mobile citizens' observatory applications users are presented with two different ways to inform AGEO about events they face

screen also allows users to add any comments related to the event if they want to.

Next, AGEO mobile observatory detects the user's location and shows it to the user and allows the users to select pins on the map as the place of occurrence for the event. Finally, users are asked to select an event type from the ones listed on the phones screen with their related icons which are also used as their pins (see Fig. 6.17). After successful users are presented with a URL the event's detail, which they just reported to the AGEO platform which is also a sharable link for social media platforms.

6.4 EZ-InSAR: A Toolkit for Automated Accessing, Processing and Visualising Multi-Temporal InSAR Data from Sentinel-1

6.4.1 EZ-InSAR: A New Open-Source InSAR Environment

Since the ESA launched the Sentinel-1A and Sentinel-1B SAR satellites (S1) in 2014 and 2016 under the European Union's Copernicus Earth Observation program, the availability of satellite SAR images free-of-charge has accelerated InSAR applications in recent years.

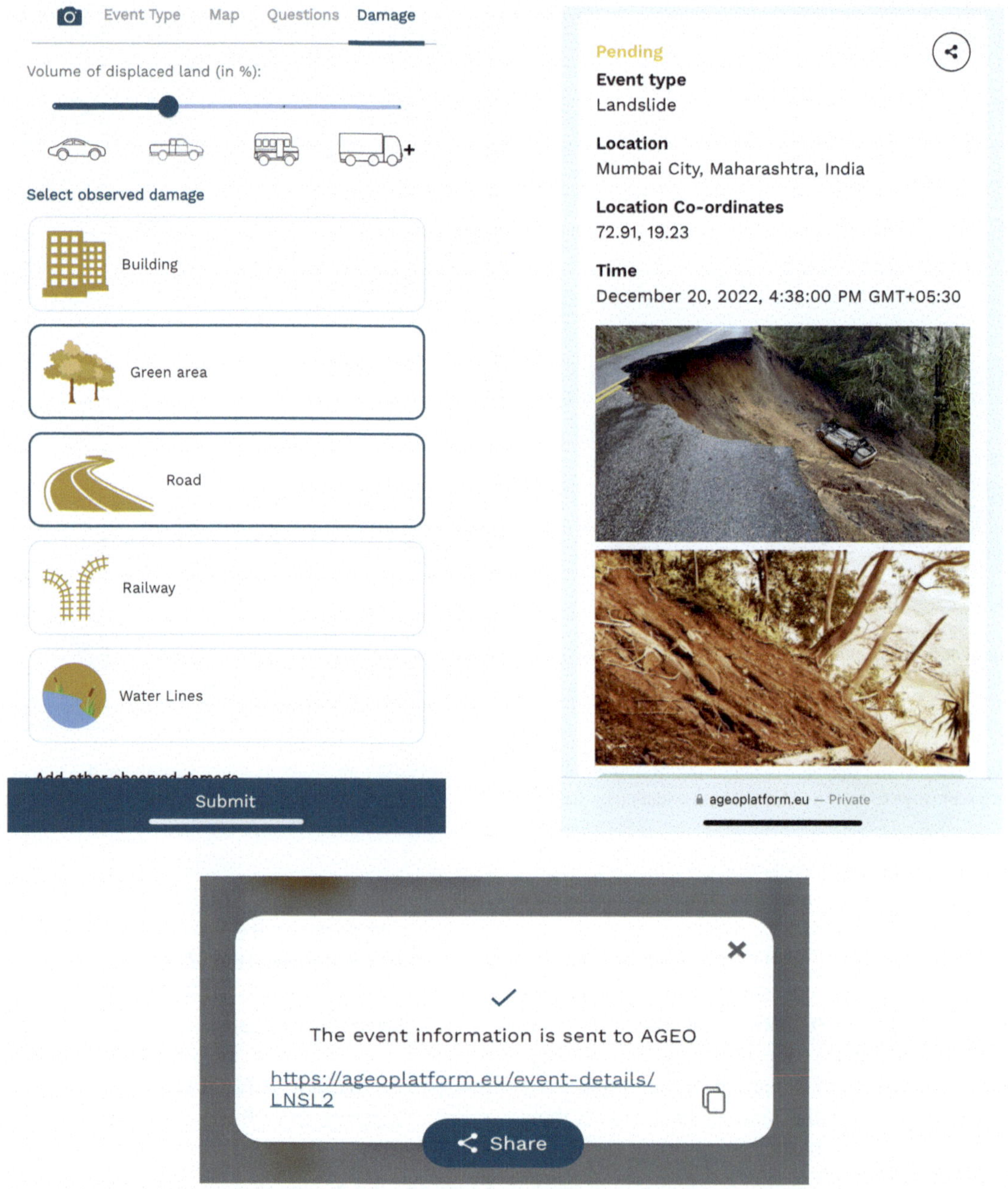

Fig. 6.15 At the starting screen of the AGEO mobile citizens' observatory applications users are presented with two different ways to inform AGEO about events they face

Today, geoscientists can use commercial or open-source software packages to process InSAR data: e.g., the commercial code from GAMMA® software (Werner et al., 2000) used in the previous section. However, open-source packages for InSAR purposes currently have one main disadvantage: they are designed for dealing with only one sub-task of InSAR analysis—i.e., for generating interferograms (InSAR stack) or generating time-series of displacement. Indeed, users must switch between the different packages to complete a full InSAR analysis. In

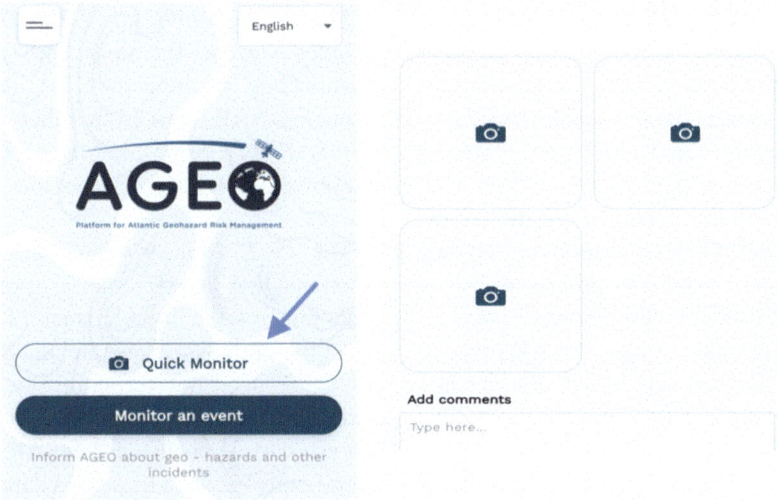

Fig. 6.16 At the starting screen of the AGEO mobile citizens' observatory applications users are presented with two different ways to inform AGEO about events they face

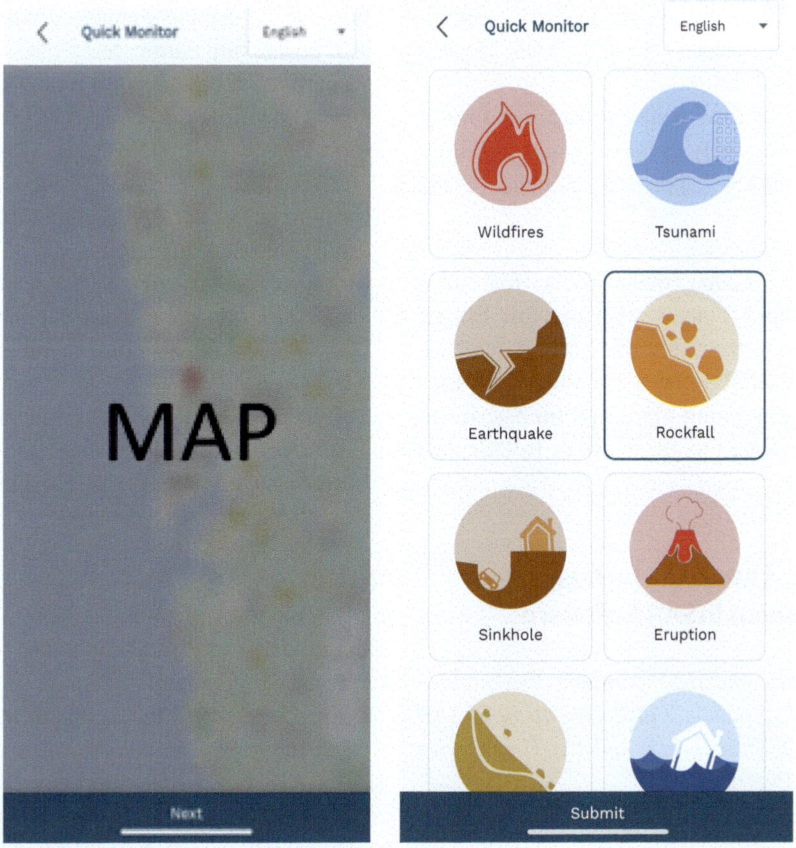

Fig. 6.17 At the starting screen of the AGEO mobile citizens' observatory applications users are presented with two different ways to inform AGEO about events they face

addition, the same open-source codes mainly are distributed for Linux or Unix operating systems and are controlled by command-line limiting the use of InSAR measurements inside the Earth-Sciences communities and blocking this use by citizens.

EZ-InSAR toolkit, developed by the AGEO project, is a MATLAB®-based toolbox that enables an open-source InSAR processing chain provided with a user-friendly graphical interface (GUI) for intuitive and seamless InSAR processing with both conventional InSAR approaches. EZ-InSAR[1] transparently integrates existing open-source codes:

- InSAR Scientific Computing Environment (ISCE) for generating InSAR stack (Rosen et al., 2000)
- Stanford Method for Persistent Scatterers (StaMPS) for time series analysis using Persistent Scatterers (PS) and Small-Baselines approaches (Hooper, 2008; Hooper et al., 2012)
- Miami InSAR time-series software (MintPy) for time series analysis using Small-baselines Subsets (SBAS) approach (Zhang et al., 2019).

Completed by management tools for SAR imagery, EZ-InSAR deals with the full InSAR processing chain, from automatically downloading Sentinel-1 SAR images for the user's area of interest to generating maps and time series of ground displacement to allow use of InSAR data for everyone.

6.4.2 EZ-InSAR: An Interface Between InSAR Software

The EZ-InSAR interface contains three modules separated to panels: (1) *Preparation of SAR Data*, (2) *ISCE Processing*, and (3) *Time Series Analysis* (see Fig. 6.18).

[1] EZ-InSAR toolkit in version 2.0.0 Beta.

The *Preparation of SAR Data* module includes various functions for searching and downloading Sentinel-1 SAR data and for managing SAR data from other sensors—i.e., TerraSAR-X, Cosmo-SkyMed. EZ-InSAR also can download DEMs from various conventional sources.

The *ISCE Processing* module includes the functions for interferometric processing and generating of the InSAR stack. The users can select the options to generate an "SLC stack" or an "Unwrapped interferogram stack", related to a StaMPS-based or a MintPy-based time series analysis, respectively. EZ-InSAR also allows conversions between these two data stacks.

The *Time Series Analysis* module provides the option of using either the StaMPS or MintPy time-series processors. The StaMPS and MintPy processors can be activated by clicking the corresponding tabs in the panel (see Fig. 6.18). Thus, the GUI panels enable control of the full set of processing parameters.

Finally, EZ-InSAR toolkit has several unique tools to facilitate the InSAR processing, for example:

- tool to select the SAR acquisition as the optimal reference image;
- tool to display to easily visualise all SAR interferograms, unwrapped phase maps, and coherence maps within an unwrapped interferogram stack;
- tool that allows the user to manually modify the SAR image connection network by removing or adding specific image pairs for Small-baselines approaches.

Regarding the future development of EZ-InSAR toolkit, the priority could be the translation of the EZ-InSAR scripts to Python language to avoid requiring commercial language and to focus on the cross-platform compatibility. Secondly, the visualisation and plotting capabilities of EZ-InSAR could be improved in the future releases. Finally, EZ-InSAR should support the new satellites with free-of-charge data policy: i.e. ESA Sentinel-1C/D SAR satellites in 2024/2025 (Spataro et al., 2021) and NASA-ISRO SAR

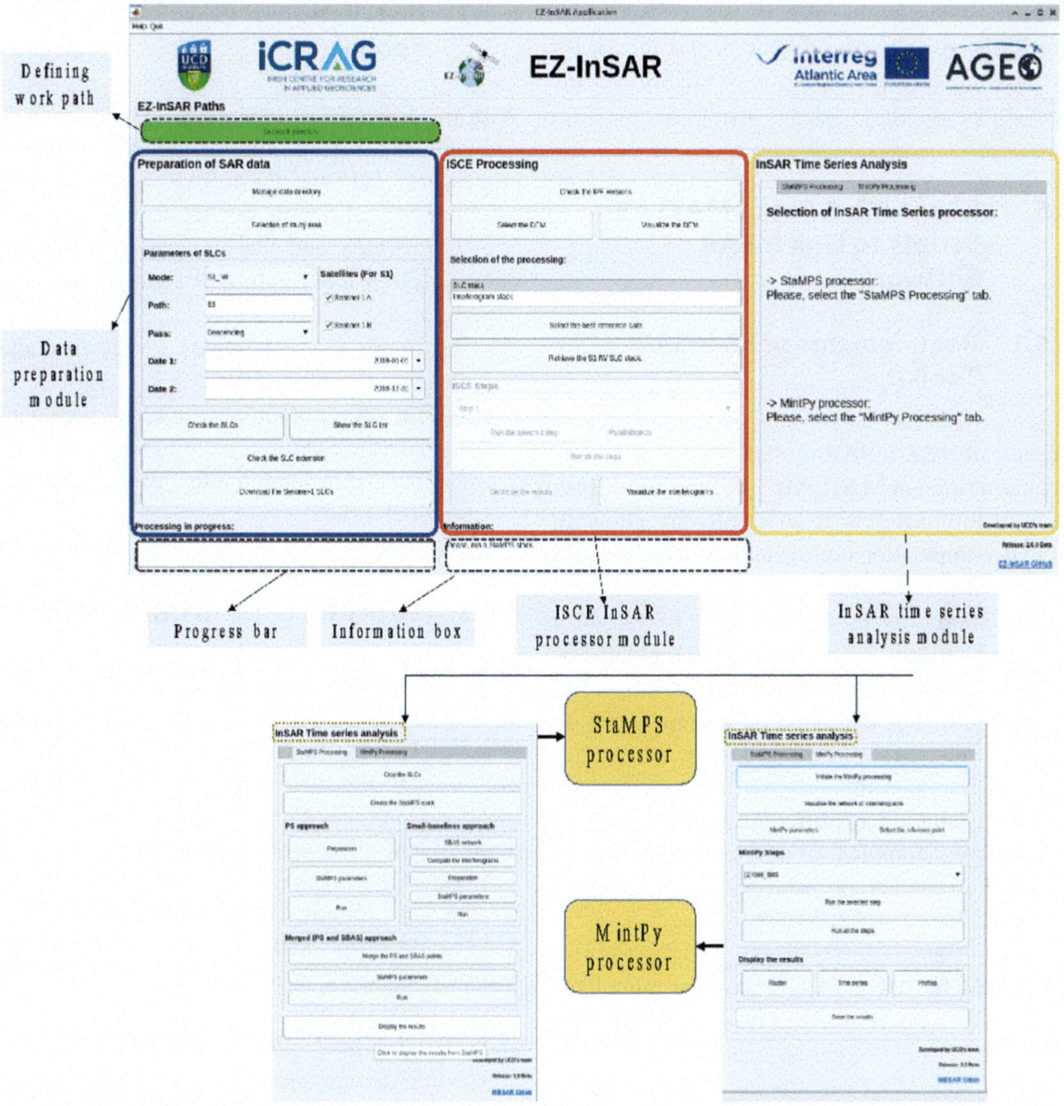

Fig. 6.18 Snapshot of the interface of EZ-InSAR

(NISAR) launched in 2025 (Rosen et al., 2015). For further details on EZ-InSAR and its use on two test sites with comparison to GNSS data for validation, see (Hrysiewicz et al., 2023).

6.5 EZ-InSAR: A Set of MATLAB® Scripts to Link InSAR Software

6.5.1 Requirements of EZ-InSAR Toolkit

Technically, EZ-InSAR toolkit is a set of scripts mainly written in MATLAB®, Python3 and Bash shell to link different InSAR software allowing InSAR computations of ground surface motions from SAR images.

The source codes are available to all via a GitHub repository[2] (or on request) under the GPL-3.0 licence. Associated to the scripts, a full documentation of the toolkit is also available on GitHub with a precise description of:

- the installation of different InSAR software (relatively complex for non-experts);
- some examples of how to use EZ-InSAR toolkit

To use EZ-InSAR toolkit,[3] each user must have MATLAB® environment (version $\geq$ 2020b), Python3 (with required packages), *ISCE*, *StaMPS* and *MintPy* software. All the scripts have been developed and tested under Linux (Ubuntu 20.04), which means that EZ-InSAR requires a UNIX-Linux environment to run.

For MacOS® and Windows® environments:
EZ-InSAR[1] is not currently compatible for systems using MacOS® and Windows® environments but will be compatible in the future.

6.5.2 Installation of EZ-InSAR Toolkit

With the EZ-InSAR documentation, users have complete instructions for installing the InSAR processors and the EZ-InSAR toolkit. After installing[4] the different InSAR software (*ISCE*, *StaMPS*, *MintPy* and the associated tools), the only steps to install EZ-InSAR are as follows:

(1) copy/paste the EZ-InSAR directories and files into the desired directory;
(2) modify the *pathinformation.txt* file with the Alaska Satellite Facilities (ASF) user account information to download the Sentinel-1 data.

6.5.3 Structure of EZ-InSAR Toolkit

As described above, EZ-InSAR links the InSAR software and provides a graphical user interface (GUI). The script directories are organised according to the main steps of MT-InSAR processing: (1) preparation of data; (2) InSAR computation and (3) time series analysis.

Figure 6.19 shows the structure of the EZ-InSAR toolkit, the names of main MATLAB® scripts and their links to the external InSAR processors:

- **"GUI scripts"**, the two main scripts of EZ-InSAR are stored: (1) the run scripts of EZ-InSAR named *EZ-InSAR.m* and (2) *GUIMIESAR.m* which are the scripts to generate the EZ-InSAR GUI.
- **"SLC module"** (*SLC_management* directory), all the SAR image management scripts are stored (selection of the study area, image selection, verification, downloading, etc.). Overall, **"SLC module"** provides a full set of tools for SAR data preparation.
- **"ISCE module"** (*ISCE_functions* directory), these scripts control the ISCE processing for

[2] https://github.com/alexisInSAR/EZ-InSAR
[3] EZ-InSAR toolkit in version 2.0.0 Beta.

[4] This part of the installation is available in the EZ-InSAR documentation.

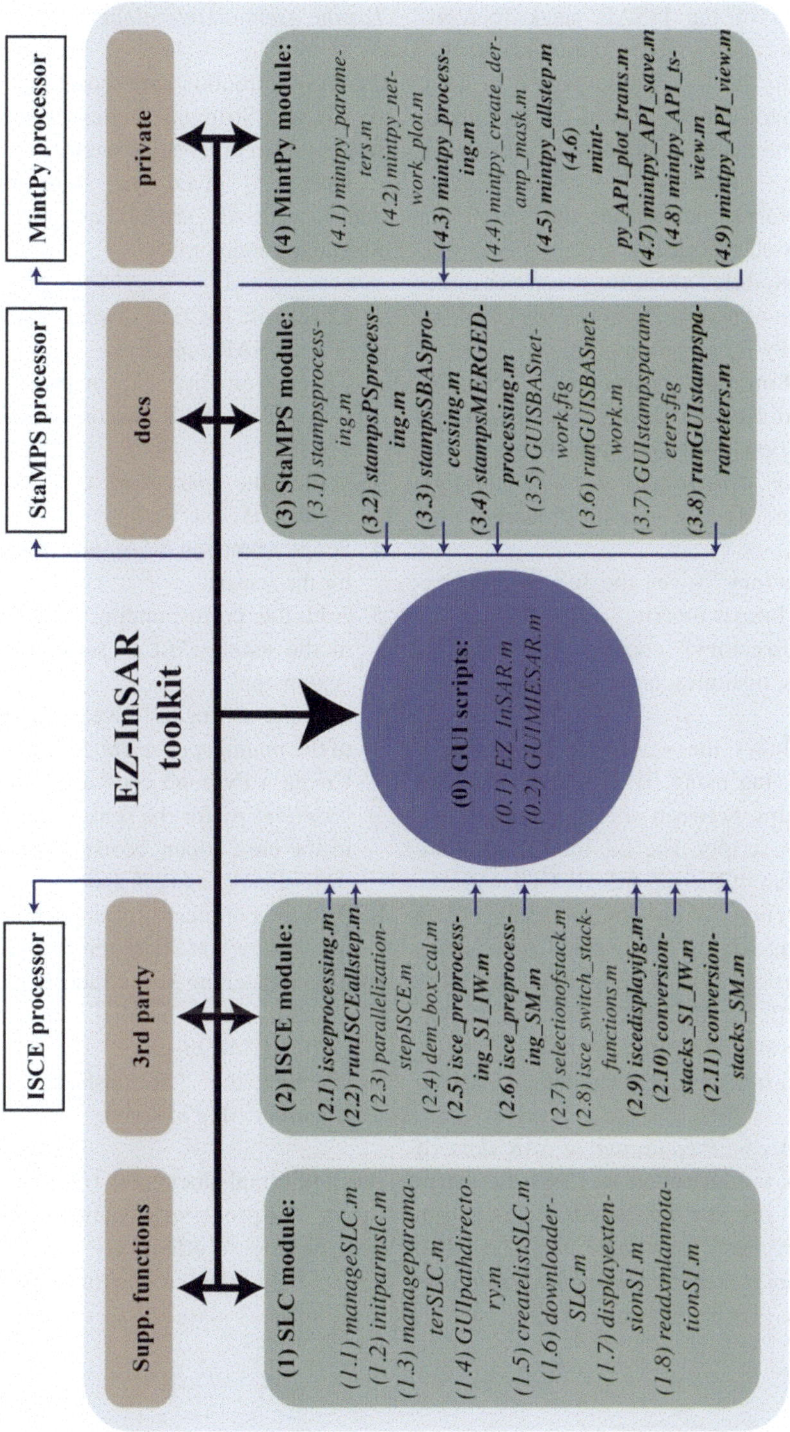

Fig. 6.19 Main structure of EZ-InSAR toolkit. The purple circle is the two main scripts of EZ-InSAR. The green boxes give the sub-directories of EZ-InSAR with names of main MATLAB® scripts in italic. The brown boxes are the other sub-directories containing complementary tools of EZ-InSAR. The blue arrows represent the links between EZ-InSAR functions and InSAR software

the creation of the InSAR stack (coregistered images and/or interferogram stack, etc.) according to the user's parameters. In addition, the management (download and verification) of the required DEM is done here.
- **"StaMPS module"** (*StaMPS_functions* directory) contains all the scripts for StaMPS-based processing.
- **"MintPy module"** (MintPy_*functions* directory) is a set of scripts for time series analysis using MintPy-based processing.
- **"Supp. Function directory"** contains some additional MALTAB®, Bash shell and Python3 scripts for the required tools.
- **"3rd party directory"** contains non-proprietary (or slightly modified) scripts for EZ-InSAR.
- **"docs directory"** gives the full documentation of EZ-InSAR toolkit.
- "**private directory"** contains the logos of universities, institutes, and funding agencies.

Figure 6.20 shows the workflow for a typical InSAR processing using EZ-InSAR toolkit and the relationships between the user inputs and the EZ-InSAR scripts. The EZ-InSAR structure is similar to that of the EZ-InSAR GUI with the three main sections of InSAR processing (data preparation, InSAR processing and time series analysis). Vertically and highlighted by blue arrows, the different user interactions (e.g., buttons) are indicated with the identification numbers of the EZ-InSAR scripts (in italic and given by Fig. 6.19).

For example, in *Preparation of SAR data*, if the user clicks on the button to check the extent of the SAR images, EZ-InSAR will launch the scripts named *manageSLC.m*, and then *displayextensionS1.m* with *readxmlannotationS1.m* function.

For the users and developers

EZ-InSAR toolkit can process Sentinel-1 data (in IW and Stripmap modes), and Stripmap-like images from other SAR satellites.

Here are given the steps to implement additional SAR sensors and new sensors (in Stripmap mode only):

1. Check if ISCE software can process the desired SAR sensor.
2. Check from line 172 in the *GUIMIESAR.m* script if the SAR sensor is available. If not, add it.
3. Write the part for this sensor in the *createlistSLC.m* script.
4. Script a function as *displayextensionTSXPAZ.m* for the sensor.
5. Add the corresponding line for the sensor in the *manageSLC.m* script (under the case "extension").
6. Add the sensor in "save" and "update" cases in the *manageparameterSLC.m* script.
7. Create a Python3 code as *coarse_TSX_PAZ_baselines.py* for the sensor and add this script in the case "open_coarse_network_check" in the *isceprocessing.m* script.
8. Add the command for unpacking in *isce_preprocessing_SM.m.* If the user has scripted a new unpacking script, add it to the *3rdparty* directory and modify line 694 of the *GUIMIESAR.m* script.
9. Change the "prerunstack" case in the *isceprocessing.m* script.

An additional document (manual) summarises the InSAR processing chain and providing tutorials on how to effectively use it. This is available in the GitHub repository (see link above).

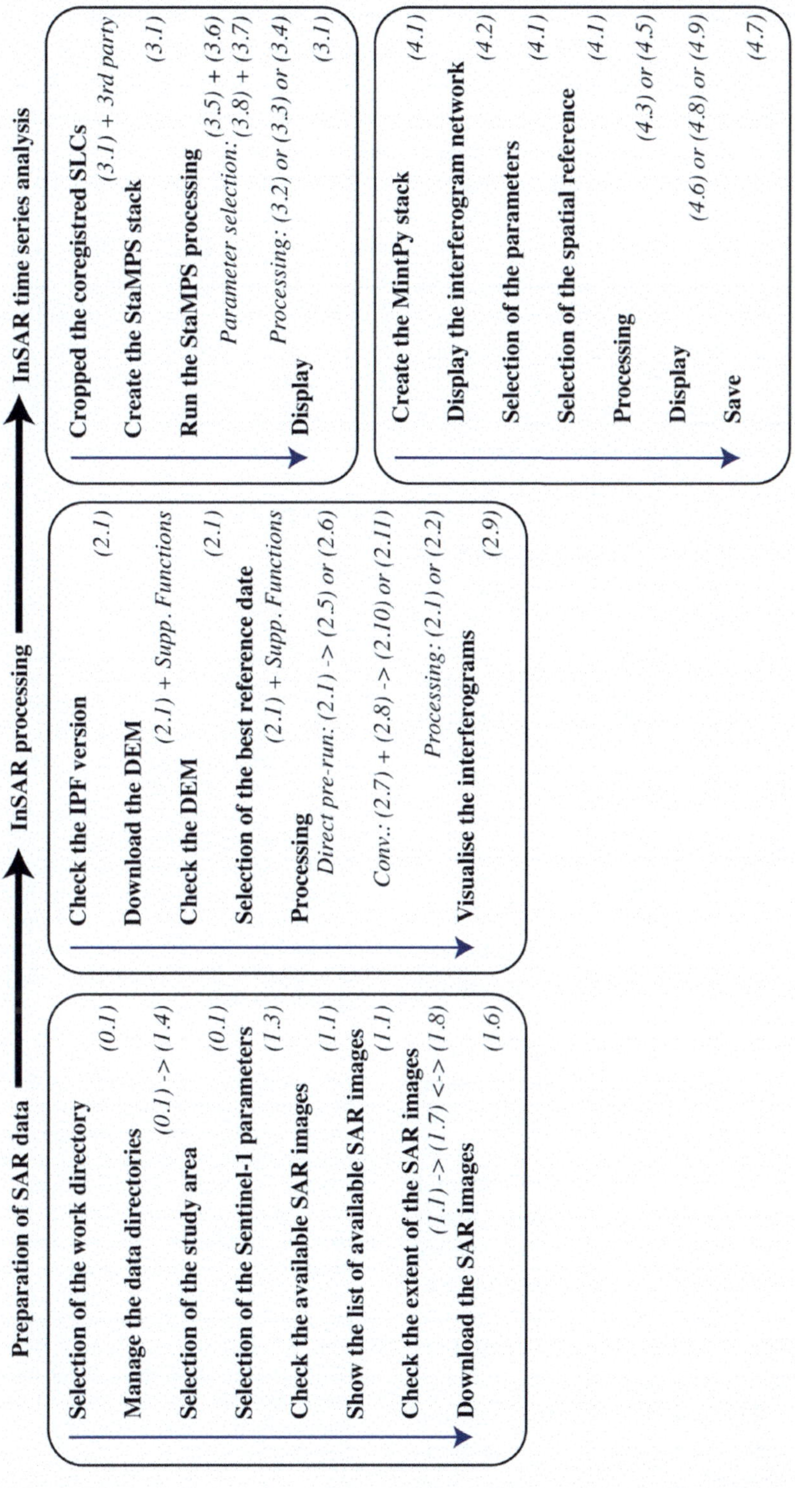

Fig. 6.20 EZ-InSAR workflow

6.6 Conclusions

Citizen science through the AGEO platform involves the active participation of people with minimal expertise in geohazards to collect and report scientific data. It has been proved that the citizens' contribution can be extremely positive in diverse scientific fields.

This chapter has presented the design and development of a set of customizable software artefacts that will support the functional needs of citizen observatories of geohazards. Especially within multi-hazard settings, the set of software resources is available through the AGEO platform for people to develop a shared understanding of geohazards. Moreover, communities that are impacted or might be impacted can contribute to geohazard forecasting and early warning systems.

The AGEO Citizens' Observatory Platform brings citizens and authorities together, and it enables communication and interaction between them. Within this platform is a map-based interface for social engagement through which citizens can observe the collective citizen-observed geohazard data.

Coupled to the AGEO platform is the AGEO Citizens' Observatory Mobile App. This tool enables citizens to monitor and report an event they are facing. Citizens are given a streamlined and intuitive scheme to record and upload pertinent digital data for a range of hazards. These hazards currently include: wildfires, tsunami, earthquakes, volcanic eruptions, flooding, sinkholes, and building settlement.

Satellite-derived multi-temporal Interferometry of Synthetic Aperture Radar (InSAR) data for the AGEO Citizen Observatory Pilots are essential for the citizens observatory. An easy-use, open-source, multi-temporal InSAR processing tool for future (post-AGEO) geohazard monitoring has been described.

The AGEO Platform can be the medium between citizens and responsible authorities, national or local, as an icebreaker to start some form of conversation and interaction. We believe collaborations that the AGEO project can produce will encourage authorities to see the opportunity in citizen science and attract regular citizens to voluntarily be more interested in their environment's status and what is happening around them.

In summary, this chapter has presented the design and development of a set of customizable software artefacts that will support the functional needs of the citizen observatories for geohazards. This set of software resources will be made available for use by other Atlantic Interreg initiatives involving citizen science or citizen engagement also.

References

Görgü, L., O'Grady, M., Mangina, E., & O'Hare, G. M. P. (2022). Participatory risk management in the smart city. *2022 IEEE International Smart Cities Conference (ISC2)*, Pafos, Cyprus, pp. 1-6. https://doi.org/10.1109/ISC255366.2022.9922040.

Schubert, A., Miranda, N., Geudtner, D., & Small, D. (2017). Sentinel-1A/B combined product geolocation accuracy. *Remote Sensing, 9*(6), 607. https://doi.org/10.3390/rs9060607

AGEO Citizens' Observatory Web Platform. Retrieved from https://ageoplatform.eu.

AGEO Mobile Observatory on Google Play Store (for Android devices). Retrieved from https://play.google.com/store/apps/details?id=eu.ageoplatform.app.

AGEO Mobile Observatory on Apple AppStore (for iOS devices). Retrieved from https://apps.apple.com/ie/app/id6444765392.

Werner, C., Wegmüller, U., Strozzi, T., & Wiesmann, A. (2000). Gamma SAR processor and interferometry software. In Proceedings of the ERS-ENVISAT Symposium, Gothenburg, Sweden.

Rosen, P. A., Hensley, S., Joughin, I. R., Li, F. K., Madsen, S. N., Rodriguez, E., & Goldstein, R. M. (2000). Synthetic aperture radar interferometry. *Proceedings of the IEEE, 88*, 333–382.

Hooper, A. (2008). A multi-temporal InSAR method incorporating both persistent scatterer and small baseline approaches. *Geophys Res Lett, 35*(16). https://doi.org/10.1029/2008GL034654.

Hooper, A., Bekaert, D., Spaans, K., & Arikan, M. (2012). Recent advances in SAR interferometry time series analysis for measuring crustal deformation. *Tectonophysics, 514*, 1–13. https://doi.org/10.1016/j.tecto.2011.10.013

Zhang, Y., Heresh, F., & Falk, A. (2019). Small baseline InSAR time series analysis: Unwrapping error correction and noise reduction. *Computers & Geosciences, 133*.

Spataro, F., Pavia, P., Roscigno, R., Torres, R., Bibby, D., & Cossu, M. (2021). AIS P/L on SAR satellite: the Copernicus Sentinel-1 solution. EUSAR 2021; 13th European Conference on Synthetic Aperture Radar.

Rosen, P. A., Hensley, S., Shaffer, S., Veilleux, L., Sagi, V. R., Satish, R., Chakraborty, M., Misra, T., & Bhan, R. A. (2015). The NASA-ISRO SAR Mission—An International Space Partnership for Science and Societal Benefit. *IEEE Int Radar Conf (Radarcon), 2015*, 1610–1613. https://doi.org/10.1109/RADAR.2015.7131255

Hrysiewicz, A., Wang, X., & Holohan, E. P. (2023). EZ-InSAR: An easy-to-use open-source toolbox for mapping ground surface deformation using satellite interferometric synthetic aperture radar. *Earth Sci Inform*. https://doi.org/10.1007/s12145-023-00973-1

Improving Risk Management for Geohazards Based on Citizens' Observatories

7

Delta Sousa e Silva, Maria Luísa Sousa,
Laura Caldeira, Joana Carreto, Maria João Coelho,
Filipe Telmo Jeremias, Rute Ramos,
António José Roque and João Bilé Serra

Abstract

Within the realm of natural hazards, geohazards are recognised as being particularly complex and often having the potential of triggering disasters. This complexity has led to increased political commitment and interest in engaging multi-stakeholder and citizens in disaster risk management. The AGEO project explored new forms to foster collaboration between civil society and authorities, using the rapidly developing field of citizen science and its innovative tool of citizens' observatories (COs). Five COs were established in different areas across the Atlantic region. This chapter summarises the lessons learned from this experience, based on the results of a comparative analysis and evaluation of AGEO's COs. The aim was to capture the benefits of creating and operating people-centred COs for geohazards monitoring, coupled with citizen education and training. The key findings of this evaluation formed the basis for a roadmap for establishing future COs, which is outlined in this chapter.

D. S. e. Silva · M. L. Sousa (✉) · L. Caldeira ·
J. Carreto · M. J. Coelho · F. T. Jeremias · R. Ramos ·
A. J. Roque · J. B. Serra
Laboratório Nacional de Engenharia Civil,
Av. do Brasil 101, Lisboa, Portugal
e-mail: Luisa.sousa@lnec.pt

7.1 Introduction

Unprecedented advances in science and technology in modern societies created opportunities for people to enjoy a safer and rewarding existence. However, despite these achievements, current trends reveal that the overall economic losses caused by natural disasters keep increasing worldwide (EM-DAT, 2022). These trends question the effectiveness of risk management policies and call for a paradigm shift. Disaster risk management (DRM) needs to be a multi-sectoral-systemic effort, top-down and bottom-up, across scales from global to local, and based on the active engagement of common people (Wisner et al., 2012).

The Platform for Atlantic Geohazard Risk Management (AGEO) has explored new ways of engaging people in geohazard risk management, taking advantage of the emerging field of citizen science, in particular, the citizens' observatories (COs) approach, to promote public participation. This chapter provides a critical review of achievements, along with a discussion of the innovations that the COs approach bring to the field of participatory methods applied to disaster risk reduction (DRR).

It is important to clarify the concept of COs as adopted in this chapter. COs are understood as a participatory approach, aligned with citizen science, aimed at fostering collaboration between citizens, scientists, decision makers and

R. C. Gomes et al. (eds.), *Citizens' Observatories on Geohazards*, Geoenvironmental Disaster Reduction,
https://doi.org/10.1007/978-3-031-53371-6_7

technical officials to support community-based risk management.

This chapter is organized into six interrelated sections. The next two sections set the discussion into context. Section 7.2 briefly outlines the policy background underlying DRM, both at international and European Union (EU) level, as well as the way in which science has approached the field. An increasing emphasis is placed on the urgency of a paradigm shift from hierarchical management systems to community-based risk management processes. Section 7.3 addresses the concepts of DRR and DRM. Afterwards, Sect. 7.4, situates COs approach in the realm of DRM participatory approaches. There is a large volume of applied research and experience on empowering local communities for more effective DRR. Following this policy and theoretical background, Sect. 7.5 presents a comparative analysis and a brief evaluation of the five AGEO pilots where COs approach has been tested. The review concludes with a summary of lessons learned based on the experiences of the pilots, highlighting the challenges of adopting this participatory approach to risk management. Finally, Sect. 7.6 sets up the basis for a roadmap for applying the COs' approach to geohazard risk management, recognising that implementing COs to support risk management is a systematic approach that requires in-depth knowledge of the risks and target areas as well as time and persistence to build solid relationships with local people.

7.2 Policy Background

A brief analysis is made of the main policy developments both at international and European level, taking into account their relevance for DRR and DRM of natural and human-induced risks, citizen science and the linkages between the two topics.

The Global Assessment Report on Disaster Risk Reduction (UNDRR, 2019) and *PreventionWeb*, (UNDRR, 2023) provides an overview of the evolution of the global policy initiatives for DRR. From this report, five policies stand out whose chronology is shown in Fig. 7.1.

The **Sendai Framework for Disaster Risk Reduction 2015–2030**, or simply called "the Sendai Framework" (UNISDR, 2015) contains four priorities of action, quoted below:

(i) understanding disaster risk,
(ii) strengthening disaster risk governance,
(iii) investing in DRR, and
(iv) improving disaster preparedness to make effective impact on response and to "Build

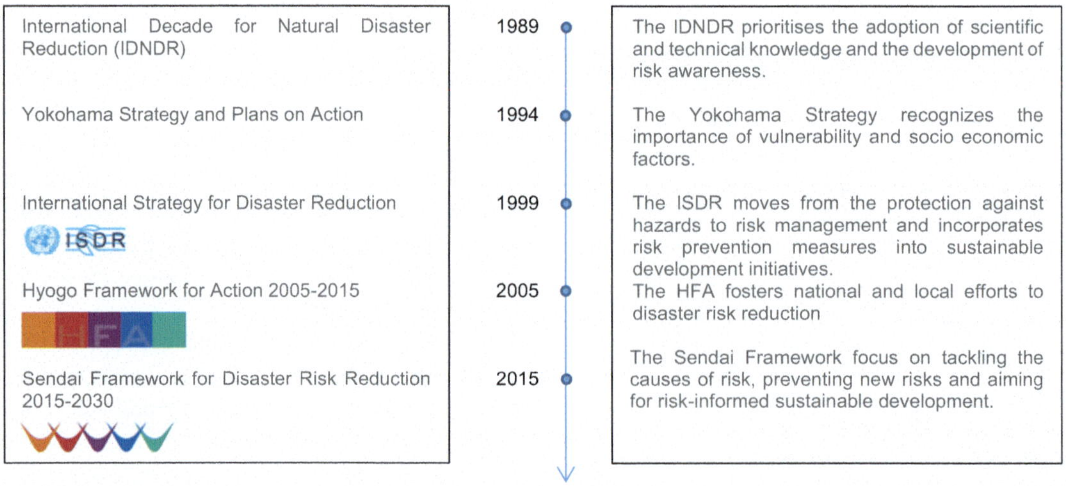

Fig. 7.1 Timeline of global policy for DRR (adapted from UN, 2020)

Back Better" in recovery, rehabilitation and reconstruction.

The Sendai Framework is one of three important Agreements that were adopted in 2015 by the United Nations (UN). The other two Agreements are the **Sustainable Development Goals of Agenda 2030** (UN, 2015) and the **Paris Agreement on Climate Change** (UNFCCC, 2016). All these initiatives share a common concern of achieving DRR. However, the Sendai Framework has a specific focus on preventing new disasters and mitigating existing risks through the implementation of various measures that aim to enhance disaster resilience. Moreover, the Sendai Framework calls for an all-of-society engagement and partnerships for DRR, as well as encourages the participation of citizens in voluntary work.

Starting in December 2019, the European Commission embarked on an ambitious roadmap known as the European Green Deal (COM(2019) 640, 2019). The initiative aims to convert the EU economy into "a modern, resource-efficient and competitive economy" that should be climate neutral by 2050. The European Green Deal underlines the EU's commitment to reaching the Paris Agreement goals. Moreover, the Green Deal has a pivotal role in the Commission's approach to implementing the UN's 2030 Agenda and achieving the Sustainable Development Goals. Regarding the Sendai framework, the European Commission supports the EU Member States in the implementation of this agreement, thus advocating for a wide range of DRM programs closely aligned with the goals of such a global initiative (DG ECHO, 2023).

The EU has formulated extensive policy and legal frameworks dedicated to DRR. These initiatives establish the basis for efforts and strategies aimed at mitigating the physical, social, economic, and environmental consequences of disasters. Additionally, they actively encourage citizen participation and foster a culture of self-protection by enhancing awareness, promoting education, and facilitating widespread dissemination of relevant information.

In this respect, it is worth noting **the Union Civil Protection Mechanism** ('Union Mechanism' or **UCPM**), established by the Decision No 1313/2013/EU (2013), aiming to "strengthen the cooperation between the Union and the Member States and facilitate coordination in the field of civil protection to improve the effectiveness of systems for preventing, preparing for and responding to natural and man-made disasters". The act was amended by Decision (EU) No 2019/420 (2019) and by Regulation 2021/836/EU (2021), which lay down the foundation for **the Union disaster resilience goals** in civil protection, materialised by the 2023 European Commission Recommendation on Union disaster resilience goals (Commission Recommendation, 2023/C 56/01, 2023). The Recommendation set out five disaster resilience goals related to disaster actions of the Civil Protection System, namely improving risk assessment, preparedness, early warning, response and ensuring robustness. In the context of the AGEO project, the most relevant Union disaster resilience goal is undoubtedly the second one, which deals with preparedness, i.e., to "increase risk awareness and preparedness of the population". This Union disaster resilience goal has specific purposes quoted in the following: (i) increase the overall level of risk awareness, (ii) improve public access to disaster risk information, (iii) increase the awareness and adoption of risk prevention and preparedness measures amongst the population, and (iv) enhance the culture of risk prevention amongst the population.

Note that the UCPM and the Union disaster resilience goals, while intending to involve citizens, do not call for a change of paradigm on DRM, i.e., abandoning classical top-down DRM systems in favour of people-centred disaster risk governance. Yet, this paradigmatic change is advocated in the European Commission Disaster Risk Management Knowledge Centre (DRMKC) report, "Science for Disaster Risk Management 2020" (Casajus Valles et al., 2020). Other initiatives also place a strong emphasis on citizens' involvement and strive to empower local communities to respond effectively to risks. For instance, Claassen et al. (2020) provide justification for the implementation of participatory risk management systems for building more resilient communities. They advocate for the inclusion of citizens' perspectives throughout all stages of the risk management process. The authors propose the development of comprehensive bottom-up approaches encompassing legislation, organisations, and technologies that facilitate citizen consultation, information dissemination, and empowerment. These measures would enhance risk awareness, preparedness, and shared responsibility for risk reduction, thereby turning it into a collective societal mission that requires contributions from everyone.

In practice, the transition towards the new paradigm of DRM, as referred above, is more noticeable when it comes to (1) environmental monitoring, (2) climate change risk management, or (3) supportive actions to achieve the UN Sustainable Development Goals, as exemplified below:

1. The potential of citizen involvement has been widely explored in environmental protection policies. Indeed, the recognition of scientific and societal benefits of citizen science had a major breakthrough in 2008, in the scope of European environmental policies, when the creation of the first CO for monitoring water quality was officially announced (Haklay, 2015 as cited in Schade et al., 2021). Another example is the European Commission Staff Working Document on "Best Practices in Citizen Science for Environmental Monitoring" (SWD 149, 2020), which exploits opportunities and provides guidelines and methods for the easy usage of citizen science data and information for environmental monitoring and reporting (EPA, 2022).
2. In what concerns climate risks, the European Climate Pact (COM(2020) 788 final, 2020), acknowledges the active engagement of individuals, communities and organisations to help deliver the European Green Deal, anchored in open science partnerships, EU research programmes, or other funding opportunities at a national level. An example is the Horizon Europe Mission on "Adaptation to Climate Change, including Societal Transformation" (Hedegaard et al., 2020). The Mission aims to combine research and innovation with an extensive engagement of citizens and stakeholders. This includes fostering communities of practice and granting access to data and knowledge to assist regions and communities to understand and manage climate risks. Lastly, the Mission is developing a unified framework for climate risk assessment, guiding and training communities and regions in adopting community-based emergency plans for climate risk management, in support of the new EU Strategy on Adaptation to Climate Change (COM(2021)82 final, 2021).
3. Fritz et al. (2019) highlight an increasing demand for detailed spatial and temporal data to monitor progress towards achieving the UN Sustainable Development Goals. The authors urge all interested parties to encourage the uptake of citizen contributions, establish the necessary framework for citizens to provide accurate data for monitoring the Sustainable Development Goals and actively participate in their implementation.

Other initiatives are exploring the potential benefits of integrating real-time local observations provided by crowdsourcing (e.g. the so-called Volunteered Geographic Information) and citizen science projects into Copernicus tools, such as the Copernicus Emergency Management Service. One notable example is the use of near real-time in situ data in Rapid Mapping or Risk Recovery Mapping activations as well as in emergency awareness situations (Arnoldi et al., 2021). Contributions put in place by citizen science practices such as verifying and validating Earth observations (the so-called "ground truth" of occurrences) and filling data gaps when official data sources are scarce, improve the success of early warning systems (WeObserve consortium, 2021).

In addition, the popularity of citizen science has increased in recent years due to the proliferation of technological tools (e.g. sensors, Apps, online activities) that allow citizens to collect and share data about hazards and their impacts with researchers or with DRM practitioners.

On the other hand, since the advent of the CO concept, the European research priorities and frameworks offer numerous opportunities to advance citizen science, for instance, under Horizon 2020 and Horizon Europe (2021–2027) research and innovation programmes, or under the Interreg Atlantic Area funding programmes (WeObserve Consortium, 2021). The Pact for Research and Innovation in Europe (COM(2021) 407, 2021) also deserves mention as it prioritises active citizen and societal involvement across all aspects of research and innovation (R&I). This approach aims to enhance public awareness of the benefits of R&I in everyday life and facilitate access to and adoption of new solutions.

In short, citizen science has multiple relationships with global and European policies and receives enthusiastic support from the EUs research and innovation programmes. Yet, as will be seen in this chapter, there are still a number of challenges and barriers that prevent citizens' views, inputs and solutions from being accepted in risk management practices.

7.3 Reduction of Disaster Risk Through Risk Management

The report of the open-ended intergovernmental expert working group on indicators and terminology concerning DRR (UNDRR, 2016) includes a comprehensive group of indicators for evaluating the advancements made in the implementation of the Sendai Framework for Disaster Risk Reduction 2015–2030. Additionally, it has updated the widely used "2009 UNISDR Terminology on Disaster Risk Reduction". The report defines **hazard** as "a process, phenomenon or human activity that may cause loss of life, injury or other health impacts, property damage, social and economic disruption or environmental degradation" and **disaster** as "a serious disruption of the functioning of a community or a society at any scale due to hazardous events interacting with conditions of exposure, vulnerability and capacity, leading to one or more of the following: human, material, economic and environmental losses and impacts".

Disasters result from the combination of hazards, vulnerability, and exposure of people and assets, but **"hazards do not have to turn into disasters"** (UNDRR, 2023). In fact, the severity levels of hazards, the value of exposed elements, and the vulnerability features of those elements may all be regarded as acceptable, if efforts to reduce risk and strengthen resilience to disasters have been made, i.e., appropriate risk management practices have been implemented.

DRR and DRM share the common objectives of preventing future risks, reducing existing risks and managing residual risks, with the former being regarded as a policy objective, while the latter encompasses the practical implementation of measures to manage risks (Hicks et al., 2019; UNDRR, 2016). To note that the concept of DRM was initially developed to address the management of risk associated with natural hazards (Surminski et al., 2017).

DRM has evolved from early works in the 1960s focusing on understanding natural hazards, to recognising the importance of vulnerability and socio-economic factors in the 1970s, illustrated by the ground-breaking paper "Taking the naturalness out of natural disasters" (O'Keefe et al., 1976). Back then it became clear, for both scientists and practitioners that people's vulnerability was not solely explained by their physical proximity and exposure to hazards. For a more thorough understanding of vulnerability, it was imperative to consider people's pre-existing social and economic conditions. They actively interfered on their vulnerability and, consequently, should form part of risk equation (Blaikie et al., 1994). This insight led to community-scale approaches that emphasised the importance of uncovering local vulnerabilities and capabilities, followed by grassroots initiatives aimed at involving communities in DRR. Today, this approach is formally developed as an alternative paradigm, to the still dominant top-down DRM paradigm, and became internationally known as *Community-Based Disaster Risk Reduction* (CBDRR) (Wisner et al., 2012).

Later on, the fifth edition of the UN "Global Assessment Report on Disaster Risk Reduction" (UNDRR, 2019) introduced the Global Risk Assessment Framework (GRAF) as a means to facilitate the implementation of the Sendai Framework. GRAF serves as an illustrative model of a risk assessment framework that understands risk as a combination of three key components, hazard(s), exposure and vulnerability.

Recently, the sixth edition of the UN "Global Assessment Report on Disaster Risk Reduction" (GAR) (UNDRR, 2022) underscores the significance of tackling systemic risk and comprehending its impact on sustainable development. It reinforces the perspective laid out in the previous GAR report (UNDRR, 2019) of integrating risk-informed disaster reduction into strategies for sustainable development. Additionally, the sixth GAR report urges the involvement of affected populations in consultations to expedite risk reduction and enhance resilience-building.

Systemic risk assessment goes beyond the traditional approach of assessing risk through a limited set of consequences and the likelihood of their occurrence (IRGC, 2017). A systemic approach recognises that the impact of an event may be exacerbated by cascading effects capable of disrupting an entire system or even spilling over to interconnected systems and sectors, crossing national borders, and culminating in severe and irreversible consequences, or even causing global impacts. Due to the increasing level of connectivity between systems, the International Risk Governance Council (IRGC), argues that systemic risks cannot be managed within a single sector and require close cooperation between several stakeholders such as policymakers, public institutions, industry, academia and civil society at large.

There are several other conceptual frameworks for risk assessment. Another example is the extensively used international standard ISO 31 000:2018 Risk management—Guidelines. This framework delineates risk by considering its sources, potential events, their resulting consequences and the likelihood of occurrence (Hagenlocher et al., 2020). The standard defines risk as "the effect of uncertainty on objectives", thus assuming an inherent connection between risk and uncertainty. In ISO 31 000, risk management is built upon a systematic process that involves activities such as communication and consultation, establishing the context, scope and risk criteria, assessing risks, implementing treatment, monitoring progress, conducting reviews, maintaining records and reporting on risk. ISO 31 000 recognises risk management as a cyclic process, emphasizing that it is not a one-time activity, but rather an ongoing and iterative process that necessitates regular review and update of risk assessment and treatment strategies.

From the broader perspective of risk governance, the IRGC framework is worth mentioning. This risk governance framework is a modular, adaptable and supports various risk management frameworks. Going beyond standard risk analysis and management, the IRGC framework supports effective risk governance strategies that take into account societal values and different perceptions of risk, and examines linkages between potentially affected parties (IRGC, 2017). Experts argue that the IRGC framework is an example of a more convenient model for engaging citizen observatories in risk management than the ISO 31 000 approach (Rodrigues, 2023).

Equally important is that the Sendai Framework focuses on risk management rather than disaster management, aiming at better managing uncertainty, addressing uncontrollable risks, and improving knowledge of dynamic interactions within systemic risk (UNDRR, 2019). Moreover, the Political Declaration of the High-Level Meeting of the Sendai Framework Midterm Review, adopted in May 2023, adverts for the need for a preventive approach to DRR that places people at the centre, aligning with the principles of the 2030 Agenda for Sustainable Development.

DRM also sits on a cyclic structure, as shown in Fig. 7.2, ensuring adaptation to changing circumstances and continuous improvement through the incorporation of risk knowledge and best practices. The illustration is one of many existing schemes to represent the disaster

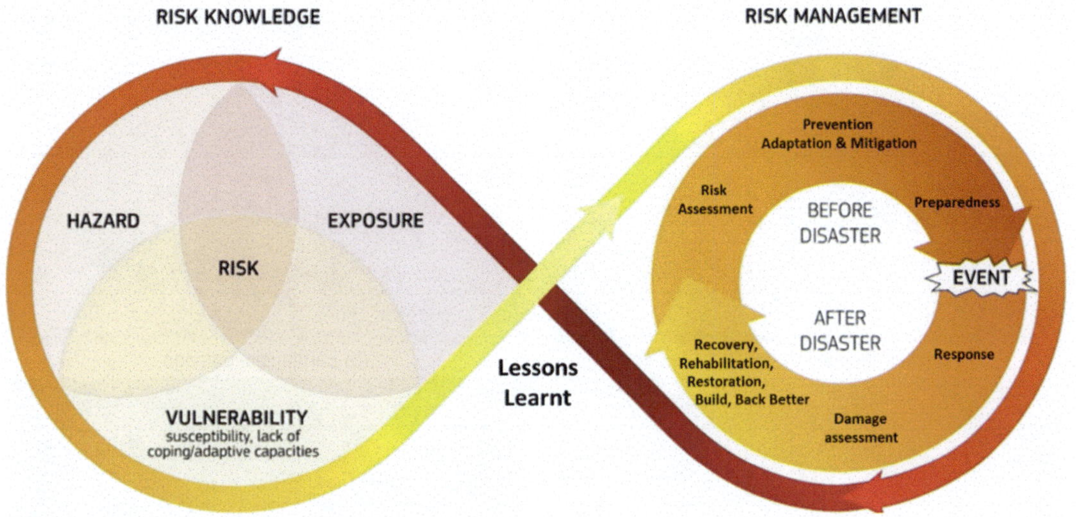

Fig. 7.2 Relationship between risk knowledge and DRM (DRMKC, 2023)

cycle (right side of Fig. 7.2). It was chosen as an illustrative example since it integrates practical actions such as damage and risk assessment into the disaster management cycle. In addition, it associates risk knowledge and lessons learnt with the risk management process, i.e., scientific, and practical dimensions, that form the base of risk-informed decision-making. Finally, it demonstrates that risk knowledge and risk management form an integrated and dynamic process that is inherently iterative. Risk knowledge encompasses the assessment of hazards, exposure, and vulnerability, and ultimately the assessment of risk, contributing to informed decisions in DRM. Risk knowledge assumes an essential role throughout all phases of the disaster cycle, i.e., the pre-event phases of mitigation, prevention and preparedness, and the post-event phases of response and recovery. For instance, after an event there is progress in the assessment of risk as a result of data and knowledge derived from the event characterisation and observed impacts, which in turn improve DRM (Casajus Valles et al., 2020).

It is important to recognize that, in practice, these phases are not distinct and discrete. Instead, they form a continuous process and overlap, or boundaries between them are rather blurred.

To finalise this section, special attention will be paid to the characterisation of geohazards. Recently, the United Nations Office for Disaster Risk Reduction (UNDRR) with the International Science Council have recognised the importance to identify a complete set of hazards, among them **geohazards**, to facilitate informed decisions on DRR and DRM practices, and report on the three international agreements mentioned in Sect. 7.2, in particular the Sendai Framework. In this context, it was identified an open-ended list of 302 specific hazards to address (UN, 2021) and the scientific knowledge base for defining these hazards. The hazards were grouped into eight types: meteorological and hydrological (60), extraterrestrial (9), geohazards (35), environmental (24), chemical (25), biological (88), technological (53), and societal (8), most of them with a UN agreed definition. The 35 specific hazards included in the geohazard type are divided into three clusters: seismogenic, volcanogenic, and other geohazards (also titled shallow geohazards). The first two relate to the Earth's internal geophysical processes, and the latter to surface or near-surface processes.

As already mentioned in previous chapters, the AGEO project addresses hazards such as coastal and urban floods, landslides and rockfalls, coastal erosion, earthquakes, liquefaction,

Fig. 7.3 Illustration of some geohazards addressed in the AGEO project

subsidence, and uplift. Figure 7.3 illustrates five of these hazards. Notice that two hazards are illustrated in one of the pictures, rockfall and coastal erosion. Some are induced by geological processes; others result from hydrological and meteorological phenomena or are driven by anthropogenic activities, making it difficult to define boundaries. For example, if an earthquake is the triggering factor, landslides (including rockfalls) are included in the geohazard type. However, landslides can also be triggered by hydrological conditions, in which case, they have been classified as a meteorological and hydrological hazard since 2021. However, taking into account the initial concept and the very name (and acronym) of the project, as well as for the sake of simplification, all hazards considered in the AGEO project, are herein referred to as geohazards.

7.4 Situating Citizen Observatories on the Realm of Participatory Approaches for Disaster Risk Management

As previously emphasised, DRM in western societies has historically been shaped by top-down and sectoral systems (Handmer et al., 2007; McEntire et al., 2002; Boersma et al., 2020). As referred in Sect. 7.3, such a classical DRM paradigm was questioned in the 1970s when it became clear that a more thorough comprehension of vulnerability to natural hazards demanded going beyond the analysis of geomorphological processes to include the social processes that generate differentiated patterns of human susceptibility to hazards. Scientific research on vulnerability and risk reduction became the basis for the above-mentioned

CBDRR new paradigm. At the policy level, governments have been encouraged to involve local communities in DRR. Delica-Wilison et al. (2012) define participatory approaches in CBDRR as based on three main principles. One principle is the emphasis on *people's participation and empowerment.* The strategy for DRR must be rooted in people's understanding of risk and their own coping capacities. CBDRR may require external resources and knowledge, but community and people's engagement in risk reduction processes is central. The second principle concerns the imperative of considering risk reduction as part of *development-oriented actions*. Once the root causes of vulnerability rely in pre-existing socio-economic conditions and political processes, development-oriented actions should integrate DRR into their strategy. The third principle concerns the importance of taking CBDRR as *a multi-stakeholder approach*. Local communities are the locus of DRM, but they are not self-sufficient and need the support of national government and external stakeholders to provide them with resources and regulatory mechanisms designed to legitimate local risk reduction activities and disaster preparedness.

CBDRR gained wide ground over these past fifty years. It has become well established at international level. It has also become a scientifically grounded approach with a large body of research on social vulnerabilities and capabilities as well as specific methodological frameworks aimed at promoting community-based DRR and risk management (Delica-Wilison et al., in ibid). One of these frameworks is the Participatory Action Research (PAR), whose relevance to this discussion comes from its points of convergence with citizen science. As described by McCall et al. (2012), PAR stands on a methodology in which risk reduction is sought through a process where researchers and inhabitants work together to examine their condition of vulnerability and transform it. Local knowledge and capacities are brought into the social arena, jointly with the formally accepted knowledge of scientists and decision-makers, with the aim of supporting the development of a risk management strategy.

As mentioned in Sect. 7.2, COs were first introduced in 2008 by environmental sciences, trusting that it was imperative to engage with citizens and motivate them to become involved in environmental monitoring, rather than taking them as silent observers.

The idea of COs was embraced as a potentially powerful tool, and several research projects involving citizen science have been developed. There is now a significant body of scientific research on the subject. Its application has gone beyond the strict domain of environmental governance and has been extended to other domains, namely DRM (Hicks et al., 2019). Despite this vitality, there is still a lack of a clear definition of what is meant by CO. This term labels a diversity of initiatives, ranging from those aimed at motivating the public in general (i.e., the crowd) to more focused activities aimed at building a network of citizens to become engaged in local environmental monitoring or risk management (Hicks, in ibid; Lanfranchi et al., 2014). Thus, what do we mean by COs and what innovations do they bring to participatory approaches in DRM?

Recalling the definition given in the introduction to this chapter, citizen observatories are a participatory approach aligned with citizen science that aims to foster collaboration between citizens, scientists, decision-makers, and technical officials to support community-based risk management. Community participation has a long tradition in the DRM field. Any effort to define the concept of citizen observatories should clarify the innovations they bring.

Participatory DRM approaches and COs share the same principle of making *people's participation and empowerment* a core goal of risk management. The COs approach emphasises citizen collaboration in scientific monitoring processes as a means of generating community-based risk management projects (Hicks et al., in ibib). They also differ in the emphasis placed on technology as a means of communication. COs approaches are conventionally technology-led initiatives, with a strong emphasis on the innovation of new digital forms of communication and information exchange.

7.5 AGEO Observatories: Evaluation and Lessons Learned

7.5.1 Aims and Methodological Approach

AGEO aimed to explore new forms of engaging people in geohazards risk management, through the development of COs in five pilot areas across the Atlantic region, respectively: Canary Islands (Spain), Brittany (France), Lisbon and Madeira Island (Portugal), and the Causeway Coast (Northern Ireland).

Project's design included an evaluation of the pilots' experience with the COs approach. As part of the learning process, the evaluation was sought as an opportunity for a better understanding of the benefits of COs for people-centred geohazard risk management processes and, also, a way of identifying *lessons learned* coming from the pilots' experience. With these two goals in mind, a comparative analysis of the five AGEO pilots was pursued, which became the basis for a brief evaluation and identification of lessons learned.

As mentioned above, scientific literature encompasses diverse definitions of CO. For analytical purposes, it was imperative to clarify the concept and operationalise it into analytical dimensions. Based on the literature review, four dimensions were identified that cut across several definitions of CO. The first dimension concerns the issue of multi-stakeholder *collaborative participation*. COs aim to become a platform for stakeholder's dialogue and engagement in risk management. The second dimension concerns the *specific role* assigned to citizens. They are stimulated to become involved in activities that traditionally fall within the realm of science and institutions, as for example data collection, risk monitoring, and hazard mapping. As highlighted by Rubio-Iglesias (2013), instead of just recipients of information, citizens are called to perform "new functions", such as *in-situ* observations and hazard mapping. This specific role of citizens sends us to the third cross-cutting dimension, *complementarity*. Both scientists and non-scientists bring different forms of knowledge to the same arena, which, once integrated, generate new understandings of risk and form the basis for a shared agenda for DRR (Hicks et al., 2019). The fourth dimension concerns the emphasis put on technology as means of multi-stakeholder communication. COs are conventionally technology-led (Hicks et al., in ibid). Strong emphasis is put on the ability of specialised monitoring devices (earth observation systems, sensors), mobile personal devices (smart phones, tablets, laptops) and *social media* as means of stakeholder communication and exchange communication. As mentioned, the innovations of COs, by comparison with previous participatory approaches, lies in the third and fourth dimensions. The "new functions" attributed to citizens along DRM cycle intends to become a way of promoting risk awareness, behavioural change, and local empowerment. The other innovative breakthrough concerns the use of technology for information exchange.

AGEO pilots' comparative analysis was structured around the above-mentioned four cross-cutting dimensions. In other words, it aimed to answer to four main questions:

- How was *collaborative* stakeholder *participation* designed in each pilot area?
- What roles were assigned to the stakeholders, in particular to citizens?
- How did the pilots deal with non-scientific information and its integration with technical knowledge?
- Which means of stakeholder communication were used and how was data sharing sought?

Methodologically, an effort has been made to follow the framework of Kieslinger et al. (2018). These authors propose a citizen science evaluation framework structured around two main axes, respectively: (a) process-based evaluation, by which the operational strengths and weaknesses along the project implementation are considered; (b) outcome-based evaluation, which assesses the achievement of the overall goals of the project and its benefits to participants. In this chapter, the first axe was privileged. The second one would demand to be pursued when all COs are working in practice.

In operational terms, the comparative analysis of the AGEO pilots rendered concrete through an iterative process with the five pilots and, also, documents consultation. A series of face-to-face semi-structured interviews were prepared and carried out with representatives from the five pilots, i.e., *Canary Islands* (Spain), *Madeira Island* (Portugal), *Lisbon* (Portugal), *Brittany* (France) and the Causeway Coast (Northern Ireland). The interviews were oriented by an interview guide (see Annex), based on predefined criteria, and recorded.[1] The interviews were subjected to a content analysis. The main results of this analysis are summarised below.

7.5.2 AGEO COs Throughout DRM Continuum

This section provides a brief overview of the five AGEO pilots (see Chaps. 8 to 12 for a detailed description), attending to: (i) the type of geohazard addressed by each pilot, (ii) the socio-environmental specificities of the target areas, and (iii) the objectives that each pilot sought to achieve through the establishment of a CO.

Globally, AGEO covered a wide range of geohazards through its five pilot areas. These pilot areas aimed to tackle one or two geohazards, with the exception of the Lisbon Pilot that covered more geohazards (landslides, earthquakes, uplift and liquefaction, soil settlement and urban flooding).

Rockfall is the type of geohazard most often covered by the pilots, namely Canary Island (Spain), Causeway Coast (Northern Ireland), and Madeira Island (Portugal). This is understandable given the geomorphology of their territories. It is, particularly, the case of Causeway Coast (Northern Ireland) that has several areas prone to frequent slope instabilities, ranging in size from shallow translational flows to large rotational landslides and rockfalls. Slope instabilities are an essential component of the site as they (i) have intrinsic importance as geomorphological features and continually create new exposures for scientific study; (ii) are essential to maintaining the overall rugged nature of the coastline and creating the spectacular landscape of the site; and (iii) regularly expose soil and rock masses, which are essential to the biodiversity of the site.

Several pilots, i.e., Britany (France), Lisbon (Portugal) and Madeira island (Portugal), also, targeted flooding. There are, however, specific features resulting from the geomorphological and social characteristics of each area. Britany pilot (France) is particularly concerned with coastal erosion. Lisbon pilot (Portugal) deals with a specific type flooding, urban floods, which results from the combination of exceptional hydrological events and the inability of drainage systems to meet its demands. Madeira Island (Portugal) is focused on flash floods phenomena associated with exceptional meteorological events and steep terrain (cf. Table 7.1).

As mentioned, Lisbon pilot (Portugal) aims to cover several types of geohazards. Besides flooding, this pilot also aimed to cover the problem of landslides and earthquakes. In fact, Lisbon is one of the regions with the highest seismic risk in the Portuguese territory. Although Madeira Island has a lower seismic hazard and risk, by comparison to the Portuguese Continent, it was found relevant to expand the seismological network, after recent earthquake events felt on the island.

The five AGEO pilots are diverse in socio-environmental terms. Lisbon pilot (Portugal) operates in a densely populated urban area. The Causeway coast Pilot (Northern Ireland) covers a rural coastal area and its World Heritage status attracts a significant number of visitors *per* year (over 1.5 million). The high level of tourist interest, due to the spectacular landscape of

[1] With exception of the Causeway Coast pilot (Northern Ireland) pilot, all the interviews were face-to-face. In the case of the Causeway Coast pilot, the interview rendered concrete through by sending the interview guide to the team and asking them to answer in a written format.

Table 7.1 Aims and type of geohazards covered by the AGEO pilots observatories

Pilots	Target-area	Type of geohazard	Aims
Canary islands (Spain)	Canarias archipelago, with three sub-pilots areas: – El Hierro island; – San Bartolomé de Tirajana Municipality (Gran Canaria); – San Cristóbal de La Laguna Municipality (Tenerife)	Rockfalls	To improve geohazards risk awareness To encourage citizens to take an active part in risk management, particularly in the prevention stage To promote participatory strategies of rockfall risk management, i.e., assessment and monitoring To increase and improve the rockfall inventory and susceptibility map
Brittany (France)	Britany coastline	Coastal erosion and marine flood	To improve geohazards risk awareness To improve coastline monitoring capabilities To stimulate the adoption of risk mitigation measures through land-use planning
Lisbon (Portugal)	Lisbon Municipality	Floods, earthquakes, liquefaction, landslides, soil settlements and uplift	To improve geohazards risk awareness To improve disaster preparedness To demonstrate that citizens can play an active role in risk management To develop a platform to gather and visualize data, namely the one coming from citizens
Causeway Coast (Northern Ireland)	Causeway Coast World Heritage Site and Carrick-a-rede on the northern coast of Northern Ireland	Landslides, including rockfalls	To improve geohazards risk awareness To enhance the monitoring and understanding of the key triggers and mechanisms for landslides To support risk management in the prevention stage with information for risk assessment
Madeira island (Portugal)	Funchal municipality	Rockfalls, earthquakes, flash floods	To improve geohazards risk awareness To encourage citizens in the prevention, mitigation, and management of geohazards To provide real-time data for risk management through meteorological and seismic stations installed in schools To collect data from social media (Facebook, Twitter, WhatsApp) and other informal citizen groups, regarding geohazard events

the site resulting from the overall rugged nature of the coastline, induces constraints to institutional actions concerned with the mitigation of rockfall events. Madeira Island pilot is the most diverse, covering urban and rural areas, inland and coastal zones, extremely variable in terms of human occupation. In the case of Brittany, its observatory aims to cover the entire coastline of the region, which comprehends urban and rural areas. Given the size of the region, it is challenging to involve all stakeholders. The Canary Islands pilot has designed an observatory for all the islands but selected three small areas as sub-pilots: El Hierro Island, San Bartolomé de Tirajana (Gran Canaria), and San Cristóbal de La Laguna (Tenerife). These three sub-pilots cover a wide range of environments, from urban to rural, some of them isolated.

All AGEO pilots share the common goal of raising awareness of geohazard risks. For each of them, CO approach was envisaged as a means of raising awareness and knowledge transfer about geohazards. They also share the same concern to contribute, directly or indirectly, to the improvement of DRM systems.

Despite common goals, the comparative analysis revealed some differences in the focus of each CO. Brittany pilot, as well as the Causeway Coast and Canary Islands pilots, revealed to be particularly focused on the issue of geohazards monitoring. Brittany pilot has established as main goal of its CO to improve coastal monitoring capabilities. Causeway Coast pilot aimed to develop tools to support a more effective monitoring of rockfalls and landslides in this World Heritage Site, contributing to improve the performance of the National Trust, i.e., the entity responsible for managing the site. The Canary Islands monitoring tool aimed to support the characterisation of geological processes, the identification of thresholds for triggering factors and the updating of mass movement susceptibility maps. Lisbon and Madeira Island pilots also seemed to enlarge their focus beyond geohazards monitoring, aiming to build a CO capable of improving disaster preparedness and response in their own areas.

7.5.3 Target Groups and Collaborative Participation Process

As emphasized, two key dimensions of *citizens' observatories* are the nature of the *collaborative participation* between citizens, scientists, elected officials and practitioners, and the *specific role* to be assigned to participants. In order to characterise these dimensions, it is important to identify who the participants were; what kind of engagement they were expected to have in the CO; and what strategies were put in place to stimulate and facilitate participation and collaboration.

With the exception of the Brittany pilot, which limited its observatory to local decision-makers and technical staff, the other AGEO pilots aimed to reach a wider variety of stakeholders. The Causeway Coast pilot was built on a strong partnership with the site managers, i.e., the National Trust, and the World Heritage Site steering group[2] and sought to involve local communities and schools. The Canary Islands pilot selected three sub-pilot areas and channelled networking efforts on elected officials, technical staff, and inhabitants of these areas. Lisbon pilot tried to take advantage of being led by the municipality and established close links with the Municipal Civil Protection Unit and with technical staff from other services. It made efforts to reach out to the citizens of Lisbon and establish connexions with specific groups of citizens, such as the scouts. The Madeira Island pilot envisaged citizens in general, from the municipality of Funchal, as potential providers of real-time geohazard data, but started its experience with two small local communities and their schools.

As mentioned above, Brittany pilot aimed to improve coastal-monitoring capacity (both in terms of the number of sites monitored and

[2] The Causeway Coast World Heritage Site' steering group includes, among others, the National Trust, the Crown Estate, national tourism bodies, the GSNI, the NI Environmental Agency, and private landowners.

the amount of data collected per site) and raise awareness on the importance of integrating geohazards into land-use planning among elected officials and staff. The relationship between scientists and local stakeholders started with the refinement of the methodological framework for vulnerability assessment, with indicators being co-designed by both parts (Le Dantec et al., 2023). Gradually, a collaborative relationship has been stablished with several municipalities in Brittany, to collect data and monitor the evolution of the coastline. In general terms, interaction between scientists, elected officials and technical staff took place on an annual basis, through technical meetings and one-day fieldwork to test monitoring tools and collect data.

In the vision of the Brittany pilot, the interaction with stakeholders was an opportunity to deconstruct the tendency to evaluate coastal erosion on an event-basis and to favour engineering hard engineering solutions, based on the false expectation that such solutions would slow down, or even stop the process of coastal retreat. The dialogue allowed scientists to emphasise that coastal erosion, as a slow-manifesting hazard, requires a long-term strategy and should be addressed through land-use planning, focusing on reducing vulnerability and exposure. In addition to providing *in-situ* data, local technical staff provided the scientists with informal local knowledge and information on the land-use normative system, both of which are fundamental to risk assessment.

Despite the ambition to reach out to a wider public, namely the tourists, the real *collaborative participation* in the Causeway Coast pilot was with the National Trust and World Heritage Site stakeholders. The collaboration between AGEO scientists and the National Trust staff started in an early stage of the project and led to the co-production of geohazard monitoring protocols and a user-friendly tool for in situ data collection. Interaction with these stakeholders throughout the project was high, occurring on average every two months, with joint fieldwork activities, technical training workshops and geological walkover surveys (Lemon et al., 2023). Following a two-way communication, the National Trust contributed a wealth of historical data to support the characterisation of geohazards. From 2022 onwards, efforts were also made to involve local communities and schools in the observatory. However, this goal proved to be difficult to achieve. Getting citizens interested in natural hazards is a challenging and time-consuming task in itself. It is important to acknowledge that AGEO had to deal with the constraints imposed by the COVID-19 pandemic. All AGEO pilots mentioned the pandemic as an obstacle to developing a network of participants around their geohazard observatory (cf. Table 7.2).

The Canary pilot has made great efforts to gain the commitment of the inhabitants. Such efforts consisted of a regular presence of the AGEO team in the target areas, with geohazard training workshops, as well as quarterly meetings to collect the opinions, suggestions, and the experience of inhabitants regarding rockfalls, aiming to improve the inventory and susceptibility maps. In addition to these two-way communication strategies, efforts were made to involve local communities through one-way communication methods, namely leaflets, information in traditional *media* (local press, TV, radio) and online channels. For the Canary pilot team, gaining the interest of citizens was one of the main challenges of building a CO. They dealt with this challenge by identifying local *allies*. For example, they highlighted the importance of a local firefighter in one of the subpilots as a bridge between the AGEO team and the community. In the literature, these internal *allies* are referred as *local champions*, i.e., individuals who have credibility among inhabitants, possibly due to the relevant role they play in the community, and who are willing to advocate for a particular policy (EERI, 1996).

The Lisbon pilot stands out from other AGEO pilots for its efforts to have a communication strategy and a visual identity for the Lisbon observatory. In this respect, it is important to highlight the effort made to develop risk information tools in a user-friendly language. The task of selecting, adapting, and transforming expert knowledge into clear and relevant

Table 7.2 Roles and patterns of interaction with target groups, according to AGEO pilots

Pilots	Target-groups	Roles	Interaction and training
Canary Island (Spain)	Citizens in general, especially inhabitants of the three sub-pilots Civil Protection local services, water supply and maintenance local services Neighbor associations, trekking groups	Citizens and civic organizations: report rockfall events in their area	Offline/co-presence: project dissemination during science-related events (summer schools, scientific holidays, technical visits); Workshops and participatory mapping in secondary schools; Workshops and training sessions with citizens of sub-pilot areas focused on geological hazards Online/co-presence: short courses with First Aid and Emergency students Offline/one-way: Information through leaflets, local press, TV and radio Online/one-way: Video promoting the CO, Social media (Facebook, Twitter, WhatsApp)
Brittany (France)	Decision-makers Technical staff from municipalities Intermunicipal government bodies and associations	Collaborative on-site monitoring of coastal retreat process and marine flooding	Online: App CoastAppli to support on-site monitoring Offline/co-presence: Technical training workshops
Lisbon (Portugal)	Citizens in general Municipal services and staff Local civic organizations: groups of scouts Schools and senior universities	Report geohazards events Participate in the preparedness phase	Offline/one-way: CO information through leaflets, posters, sugar packets, tray napkins, bookmarks and rulers Offline/ co-presence: Meetings, training workshops, dissemination actions Online/one-way: Educational and call-to-action videos, tutorials; Web portal on Lisbon CO; Social media information (Facebook, Instagram)
Causeway Coast (Northern Ireland)	Staff from The National Trust, (responsible for site management) Giant's Causeway and Causeway Coast World Heritage Site' Steering Group Local communities in the surrounding area Schools	National Trust: site geohazard monitoring in a daily basis Steering group and local communities: future participation in geohazard management activities	Offline/co-presence: Technical training workshops; one-to-one interviews; guided walks and public talks Offline/one way: Educational resources Online/one-way: Zoom webinars; Social media information (Facebook, Twitter)
Madeira Island (Portugal)	Citizens in general Two local communities Schools	Citizens and schools: to report geohazards events Local communities and schools: to send real time data from meteorological and seismic stations Local communities: to participate in the preparedness phase	Offline/co-presence: Annual program of activities in schools, including lectures, practical visits and games Technical workshops, training and participatory mapping with local communities

information for the public is very challenging. This effort was made in the Lisbon pilot and was evident in the variety of risk information tools produced, from leaflets to the observatory web portal and call-to-action videos. The observatory was built on the assumption that common citizens of Lisbon could be data providers and report an event. Consequently, a huge effort was made to disseminate the AGEO observatory, and the App tool was created for citizens to report events. Reaching common citizens in Lisbon proved to be difficult. To overcome this problem, the Lisbon pilot opted to focus on specific civic associations (*e.g.* scouts) or institutions (*e.g.* schools and senior universities) and to extend the networking to other municipal services, in addition to Civil Protection. In this respect, it should be emphasising the importance of considering the specificities of the social arena targeted by a CO. Arenas that are tightly knit, with frequent and sustained interaction between inhabitants, are potentially easier to implement participatory approaches. Social arenas, such as the city of Lisbon, that are more diverse and fragmented are more challenging. In such contexts, social relations tend to be infrequent, transient, and loosely knit, making it difficult to engage people in participatory action (Berke et al., 1992).

On the contrary, the Madeira Island pilot launched its CO with inhabitants of two small and remote communities in Funchal and promoted proximity between the inhabitants and the technical officers of the local civil protection organisation through regular meetings. The main objective of these meetings was to discuss geohazard-related issues of concern to the neighbouring communities. Some of the topics discussed were emergency training, preparedness, and risk mapping. An early warning system was installed in the schools of these communities, using meteorological stations and seismometers, and a series of activities were developed, involving both students and teachers. This was used as a pretext to start raising awareness of geohazards. More recently, students have been asked to use the existing Funchal Municipality App and, then, the AGEO App in a test phase of event reporting. Future plans for the Madeira Island pilot include the use of the App by all residents of Funchal to report events.

7.5.4 Means of Communication and Information Exchange

A distinctive dimension of the COs approach, compared to other DRM participatory approaches, is its reliance on technology as a means of sending data and reporting an event. The characterisation of this dimension among the AGEO pilots focused on aspects related to: (i) means of data collection/event reporting by pilots, (ii) feedback to participants and visualisation of data, and (iii) data validation.

In general, all AGEO pilots favoured technology as a means for stakeholders to send data or report an event. Some pilots developed a web platform and a mobile application for this purpose. The AGEO project envisaged the development of a unique App to be tested by the five observatories. However, this goal was not compatible with the needs of some AGEO pilots. Some of them felt that they needed such a technological base earlier in the process and decided to develop their own web and mobile application (cf. Table 7.3).

In Brittany, data collected by the stakeholders participating in the observatory was sent by means of a mobile application so-called *CoastAppli*. This data was stored and analysed through two web platforms (Le Dantec, 2023). The other four pilots followed the same pattern, sending and processing data around an App and a web platform. The Canary pilot created alternatives to the mobile App, by setting up a social media account (Twitter, Facebook and WhatsApp) for citizens to report events. Madeira Island pilot also took advantage of social media as a means for citizens to report events. All pilots referred to face-to-face interactions as a way of testing the data submission process and confirming its accuracy.

Table 7.3 Ways of data sending and feedback, according to the AGEO pilots

Pilots	Ways of sending data	Feedback	Informal knowledge
Canary sland (Spain)	Virtual: via mobile phone App and platform Virtual: social media (Facebook, Twitter and WhatsApp for the CO profile/number) Co-presence: Phone calls; face-to-face interaction activities	Automatic responses are always sent out. After data validation another response is provided Feedback is presumably provided at the trimestral meetings as well	The CO takes advantage of experiential knowledge coming from citizens to improve rockfall inventory and susceptibility map
Brittany (France)	Virtual: via mobile phone App and platform Co-presence based on annual meetings with stakeholders	Feedback through meetings with stakeholders (implicit)	Stakeholders have informal knowledge of both the location's geography and the population, which is used in the coastal monitoring process
Lisbon (Portugal)	Virtual: via mobile phone App and platform Co-presence: meetings with all target groups	No feedback yet	-
Causeway Coast (N. Ireland)	Virtual: via mobile phone App and platform Co-presence: meetings with the National Trust	Regular updates through the training sessions and in community workshops	-
Madeira sland (Portugal)	Virtual: via mobile phone App and platform Virtual: social media (Facebook, Twitter, Instagram) Co-presence: meeting with local communities	Schools may receive feedback from the Civil Protection Center Feedback through meetings with communities (implicit)	-

Providing feedback to citizens after their participation is important to keep them engaged with the observatory. In fact, this is one of the ten principles of the European Citizen Science Association (ECSA, 2015), which states that "citizen scientists [should] receive feedback from the project. For example, how their data are being used and what the research, policy or societal outcomes are". This dimension needs to be further developed in the AGEO pilots. Most of the pilots mentioned the regular meetings as a way of providing feedback to stakeholders and, in some cases, their mobile App was designed to automatically respond to the user after sending data. The issue of feedback should be considered more broadly. ECSA (in ibid) emphasises that in citizen science projects, participants need to feel that they benefit from their participation. Benefits "may include (…) learning opportunities, personal enjoyment, social benefits, satisfaction through contributing to scientific evidence e.g., to address local, national and international issues and through that, the potential to influence policy". Feedback should be defined in such broader terms to show that citizen participation is beneficial, namely for better risk management, and that it is worthwhile to be part of a geohazard observatory.

The data quality control and validation process is one of the most critical aspects of the COs approach (Kieslinger et al., 2018). As highlighted by Liu et al., (2014), part of the scepticism towards this participatory approach stems from questions about "the quality of the data collected and its usefulness for environmental policy". Among the AGEO pilots, there is concern about the quality of the data and the procedures to ensure it. As mentioned above, the Brittany observatory relies on a network of stakeholders that are mainly technical officials, which in a way guarantees the quality of the data. The Canary Island pilot also established a protocol for quality control of data coming from the App or social media, which consists of a technical visit to the site to confirm the reported event/situation.

In summary, although the AGEO pilots use two-way communication strategies, the approaches were mainly technology-based. A lot of faith was placed in the ability of web and mobile technologies to facilitate stakeholder exchange and engagement. Such tools may be useful for initiating dialogue but are not sufficient to sustain collaborative stakeholder participation on a giving issue. In the specific context of DRM, Hicks et al. (2019) highlight that *web* technologies do not always guarantee high data quality and participant engagement. Such limitations have motivated calls for citizen science scholars and practitioners to adopt tools other than technology alone.

7.5.5 A Balance of the Challenges of Creating Five COs. Lessons Learned

As part of the learning process, a self-assessment[3] of each pilot was sought focusing on the key challenges faced by them and Lessons learned.

Each pilot was tasked to evaluate the strengths of the COs and identify the potential opportunities that have arisen or may arise. They were also asked to identify the areas for improvement or weaknesses that, once addressed, would facilitate to deal with potential threats.

The self-assessment *per* AGEO pilot is systematized in Tables 7.4, 7.5, 7.6, 7.7 and 7.8 organized into strengths and weaknesses, i.e. internal factors over which the CO has influence, and opportunities and threats, i.e., external factors beyond CO control that can impact its development and operation. The ones more relevant are presented next.

Almost all AGEO pilots referred the expansion of data collection capabilities for monitoring as one *opportunity* of their CO. Causeway Coast Pilot highlighted that geohazards

[3] Self-assessment is an ongoing reflective process designed and conducted by those who implement the project. By contrast, evaluation provides an independent external view of the project.

Table 7.4 Synthesis of the Canary Island pilot self-assessment

Strengths	Weaknesses	Opportunities	Threats
Development of new protocols for emergency management of rockfalls at a regional level. Application agreements of these protocols established with emergency authorities Data produced by the CO used to improve hazards assessment, e.g., updating mass movements susceptibility maps and inventories The CO identified and benefited from the existence of local champions Rockfall prevention and awareness among citizens were improved A mobile App and platform proved to be useful tools for rockfall reporting and monitoring	Sustainability of citizens-participation	Integrating of data produced by the CO in the *National Database for Mass Movements*. The database is available to public and risk management practitioners CO sustainability may be achieved through its integration into the existing emergency management structure	Inexistence of a legal framework to integrate CO outputs, into the risk management process Lack of support from external stakeholders to ensure CO sustainability

Table 7.5 Synthesis of the Brittany pilot self-assessment

Strengths	Weaknesses	Opportunities	Threats
Close collaboration between academics and local authorities Development of a network of local connections for observation purposes through the allocation of a local technician to the CO A mobile App and platform proved to be useful tools for coastal retreat monitoring	A long period is needed to persuade all stakeholders to participate as well as to establish the network that will feed the CO	The active participation of local authorities and technical staff in the monitoring of the coastal retreat enables increased awareness of the risk of coastal erosion and local accountability for its mitigation Pre-existing environmental awareness among stakeholders Previous history of successful collaboration between academia and local authorities	Lack of local political support greatly hampers CO penetration into more areas of Brittany Instability of financial and human resources preventing the CO from growing and expanding

Table 7.6 Synthesis of the Lisbon pilot self-assessment

Strengths	Weaknesses	Opportunities	Threats
Data received through the AGEO App contributes to a proactive response to manage identified geoahazards Development of a network of local connections for data collection through the participation of municipal technicians The communication strategy and strong visual identity of Lisbon pilot facilitated the delivery of technical information	Lack of understanding by CO' stakeholders on geohazards hinders the use of the App Need to validate the information produced by the CO and streamline its practical use Feedback not provided to the App's users	Awareness raising in schools was extended to the community	Lack of awareness of decision makers to geohazards and their induced risks Legal restrictions to integrate COs outputs into risk management systems Disaster response agencies primarily focus on response capability, hindering CO involvement in other phases of DRM

Table 7.7 Synthesis of the Causeway Coast pilot self-assessment

Strengths	Weaknesses	Opportunities	Threats
Citizen participation in geohazard monitoring increased the number of monitored areas and monitoring frequency, reducing costs The mobile App proved to be useful tool for landslide reporting and monitoring	The delay of the publication of AGEO App postponed the operations of the CO The closure of the AGEO project is a risk for the established scientific partnerships and associated knowledge production	The CO benefited from the experience of other AGEO partners in risk management activities in the Atlantic Area A strong partnership was established between the CO and its main stakeholder, the National Trust. Such partnership CO generated an opportunity to further develop AGEO and expand the acquired experience to other sites in Northern Ireland	The COVID-19 pandemic delayed the start of operations of the CO

Table 7.8 Synthesis of the Madeira Island pilot self-assessment

Strengths	Weaknesses	Opportunities	Threats
In the occurrence of an event, the CO can produce a large amount of data in a short period, filling geographic information gaps and accelerating response	Lack of procedures to manage big data Competition between AGEO App and other existing Apps	Use of citizens collected information with personal sensors Integration of information posted on social networks for risk management processes	Inexistence of a legal framework, to integrate CO outputs into risk management systems The top-down nature of the civil protection structure is a barrier to leverage CO benefits Absence of a contact point in the civil protection authorities

monitoring through their CO enabled "more data to be collected numerous times and locations" (Lemon et al., 2023). Canary pilot also highlighted this *opportunity*, referring that data collected by their CO was included in the National Database of Mass Movement, a tool that is available to the public, experts and technical staff with responsibilities in risk management. Such emphasis that pilots put on data collection capabilities (a *strength*) was associated with the potentialities of Information and Communication Technologies (ICT). AGEO pilots heavily rely in the ability of web platforms, mobile technologies and social media for data sending, event reporting and information exchange.

Another *opportunity* highlighted by the AGEO pilots was the potential for interaction between scientists, decision-makers technical officials and citizens. All the pilots referred to these interactions as valuable opportunities to raise geohazards risk awareness. Brittany pilot emphasised this *strength* of the CO tool, recalling that the active participation of local authorities and technical staff in the CO increased awareness of coastal erosion problem and local accountability for its mitigation. Canary pilot exemplified the opportunities to raise awareness with the multiple training sessions that they promoted along the project.

AGEO pilots highlighted two main *weaknesses*. One of them concerned the issue of motivating and maintaining citizens' interest in the CO. Most of the pilots reported difficulties in persuading citizens and decision-makers to participate in the CO. The pandemic COVID-19 was referred as one barrier to promote such engagement. The other *weakness*, also salient among some AGEO pilots, concerns the availability of resources to assure the continuity of established partnerships beyond AGEO project.

Besides the above-mentioned weaknesses, Lisbon pilot manifested concerns with the reliability of information provided by non-scientists. They believe that more investment is needed on the subject of data validation and verification.

The *threats* (i.e. external factors that may affect the continuity of COs) identified by the AGEO pilots concern the nature of disaster risk management systems. Madeira, Lisbon and Canary pilots highlight the inexistence of a legal framework capable of legitimating CO tool and outputs in disaster risk management. Brittany and Lisbon pilots also highlighted the lack of political support for the implementation of participatory approaches and the low level of geohazards risk awareness among decision-makers.

7.6 Basis of a Roadmap for Including COs in Risk Management Systems

7.6.1 Overview

This section outlines the main stages of a roadmap for applying the citizen observatory approach to geohazard risk management. By roadmap we mean a guide that defines the goals or desired outcomes, and the key steps to achieving them.

In our view, COs can be a valuable tool to stimulate participatory risk management processes in line with UN policy guidelines and EU disaster resilience goals. As discussed in Sects. 7.2 to 7.4, both of them converge on the assumption that, to achieve DRR, governments and communities should invest in raising the general levels of risk awareness and involve citizens in all stages of risk management process. COs seek to achieve citizen engagement by motivating individuals to take on roles traditionally reserved for experts and risk management institutions. This is an aspiration that is not compatible with ad hoc actions. On the contrary, it requires a systematic approach that is: (i) based on a thorough knowledge of both the risk and the target area; (ii) guided by a planned vision of the observatory to be implemented; (iii) founded on stable linkages with key local stakeholders; and (iv) monitored by ongoing evaluation processes, together with the flexibility to adjust along the way.

Next, we will deconstruct each of those four stages into guiding principles or recommendations identified from the literature review and from the experiences and lessons learned by the AGEO pilots.

7.6.2 Building the Knowledge Base for a CO

The adoption of a CO tool requires a prior understanding of both the risk and the target area. Knowledge of the geohazards and social characteristics of the target area, as well as the different stakeholders and their mindsets, will help to adapt the observatory to its context.

Understand risk—As a scientific-led process, the first steps of a CO should include an analysis of the geohazard processes underlying the risks in the target area and how they can be monitored or even managed. The Brittany pilot is a good example of this. Based on previous research on coastal erosion and submergence risks, the Brittany observatory developed a model of vulnerability. This model identified the key components of vulnerability that should be monitored. The model was then completed through experimentation with the local stakeholders of the pilot sites (Lendre, 2020), who contributed to the definition and selection of the indicators in which coastal monitoring should take place.

Enquire people—In their essay on risk communication, Morgan et al. (2011) note that communicators typically rely on technical experts to decide what people should be told about risk, rather than listening to people and asking them what they need to know. Therefore, in parallel with an overview of the geohazard process underlying risk, it is important to conduct a systematic analysis of how target-groups perceive risk, their opinions on the most appropriate risk management strategies and their willingness to engage in risk management.

Identify and involve stakeholders early in the process—One of the challenges highlighted by the AGEO pilots was the time needed to persuade stakeholders to engage in a CO. It is therefore essential to identify all relevant stakeholders and bring them together early in the process. Relevant stakeholders are diverse and may vary from one context to another. They may include members of civil protection organisations, local authorities, and their technical staff, but also members of non-governmental organisations, community groups, experts and common citizens. Rather than sidelining, it is important to identify them, take their views into account and encourage them to play an active role in the observatory. Canary Island and Lisbon pilots are good examples of early involvement of stakeholders in the process, the former by engaging individuals with credibility among local inhabitants, and the latter by involving technical staff from the municipality.

7.6.3 Create a *Vision* for the Observatory

Based on the diagnosis of the target area, it is important to refine a *vision* of the observatory and its goals. In abstract terms, a *vision* is a description of a desired future that helps identify goals and the best strategy to achieve them (WeObserve consortium, 2021). In the case of the AGEO pilots, an overarching goal was to establish a CO that would enable citizen engagement in geohazard risk management and ultimately advance efforts to promote DRR. However, there are specific objectives to be defined and, more importantly, it is imperative to outline a realistic vision of the observatory. As mentioned above, one of the innovations of the COs approach is its focus on links between lay people and experts through their involvement in activities that are traditionally the domain of science and institutions. In the case of geohazards, such involvement can be made concrete in a number of ways along the DRM *continuum*, for example, by involving citizens and local officials in hazard surveillance, hazard mapping, risk monitoring and decision-making on risk mitigation and early warning. Whatever the options, the vision and objectives of the observatory must be well adapted to the specificities of the target area and the social environment, and bring active benefits to all participants, even if this means being less ambitious.

Align the observatory's objectives with local needs—Some AGEO pilots acknowledged difficulties in articulating with the DRM system in their own target areas. They attributed such difficulties to a lack of institutionalisation of active citizen involvement in DRM. As noted elsewhere, DRM systems are still dominated by top-down organisational cultures that downplay the active participation of communities at risk. Changing this dominant culture in a new direction takes time. It requires the development of new practices that can hopefully become standard operational procedures in the medium-term. There is no recipe for unlocking such change. However, an important condition is to ground the *vision* of the observatory in a solid knowledge base of the target area and its needs, and to explore ways to meet local needs. One of the strategies used by the Brittany pilot to establish links with local authorities was to demonstrate willingness to provide expert support in land-use decision-making processes. In other words, they went beyond the strict objectives of the observatory to meet other local needs for expertise and to forge closer links with local stakeholders.

Define stakeholder roles—As part of refining the *vision*, it is important to clarify the roles that local stakeholders will be asked to play. Stakeholders are likely to vary according to their role in the community and their skills, risk perceptions, priorities, and motivation to participate. This diversity will certainly have an impact on the type and degree of their involvement in the observatory. It is therefore important to clarify, with the stakeholders themselves, the valuable contributions that each of them can make to DRM and the mutual benefits of such engagement.

Design robust data validation mechanisms—The importance of local knowledge and skills has been highlighted earlier in this chapter. There is solid scientific evidence of the central role that informal knowledge existing within the communities at risk can play in DRM (Bankoff, 2015; Mercer, 2012). Several studies developed following the 2004 tsunami (Gaillard et al., 2008; Shaw et al., 2008) showed how informal knowledge, passed down from one generation to another, was used by some groups of small islands in the Indian Ocean and saved lives. However, it should not be overlooked that individuals' views and informal assessments are prone to bias and misinterpretation. In today's World, where new and competing forms of online communication are constantly emerging, the dangers of biased judgements and misinformation take on a new dimension. It is therefore important to provide the CO with resources and mechanisms to ensure data quality.

Go beyond technology-based communication—As mentioned above, one of the main innovations of the COs approach is its reliance on information and communication technologies (ICT). Hicks et al. (2019) noted that ICT do not always guarantee high data quality and participant engagement. Indeed, if the goal of a CO is to open up DRM to non-experts and motivate them to become engaged, ICT are not sufficient to achieve this goal. Social media, mobile applications and web platforms may be useful for disseminating information and initiating or complementing communication, but they are not sufficient to ensure lasting partnerships. Furthermore, as pointed out by the Madeira Island pilot, geohazard mobile applications and web platforms compete with a variety of other Apps, making long-term adoption a challenge. The current media environment is overloaded with many voices competing for the attention of different audiences on all topics, including science. Previous research has been inconclusive about the degree of accuracy of social media information and messages (National Academies of Sciences, Engineering, and Medicine, 2017). Given this, it is prudent to ensure that the CO stands in a communication infrastructure that goes beyond technology and includes stakeholders' dialogues based on co-presence.

Consider the availability of resources— Resources can be monetary or non-monetary. Monetary resources can be translated into additional staff, studies, consultants, and technology. Of the five administrative subregions (known as "departments") covered by the Brittany pilot, the Dépai tment of Finisterre was identified as the one where the observatory became more active. Three factors were mentioned as explaining this. One of these factors is staff time. The Dépai tment of Finisterre has a senior official responsible for coastal monitoring who has been given time to become involved in the observatory. The other two factors concern the existence of a well-established collaboration between the University of Brittany and The Dépai tment of Finisterre, and a high-level environmental awareness among decision-makers. Indeed, all AGEO pilots identified the availability of resources as a critical factor in ensuring the sustainability of their observatories beyond the project. As they were led by universities or research institutes, some of these pilots need to have well-established links with local authorities and stakeholders to ensure their continuity. Lisbon pilot is the exception. The fact that is led by the municipality of Lisbon, together with its efforts to involve municipal staff, in particular the civil protection, puts Lisbon pilot in an advantageous position.

7.6.4 Develop a Task Force of Stakeholders and Invest in Training

A key factor for the success of a participatory approach and COs to DRM is the ability to achieve a shared commitment between those promoting it and local stakeholders from the target areas. It may be helpful to select key-stakeholders and bring them together.

Create a stakeholder task force—A small working group should be set up with people who, by virtue of their role in the community, are key players able to define a realistic *vision* for the observatory and help to implement it. These can be representatives of local communities or affected parties, elected officials, technical staff, experts and the so-called local champions. Alesch and Petak (1986) define local champions as individuals who emerge from the community and act as advocates from within the system. Local champions can be elected officials, local experts, civil protection officers or any other person who has credibility within the community and ability to build alliances. Besides helping to design an observatory that is well adapted to the specificities of the target area, this stakeholder task force can actively contribute to ensure its continuity.

Invest in training and awareness—As an approach that encourages non-scientists to participate in tasks across the DRM *continuum*, it is important to invest in stakeholders training and risk awareness programmes. This investment has several advantages, three of which stand out. Firstly, it is an opportunity to establish or deepen links with local stakeholders and citizens. Second, it allows scientists, or those promoting the observatory to interact with non-scientists and benefit from their experimental knowledge of their own territory. Thirdly, it can prevent situations of poor data quality.

7.6.5 Evaluate the Process and Be Flexible to Make Adjustments

Systemic change requires regular evaluation of the process. Without evaluation, there is no concrete idea of how far the vision and goals are being achieved, or whether the strategy adopted is effective (WeObserve consortium, 2021). Given this, along with the several tasks involved in the design of an observatory, it is advantageous to adopt an evaluation procedure.

Evaluate the process and be willing to adapt—Monitoring the process is useful for several reasons. It provides information on how actions have progressed and whether adjustments need to be made. It can be used as a basis for setting new priorities or, if necessary, new goals. Finally, it provides learning opportunities for observatory leaders and stakeholders on the whole process of building a CO. As previously stated, an evaluation process should be twofold, including: (a) a process-based evaluation, focusing on the process of setting up the observatory, and (b) an outcome-based evaluation, with a focus on impacts and objectives achieved.

7.7 Conclusions

The AGEO project called for a paradigm shift in geohazard risk management by advocating participatory approaches through COs.

Five pilot COs have been set up in the Atlantic Area, representing a small-scale experimental initiative. Each CO developed a strategy to get citizens involved in geohazards risk management. Their experience led to a deeper understanding of the benefits and challenges posed by COs.

Based on their experience, three main benefits of the CO tool were identified, respectively: the potential for raising risk awareness among communities at risk; the opportunity for knowledge exchange between stakeholders; and the expansion of monitor capabilities. The main challenges concerned the demanding task of engaging citizens in the CO, data quality issues and CO longevity.

This chapter culminates with the identification of the key stages for a roadmap to incorporate the CO tool in risk management systems. First, COs must stand on a knowledge base of both risk and target-areas. Second, a clear vision for the observatory should be formed, and strategic goals should be defined in line with local needs. At this stage, it is important to assess the available resources, define stakeholders' roles, create reliable procedures for data validation, and go beyond technology-driven communication techniques. Although useful to spread information, initiate or complement communication, social media and web platforms are insufficient to assure enduring partnerships. The third key stage highlights the importance of creating a task force of local stakeholders with motivation to become involved in the observatory. At this level, we also emphasise the importance of multi-stakeholder training for a more accurate understanding of geohazards and related risks. Finally, it is important to endow the process of an on-going evaluation procedure to assess the extent to which *vision* and *goals* we set up are being fulfilled and if there are adjustments to make.

To sum up, this chapter examined how COs can contribute to shifting the prevailing dominant top-down approaches of DRM in favour of a people centred DRM paradigm, formulating guiding principles and recommendations to inform better CO initiatives in the future.

ANNEX: Pilots Interview Guideline

1. **Aims**—What were the aims for the creation of your citizen observatory (CO) in terms of supporting risk management processes?
2. **Scientific aims**—Thinking about them, which one would you classify as scientific aims?
3. **Role of CO in DRM**—Is your CO focused on one or more components of Disaster Risk Management (DRM) (prevention, preparedness, warning/monitoring, response and recovery)?
4. **Target-groups**—Who are the target-groups of your CO (governance-related organisations; civic associations & collectivities; citizens in general)?
5. **Citizens' role (i)**—What role does the CO assign to the target-group?
6. **Citizens' role (ii)**—Are different roles for different target groups?
7. **Data**—Does your CO require citizens to collect data and send it?
8. **Channels to send data**—Which channels do citizens have to send data and participate in CO? (Interaction virtual or based in a co-presence)?
9. **Innovation tool**—Did your team create, during AGEO, any tool aimed at facilitating the relationship and exchange between citizens and CO? Is the tool used to improve risk management systems?
10. **Communication**—Which channels does your CO provide to communicate/interact with citizens? Are they essentially virtual or based in a co-presence?
11. **Stage of engagement**—In which stage of CO creation did you start to interact with

your target groups and motivate them to participate?

12. **Recruitment**—Did/does your CO implied/ imply the 'recruitment' of citizens? If is, how and who did you recruit?
13. **Frequency of interaction**—How frequent is the interaction between CO managers and target groups? Continuous, monthly, periodically, once a year, or casual?
14. **Training**—Does your CO preview any type of citizens' training namely on risk issues or DRM?
15. **Opportunities of learning**—Is the CO an opportunity for citizen learning namely to contribute to disaster risk reduction? Examples:

Prevention: (a) increase awareness; (b) monitor hazards

Preparedness: prepare emergency plan

During the event: (a) self-protection measures; (b) warning

Response: report hazards to authorities

Recovery: implementation of vulnerability reduction measures

1.6 **Informal knowledge**—Does the CO take advantage of experiential knowledge coming from citizen? Is it useful for scientists/ practitioners?

17. **Co-created knowledge**—Is citizen information and co-created knowledge used to improve risk management systems? Who uses the information/knowledge? How is the information used in DRM?
18. **Data quality**—How do you evaluate data quality, or information accuracy coming from citizens? Is there any procedure previewed?
19. **Openness**—Do citizens know what data/ information/contribution is used for?
20. **Feedback**—Do they receive any kind of feedback for their participation?
21. **Openess**—Is data/information produced by CO publicly shared and available for the public in general?
22. **Property rights**—Does your CO raise issues of property rights? How do you deal with this (ethical guidelines for data use—clear, transparency)?
23. **Auto-evaluation (i)**—Looking back, how do you and your team evaluate this challenge of creating a CO (main challenges, difficulties, lessons learned)?
24. **Auto-evaluation (ii)**—Do you have any criteria to compare the performance of risk management systems (local/regional) before the creation of the CO and after its integration in a risk management structure?
25. **Auto-evaluation (iii)**—Can you identify the CO' strengths, weaknesses (internal), opportunities and threats (external, no control) (e.g. opportunities: provide new ideas to the DRM; threats: large return period of hazards)?
26. **Best practices (i)**—Can you identify barriers and best practices on the use of services provided by European spatial data infrastructures, such as Copernicus or EGDI (European Geological Data Infrastructure), by the pilot and risk management infrastructures?
27. **Best practices (ii)**—Can the pilot provide examples of best practices (or success stories) of using CO outputs to improve risk management systems with respect to geohazards?
28. **Recommendations**—Can the pilot and the CO provide recommendations to improve risk management systems with respect to geohazards?
29. **Sustainability (i)**—What is your perspective of this CO after AGEO ending (risks & opportunities)?
30. **Sustainability (ii)**—Do you have any strategy in place to ensure its continuity?

References

Arnoldi, C. N., Luzietti, L., & Gasperis, P. D. (2021) Actionable Proposal 5. Create a detailed overview of and analyze the Entities' current use of In Situ data delivered via crowdsourcing initiatives, social media and similar. e-GEOS, ITHAC and GAFAG. EG-RPT-EEA-SC1–0011 Issue 3.1.

Bankoff, G. (2015). Design by disasters: seismic architecture and cultural adaptation to earthquakes, in Fred Krüger, Greg Bankoff, Terry Cannon, Benedict Orlowski, E.L.Schipper, Cultures and disasters:

seismic architecture and cultural adaptation to earthquakes, Routledge, New York.

Berke, P., & Beatley, T. (1992). Planning for Earthquakes, Risk, Politics, and Policy. Johns Hopkins University Press, Baltimore (U.S.A).

Boersma, K., Martelli, C., Bruinen de Bruin, Y., Vetere Arellano, A. L., Claassen, L., & Pickl, S. (2020). Communicating risk among all, in: Casajus Valles, A., Marin Ferrer, M., Poljanšek, K., & Clark, I. (eds.), Science for Disaster Risk Management 2020: acting today, protecting tomorrow, EUR 30183 EN, Publications Office of the European Union, Luxembourg, ISBN 978–92–76–18182–8, https://doi.org/10.20/571085, JRC114026.

Casajus Valles, A., Marin Ferrer, M., Poljanšek, K., & Clark, I. (eds.) (2020). Science for disaster risk management 2020: acting today, protecting tomorrow, EUR 30183 EN, Publications Office of the European Union, Luxembourg, ISBN 978-92-76-18181-1, https://doi.org/10.2760/438998, JRC114026.

Claassen, L., Sapountzaki, K., Scolobig, A., Perko, T., Górski, S., Kaźmierczak, D., Anson, S., Carnelli, F., Bossu, R., Sousa Oliveira, C., & Laurikainen, H. (2020). Citizen participation and public awareness. In Casajus Valles, A., Marin Ferrer, M., Poljanšek, K., & Clark, I. (eds.), Science for Disaster Risk Management 2020: acting today, protecting tomorrow, EUR 30183 EN, Publications Office of the European Union, Luxembourg, 2020, ISBN 978–92–76–18182–8, https://doi.org/10.2760/571085, JRC114026.

COM(2019) 640 (2019). Communication from the commission to the European Parliament, the European Council, the Council, the European Economic and Social Committee and the Committee of the Regions. The European Green Deal. COM(2019) 640 final. European Commission. Retrieved 28 March, 2023, from <https://ec.europa.eu/info/strategy/priorities-2019-2024/european-green-deal_en#actions.

COM(2020) 788 (2020). Communication from the Commission to the European Parliament, the Council, the European Economic and Social Committee and the Committee of the Regions. European Climate Pact. European Commission, COM/2020/788 final.

COM(2021) 407 (2021). Proposal for a Council Recommendation on a Pact for Research and Innovation in Europe.

COM(2021) 82 (2021). Communication from the Commission to the European Parliament, the Council, the European Economic and Social Committee and the Committee of the Regions. Forging a climate-resilient Europe—the new EU Strategy on Adaptation to Climate Change. European Commission, COM/2021/82 final.

Commission Recommendation 2023/C 56/01 (2023). European Commission Recommendation of 8 February 2023 on Union disaster resilience goals (2023/C 56/01).

Decision (EU) 2019/420 (2019). Decision (EU) 2019/420 of the European Parliament and of the Council of 13 March 2019 amending Decision No 1313/2013/EU on a Union Civil Protection Mechanism. Official Journal of the European Union, L 77 I/1.

Decision No 1313/2013/EU (2013). Decision No 1313/2013/EU of the European Parliament and of the Council of 17 December 2013 on a Union Civil Protection Mechanism. Official Journal of the European Union, L 347/924.

Delica-Willison, Z., & Gaillard, J. C. (2012). Community action and disaster. In B. Wisner, J.C. Gaillard, & I. Kelman (eds) The Routledge Handbook of Hazards and Disaster Risk Reduction. Routledge, New York.

DG ECHO (2023). European Disaster Risk Management Factsheet. Retrieved 16 March, 2023, form https://civil-protection-humanitarian-aid.ec.europa.eu/what/civil-protection/european-disaster-risk-management_en.

DRMKC (2023). Disaster Risk Management Knowledge Centre. Retrieved 01 March, 2023, from https://drmkc.jrc.ec.europa.eu/about/posters.

EERI—Earthquake Engineering Research Institute. (1996). Public policy and building safety. EERI Endowment Fund White Paper, Oakland (U.S.A).

EM-DAT. (2023). 2022 Disasters in numbers. Center for Research on Epidemiology of Disasters, Université Catholique de Louvain and USA AID, Brussels.

EPA. (2022). Recommendations to Heads of EPAs following the European Commission's Best Practices in Citizen Science for Environmental Monitoring (Staff Working Document SWD (2020) 149 final). Paper 5. 37th EPA Network plenary meeting hosted by the Ministry of ecological transition and ADEME. 12–13 May, Paris.

European Citizen Science Association (ECSA). (2015). ECSA 10 principles of Citizen Science. ECSA, European Union, Brussels.

Fritz, S., See, L., Carlson, T., Haklay, M., Oliver, J. L., Fraisl, D., Mondardini, R., Brocklehurst, M., Shanley, L.A., Schade, S., Wehn, U., Abrate, T., Anstee, J., Arnold, S., Billot, M., Campbell, J., Espey, J., Gold, M., Hager, G., He, S., Hepburn, L., Hsu, A., Long, D., Masó, J., McCallum, I., Muniafu, M., Moorthy, I., Obersteiner., M., Parker, A.J., Weisspflug, M., & West, S. (2019). Author correction: Citizen science and the United Nations sustainable development goals. *Nature Sustainability, 2*(11), 1063. https://doi.org/10.1038/s41893-019-0426-8.

Gaillhard, J. C., Clavé, E., Vibert, O., Azhari, D., Denain, J. C., Efendi, Y., Grancher, D., Liamzon, C., Sari, D. S. R., & Setiawan, R. (2008). Ethnic groups' response to the 26 December 2004 earthquake and tsunami in Aceh. *Indonesia, Natural Hazards, 47*(1), 17–38.

Hagenlocher, M., Thieken, A., Schneiderbauer, S., Aguirre Ayerbe, I., Dobes, P., Donovan, A., Morsut, C., Paris, N., Pedoth, L., & Tonmoy, F. (2020). Risk Assessment, in: Casajus Valles, A., Marin Ferrer, M.,

Poljanšek, K., Clark, I. (eds.), Science for Disaster Risk Management 2020: acting today, protecting tomorrow, EUR 30183 EN, Publications Office of the European Union, Luxembourg, ISBN 978–92–76–18182–8, https://doi.org/10.2760/571085, JRC114026.

Handmer, J., & Dovers, S. (2007). *Handbook of disaster and emergency policies and institutions*. Earthscan.

Hedegaard, C., Mysiak, J., Lera, A., Clair, S., Bartoli, M. S., Cornieti, M., Freitas, H., Holy, M., Jacob, D., Murray, V., O'Connor, K., Pieper, H., Rockström, J., Runnel, A., Stoknes, P. E, & van Ypersele, J. -P. (2020). Proposed mission: A climate resilient Europe: Prepare Europe for climate disruptions and accelerate the transformation to a climate resilient and just Europe by 2030. Report of the Mission Board for Adaptation to Climate Change, including Societal Transformation. Publications Office of the European Union, Luxembourg, ISBN 978–92–76–21685–8 https://doi.org/10.2777/69766 KI-04–20–482-EN-N.

Hicks, A., Barclay, J., Chilvers, J., Armijos, M. T., Oven, K., Simmons, P., & Haklay, M. (2019). Global mapping of citizen science projects for disaster risk reduction. *Frontiers in Earth Science, 7*, 226. https://doi.org/10.3389/feart.2019.00226

IRGC. (2017). Introduction to the IRGC Risk Governance Framework, revised version. Lausanne: EPFL International Risk Governance Center. https://doi.org/10.5075/epfl-irgc-233739.

ISO 31000:2018 (2018). ISO 31000 Risk management—Guidelines. International Organization for Standardization.

Järvinen, L., & Sinervo, R. (2020). How to create a national circular economy road map—A guide to making the change happen. Sitra studies 170, ISBN 978–952–347–191–7 (PDF) www.sitra.fi, ISSN 1796–7112 (PDF)

Kieslinger, B., Schäfer, T., Heigl, F., Dörler, D., Richter, A., & Bonn, A. (2017). The challenge of evaluation: An open framework for evaluating citizen science activities. https://doi.org/10.31235/osf.io/enzc9.

Lanfranchi, V., Ireson, N., When, U., Wrigley, S.N., & Ciravegna, F. (2014). Citizens' observatories for situation awareness in flooding. In S. R. Hiltz, M. S. Pfaff, L. Plotnick, & P. C. Shich (Eds.), Proceedings of the international ISCRAM Conference, University Park, Pennsylvania.

Le Dantec, N., Jaud, M, Hénaff A., et al. (2023). OSIRISC: co-design of a collaborative observatory to monitor vulnerability to coastal risks of erosion and flooding in Brittany. Poster, AGEO Final Conference, Lisboa, LNEC, April 19–20, 2023.

Lemon, K., & Parker, K. (2023). Causeway Coast Citizens' Observatory. Poster, AGEO Final Conference, Lisboa, LNEC, April 19–20, 2023.

McCall, M., & Peters-Guarin, G. (2012). Participatory action research and disaster risk, in B. Wisner, J.C. Gaillard, I. Kelman (eds) The Routledge Handbook of Hazards and Disaster Risk Reduction. Routledge, New York.

McEntire, D., Fuller, C., Johnston, C. W., & Weber, R. (2002). Comparison of disaster paradigms: The search for a holistic policy guide. *Public Administration Review, 62*(3), 267–281.

Mercer, J., 2012—Knowledge and Disaster Risk Reduction, in B. Wisner, J.C. Gaillard, I. Kelman (eds) The Routledge Handbook of Hazards and Disaster Risk Reduction. Routledge, New York.

Morgan, M. G., Fishhoff, B., Bostrom, A., & Atman, C. J. (2011). Risk Communication, a Mental Models Approach. Cambridge University Press, New York (U.S.A).

National Academies of Sciences, Engineering, and Medicine (2017). Communicating Science Effectively: A Research Agenda. Washington, DC:The National Academies Press. https://doi.org/10.17226/23674.

O'Keefe, P., Westgate, K., & Wisner, B. (1976). Taking the naturalness out of natural disasters. *Nature, 260*(5552), 566–567. https://doi.org/10.1038/260566a0

Regulation (EU) 2021/836 (2021). Regulation (EU) 2021/836 of the European Parliament and of the Council of 20 May 2021 amending Decision No 1313/2013/EU on a Union Civil Protection Mechanism. Official Journal of the European Union, L 185/1.

Rodrigues, D. (2023). Madeira pilot. Personal communication, AGEO Final Conference, Lisboa, LNEC, April 19–20, 2023.

Rubio-Iglesias, J. M. (2013). Citizens' observatories for monitoring the environment: A commission perspective. In Proceedings of the Workshop on Citizen's Involvement in Environmental Governance, Arlon, Belgium, 7 October 2013; Directorate General Research and Innovation, European Commission: Brussels.

Schade, S., Pelacho, M., van Noordwijk, T., Vohland, K., Hecker, S., & Manzoni, M. (2021). Citizen science and policy. In Vohland, K., Land-Zandstra, A., Ceccaroni, L., Lemmens, R., Perelló, J., Ponti, M., Samson, R., & Wagenknecht, K. The Science of Citizen Science. Springer Nature.

Shaw, R., Uy, N., & Baumwoll, J. (Eds.). (2008). *Indigenous knowledge for disaster risk reduction: Good practices and lessons learnt from Asia-Pacific region*. UNISDR.

Surminski, S., Aerts, J., Alexander, D., Di Bucci, D., Mechler, R., Mysiak, J., & Wilkinson, E. (2017). Prevention and mitigation: avoiding and reducing the new and existing risks. In: Poljanšek, K., Marín Ferrer, M., De Groeve, T., & Clark, I. (Eds.). Science for disaster risk management 2017: knowing better and losing less. EUR 28034 EN, Publications Office of the European Union, Luxembourg, Chapter 5.1, https://doi.org/10.2788/688605.

SWD 149 (2020). Best Practices in Citizen Science for Environmental Monitoring. European Commission Staff Working Document, SWD (2020) 149 final.

UN. (2015). Transforming our world: the 2030 Agenda for Sustainable Development. A/RES/70/1, United Nations. sustainabledevelopment.un.org.

UN. (2020). Hazard Definition & Classification Review. Technical report. Sendai framework for Disaster Risk Reduction 2015–2030, International Science Council (ISC), UN Office for disaster Risk Reduction (UNDRR). United Nations.

UN. (2021). Hazard information profiles. Supplement to UNDRR-ISC Hazard Definition & Classification Review. Technical report. Sendai framework for Disaster Risk Reduction 2015–2030, International Science Council (ISC), UN Office for disaster Risk Reduction (UNDRR). United Nations.

UNDRR. (2016). Report of the open-ended intergovernmental expert working group on indicators and terminology relating to disaster risk reduction. United Nations General Assembly Seventy-first session, A/71/644. https://www.undrr.org/we/inform/publications/51748.

UNDRR. (2019) *Global assessment report on disaster risk reduction*. Switzerland, United Nations Office for Disaster Risk Reduction (UNDRR).

UNDRR. (2022). *Global assessment report on disaster risk reduction 2022: Our world at risk: Transforming governance for a resilient future*. Switzerland, United Nations Office for Disaster Risk Reduction (UNDRR).

UNDRR. (2023). PreventionWeb. Understanding disaster risk. Retrieved 30 March, 2023, from https://www.preventionweb.net/understanding-disaster-risk.

UNFCCC. (2016). The Paris Agreement. United Nations Framework Convention to Climate Change (UNFCCC).

UNISDR. (2009). UNISDR Terminology on Disaster Risk Reduction. United Nations International Strategy for Disaster Reduction (UNISDR), Switzerland.

UNISDR. (2015). Sendai Framework for Disaster Risk Reduction 2015–2030. United Nations Office for Disaster Risk Reduction. Retrieved 30 March, 2023, from https://treaties.un.org/Pages/ViewDetails.aspx?src=TREATY&mtdsg_no=XXVII-7-d&chapter=27&clang=_en.

WeObserve consortium. (2021). Roadmap for the uptake of the Citizen Observatories' knowledge base. Report submitted to the European Commission. https://doi.org/10.5281/zenodo.4646774.

Wisner, B., Gaillard, J. C., & Kelman, I. (Eds.). (2012). *The Routledge handbook of hazards and disaster risk reduction*. Routledge.

8 Presentation of the Citizens' Observatory Pilots

Marion Jaud, Nicolas Le Dantec, Caroline Lummert, Lucie Cocquempot, Isabel Montoya-Montes, Claudia Pinto, Mónica Amaral Ferreira, Kieran Parker, Kirstin Lemon, Domingos Rodrigues, Fatima P. Gouveia, Eoghan Holohan, Vitor Correia and Rui Carrilho Gomes

8.1 Introduction

The AGEO project aims to implement five pilots of citizen observatories to improve the monitoring and management of natural hazards in the Atlantic Arc Area, through active participation of stakeholders, local communities and citizens in multiple aspects of risk assessment and prevention. These will be referred to as "pilots" in the following.

The five pilot observatories are located along the Atlantic Arc (Fig. 8.1):

- Citizen's observatory of vulnerability to coastal risks of erosion and marine flooding in Brittany, France
- Multi-hazard Citizens' Observatory in Lisbon, Portugal
- Citizens' Observatory on Rockfall and Rockfall Triggers in the Canary Islands, Spain

M. Jaud (✉) · N. Le Dantec · C. Lummert
UMR 6538 GeoOcean, Institut Universitaire Européen de La Mer, University of Brest-CNRS, Rue Dumont d'Urville, 29280 Plouzané, France
e-mail: marion.jaud@univ-brest.fr

N. Le Dantec
e-mail: nicolas.ledantec@univ-brest.fr

L. Cocquempot
Oceanography and Ecosystem Dynamics, IFREMER, 29280 Plouzané, France
e-mail: lucie.cocquempot@ifremer.fr

I. Montoya-Montes
Department of GeoHazards and Climate Change, IGME-CSIC, 35003 Las Palmas, Spain
e-mail: isabel.montoya@ulpgc.es

C. Pinto · M. A. Ferreira
Lisbon Municipality, Campo Grande 25, Lisbon, Portugal
e-mail: claudia.pinto@cm-lisboa.pt

M. A. Ferreira · R. C. Gomes
Instituto Superior Técnico, Universidade de Lisboa, Av. Rovisco Pais 1, 1049-001 Lisboa, Portugal
e-mail: rui.carrilho.gomes@tecnico.ulisboa.pt

K. Parker · K. Lemon
Geological Survey of Northern Ireland, Dundonald House, Upper Newtownards Road, Ballymiscaw, Belfast BT4 3SB, UK
e-mail: kiepar@bgs.ac.uk

K. Lemon
e-mail: klem@bgs.ac.uk

D. Rodrigues · F. P. Gouveia
University of Madeira, Funchal, Portugal
e-mail: domingos@uma.pt

E. Holohan
School of Earth Sciences, UCD, University College Dublin, Dublin 4, Ireland
e-mail: eoghan.holohan@ucd.ie

V. Correia
Portuguese Association of Geologists, Lisbon, Portugal
e-mail: vcorreia@intraw.eu

R. C. Gomes et al. (eds.), *Citizens' Observatories on Geohazards*, Geoenvironmental Disaster Reduction,
https://doi.org/10.1007/978-3-031-53371-6_8

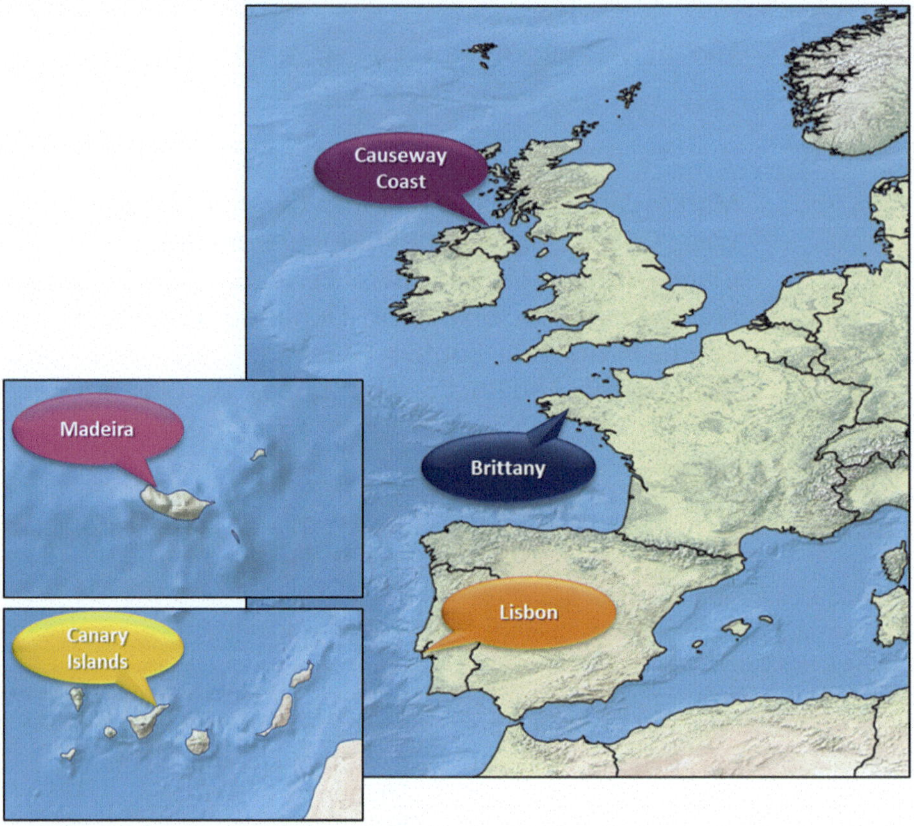

Fig. 8.1 Location of the five pilot observatories of AGEO

- Citizens' Observatory on Rockfall and Rockfall Triggers in the Causeway Coast, Northern Ireland
- Multi-hazard Citizens' Observatory on Madeira Island, Portugal.

These pilots constitute a representative panel of different hazards, stakes and risk situations for the populations at the scale of the Atlantic Arc, responding in this way to the priorities of the INTERREG program. This set of pilot observatories also includes remote or isolated territories where data is typically lacking due to the absence of systematic collection of reference data, and despite the high occurrence of natural risks. AGEO aims to involve local communities through participatory science tools in order to improve the collection of scientific data for decision-aid and awareness raising purposes. The use of European Copernicus services, based on satellite imagery, has been examined, both to assess their potential for hazards monitoring and to improve risk management methods (see Chap. 4). Above all, the main objective of pilots of AGEO is to adopt common approaches to share information between local citizen communities, natural hazard managers, emergency services and decision-makers in order to strengthen risk monitoring and risk management systems.

8.2 Methods for the Self-Characterization of the Pilot Observatories

Definition of the pilots' objectives: workgroups, brainstorming, mind mapsThe observation of geohazards is a vast domain. It is therefore not surprising that there is great diversity among the pilot observatories of AGEO. In order to be able to identify relevant axes of harmonization and to structure a collaborative roadmap while respecting the specificities of each pilot, all partners have engaged in collaborative work, including through hazard-specific working groups. Following several approaches for self-characterization has allowed each pilot could to define its objectives, means, constraints and ambitions in a homogeneous framework. Brainstorming sessions on this topic have taken place at multiple occasions throughout the beginning of the project, during consortium meetings and conference calls.

The first step was to agree on a broad, **inclusive definition of observatories** and to understand the expectations and needs of each of the partners in relation to AGEO. In the framework of AGEO, geohazard observatories aim to:

(i) improve and capitalise on **knowledge of hazards and exposure** through long-term monitoring and **data dissemination**;
(ii) improve **risk management and prevention** by involving different stakeholders and providing tools to help **decision-making**;
(iii) contribute to the improvement of the **risk culture** by involving citizens in monitoring and/or by providing them with information materials.

Then, during an online collaborative exercise of mind mapping, AGEO partners identified more specific objectives shared among all pilots:

- monitoring hazards and defining measurement protocols (data collection)
- use of Copernicus services and other Earth Observation tools (development of product and services)
- building/maintaining partnerships with local authorities and other stakeholders
- providing standardised data for the public authorities responsible of risk management
- providing access to documentary resources and/or results through a dedicated web portal
- leading/animating groups of natural risk management professionals and citizens
- providing expertise in civil engineering or support for the risk management policies at local scale
- proposing training sessions and awareness-raising measures
- setting up warning systems

For each of these points, the pilots were able to specify the action plan they intended to deploy during AGEO.

8.2.1 Survey and Semi-Directive Interviews

Taking into account the differences among pilots regarding their development stage, stakeholders configuration, and even the approach envisioned for citizen participation, a survey consisting of 56 questions was proposed to the pilot observatories' teams, to better identify the specificities of each observatory. It was structured along the following lines:

- Governance
- Types of territory concerned (area, population, environment, etc.)
- Types of risk and the dynamics of hazards,
- Main objectives of the observatory,
- Means of observation,
- Tools for data dissemination,
- Modalities of citizens' involvement,
- Target audiences (managers, inhabitants, tourists, emergency services),

These topics are common to all the pilot observatories of AGEO, therefore making inter-comparisons possible when analysing the answers to the survey.

Based on the feedback from the mind map session and the survey, a series of semi-directive individual interviews was conducted with each of the pilots, lasting approximately one hour and consisting of 16 questions. The goal was to expand on the results of the survey by clarifying certain answers and getting quantitative information, in order to gain a better understanding of the goals, functioning, and activities of the pilot observatories.

8.2.2 Classification Matrix of Observatories

The analysis of the information collected through these various means confirmed the differences in configuration and operating modes among AGEO pilots. The observation strategies and the means deployed vary from one pilot to another, which stems from the difference in their goals and also in the type of stakeholders they preferentially interact with.

In order to synthetize these results on self-characterization, the matrix initially elaborated by OSIRISC team (Philippe & Hénaff, 2021) is adapted to suit the wider scope of AGEO's observatories. The underlying classification scheme and the matrix representation offers a very visual approach, similar to a QR-code, to identify the preferred lines of action and operating methods of an observatory (Fig. 8.2).

The matrix is of dimensions 4×4. It is not intended as a rating of the quality or efficiency of an observatory, but as a way of highlighting priority areas and operating methods at a given time. Cells in the matrix describe possible actions the observatory may undertake. Results of the self-characterization study are capitalized by colouring the cells corresponding to the objectives of each observatory. Checked cells will draw a pattern, allowing a visual identity to emerge, which can eventually lead to identifying different types of observatories. Short text descriptions provide details on the activity undertaken in the coloured cell. The columns represent different operational areas of the observatory. The "geo-risks" column contains actions related to hazards monitoring and risk assessment. The columns "Risk managers", "Citizens of the territory" and "Emergency service" are meant to describe the involvement of the observatory with the different stakeholders, and whether the observatory is engaged in actions relevant for these stakeholders. The more coloured cells a column contains, the more this axis of activity is favoured by the observatory.

- Increasing involvement +	Who/what does your pilot interact with?			
	Geo-risks	Risk management stakeholders	Citizens of the territory	Emergency services
	Collecting data	Training	Informing	Monitoring risks, setting up warning systems
	Standardizing & validating data	Advising on monitoring and risk management projects and policies	Participating in projects and events	Conducting drills for a general audience
	Developing products and services	Expertising risk management projects and policies (local authorities or State)	Following territory projects	Providing support regarding decisions to organize crisis management
	Assessing geodynamics / risks assessment	Risk management planning (works, restoration...)	Leading / coordinating a community, citizen networks	Managing the deployment of protection and emergency response resources

Fig. 8.2 Standard classification matrix of the citizen observatories of AGEO

The rows define how far toward risk management the pilot is going with respect to each operational area:

- The upper rows of the matrix correspond to observation goals and awareness-raising among the various audiences,
- The further down in the matrix, the more the pilot goes beyond data collection and dissemination, by exploiting the data into products and services and engaging further with stakeholders, eventually to carry out operational missions to assist in decision-making and risk management.

The matrices therefore summarise the main actions carried out by the pilots, their working partners, and the relations with stakeholders. The cells left blank may appear as possible evolutions of the pilot observatory.

8.2.3 Identification of Action Levers

Building upon the results of the self-characterization, action levers were proposed to the AGEO pilot observatories in order to define common tools to strengthen citizen involvement and improve outputs for risk management.

Actions that emerged as priorities are:

- participative science tools (including the design of the AGEO smartphone application),
- raising public awareness,
- data sharing (development of AGEO platform) and eventually considering early-warning alert procedures
- the definition of indicators to quantify the risk.

As the project progressed and through ongoing discussions within the consortium, the pilots have updated their roadmaps, adding complementary actions.

8.3 Presentation of the Citizens' Observatory Pilots of AGEO Project: A Diverse Set of Pilots

The mind map produced less than a year after the start of AGEO project gave an overview of the aspirations any initial steps of each pilot (the formatted original mind map is shown in Fig. 8.3). The pilots' objectives have evolved since then, adapting to the implementation constraints and drawing inspiration from other AGEO pilots.

In order to have a synthetic presentation of the characteristics of each pilot observatory, the following "identity cards" have been produced from the results of the interviews and surveys (Fig. 8.4).

The pilots address different types and intensities of hazards, some pilots being "multi-hazards". Their spatial coverage, configurations and the degree of exposure of the infrastructures and human lives vary from one territory to another. From the surveys and interviews, it also appears that different types of structures are in charge of the implementation and coordination of the observatories. The five pilots also have various degrees of maturity (Fig. 8.5), the observatories in Brittany and Lisbon being more than 10 years old, while the pilots in the Canary Islands, Causeway Coast and Madeira have been initiated as observatories during the project.

With regards to the geohazards monitored by the AGEO pilots, the needs for measurements and the approaches for managing the associated risks will depend on the kinematics of these hazards. The Fig. 8.6 classifies these geohazards along two axes: one axis for the sudden nature of the process (from slow to rapid) and one axis for the duration of the event (from transient to long term). For example, the more sudden the hazard, the higher the need for an early-warning system, while the duration of the process will have a greater influence on the

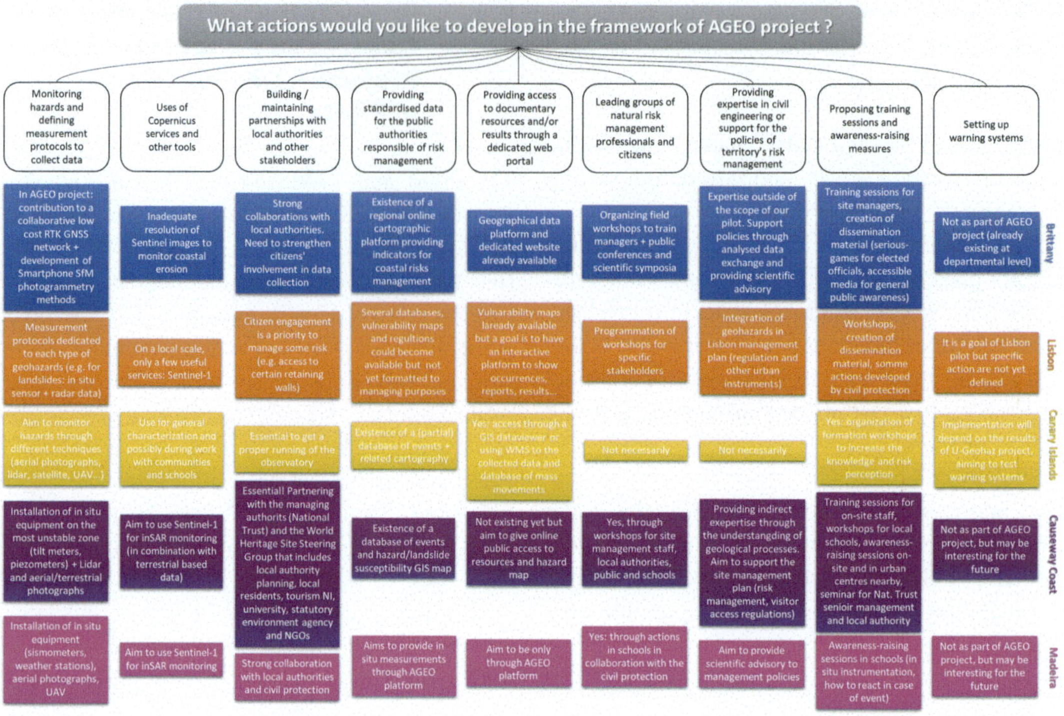

Fig. 8.3 Formatted results of the mind map exercise identifying the actions that the pilots intended to developed

timeframe of application of envisioned management policies.

The strategy of the pilot observatory will vary according to the size of the area affected by the hazards, but also the type of environment at risk: rural, natural or densely built-up zone, presence of natural or cultural heritage among the exposed elements on the territory (Fig. 8.7). The AGEO pilots again show a great diversity in the types of environments exposed, from very densely populated and with a rich cultural heritage as Lisbon to almost uninhabited but with a very high number of visiting tourists attracted by the natural heritage as the Causeway Coast.

Given these disparities in the size of the territory, the type and kinematics of the hazards and the type of environment, AGEO pilots obviously use different methods, tools and protocols to monitor their risks (Fig. 8.8). The survey showed that all pilots use (or are about to use) satellite imagery, whether from Copernicus or other Earth Observation (EO) services. The use of aerial or terrestrial photographs is also common to all pilots for the collection of in situ data. This information has contributed to guiding the development of the AGEO citizen science application around a reporting system based on taking geolocalized pictures.

The goals, operating methods and resources dedicated to risk observation are different from one pilot to another because they depend mainly on the exposed elements (human beings, economic assets, infrastructure, natural or patrimonial sites, etc.) and the stakeholders (civil society, local communities, civil protection, etc.) working with the AGEO pilot teams.

The interest of different stakeholders in the results of the AGEO pilots was also the subject of a survey sent to the pilots (Fig. 8.9). This survey shows that local authorities and communities are very concerned about these geo-monitoring issues. The role of civil protection and the interest of private structures are less significant for the more rural territories or natural areas, whereas they increase in the most urbanised areas.

Citizens' Observatory of vulnerability to coastal riks in Brittany, France	
Year of launch	2010
Governance	University of Brest
Emprise	~ 5000 km²
Observed hazards	Cliff retreat, slope instability, rockfalls, marine flooding, beach erosion
Exposed environments	Urban - medium density urbanization Rural - low density urbanization Natural Environments
Human resources	3 persons
Annual financial resources	85 000 €
Data collection (approach, revisit frequency...)	In situ instrumentation Regular field measurement campaigns One-off field measurement campaigns Post-event survey Geolocalized photo monitoring
Target audience	Managers in charge of risk management, civil society

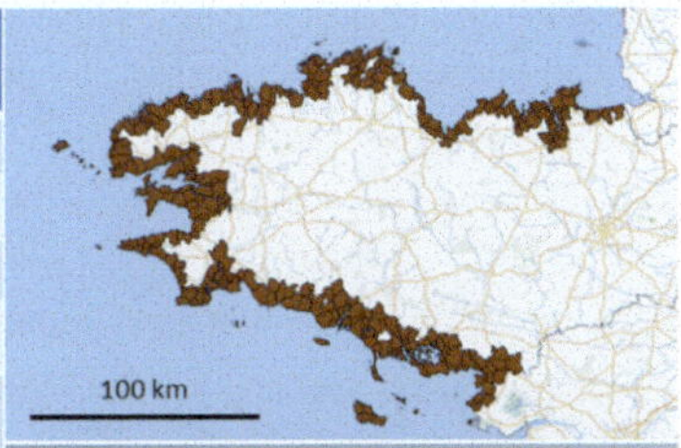

Multihazard Citizen's Obervatory in Lisbon, Portugal	
Year of launch	2010
Governance	Lisbon municipality
Spatial coverage	85 km²
Observed hazards	Landslides, earthquakes, flooding, marine flooding and geotechnical risks
Exposed environments	Urban - high density urbanization
Human resources	7 persons
Annual financial resources	Only from research projects
Data collection (approach, revisit frequency...)	No field measurements to monitor hazards
Target audience	Managers in charge of risk management, civil society/residents

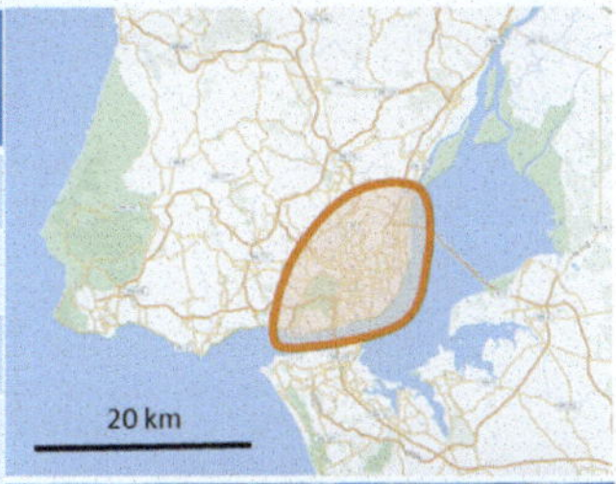

Canary Islands Citizens' observatory on rockfalls and rockfalls-triggers	
Year of launch	2020
Governance	National public organism
Spatial coverage	616 km²
Observed hazards	Cliff retreat, slope instability, landslides, rockfalls
Exposed environments	Urban - high density urbanization Urban - medium density urbanization Rural - low density urbanization Natural Environments
Human resources	1 person
Annual financial resources	100 000 €
Data collection (approach, revisit frequency...)	Regular field measurement campaigns One-off field measurement campaigns Post-event survey Geolocalized photo monitoring
Target audience	Risk managers and emergency services, civil society/residents

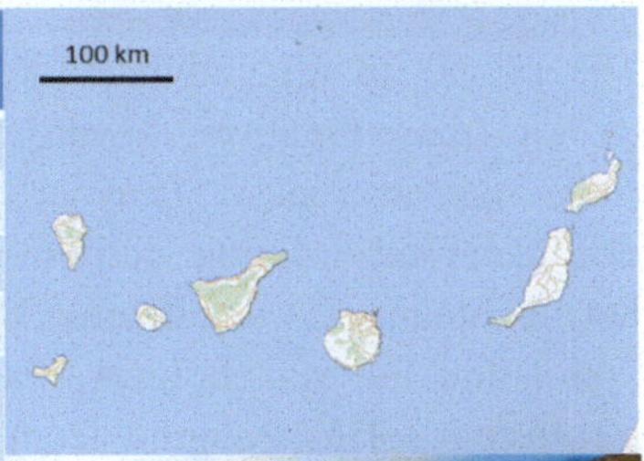

Fig. 8.4 Identity cards of AGEO pilot observatories in Brittany, Lisbon, Canary Islands, Causeway Coast and Madeira

Citizens' observatory on rockfalls in Causeway Coast, Northern Ireland	
Year of launch	2020
Governance	National public organism
Spatial coverage	0.7 km²
Observed hazards	Cliff retreat, slope instability, landslides, rockfalls
Exposed environments	Rural - low density urbanization Natural Environments
Human resources	2 persons
Annual financial resources	95 000 €
Data collection (approach, revisit frequency...)	In situ instrumentation Regular field measurement campaigns Post-event survey Geolocalized photo monitoring
Target audience	civil society, tourists

Multihazard Citizens' Observatory in Madeira Island, Portugal	
Year of launch	2020
Governance	University of Madeira
Spatial coverage	741 km²
Observed hazards	Cliff retreat, slope instability, landslides, rockfalls, flooding
Exposed environments	Urban - high density urbanization Urban - medium density urbanization Rural - low density urbanization Natural Environments
Human resources	5 persons
Annual financial resources	To be defined
Data collection (approach, revisit frequency...)	In situ instrumentation Geolocalized photo monitoring Weather stations
Target audience	Civil society/residents

Fig. 8.4 (continued)

The pilots are all convinced of the importance of dissemination toward citizens to increase risk awareness. They are all engaged in outreach and training activities for public authorities, managers or citizens. Depending on the pilot and the stakeholders, these activities take different forms, ranging from an advisory role for risk management authorities to communication via social networks.

Finally, it can be noted that the financial and human resources available are different from one pilot to another, but can also change over time. Indeed, depending on the governance of the pilot observatories, some may have difficulty in finding recurrent funding. It is essential to establish formal partnerships with public authorities, to ensure the sustainability of the pilots through political support.

A comparative analysis of the matrices obtained for the various pilot observatories makes it possible to identify those whose missions are identical or very similar (Fig. 8.10).

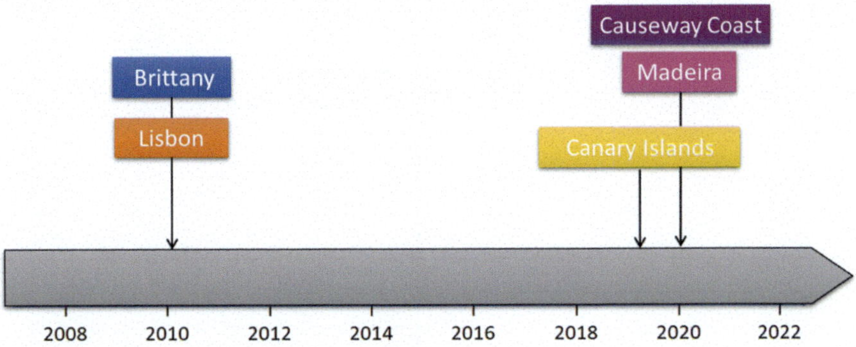

Fig. 8.5 Date of creation of the different pilot observatories

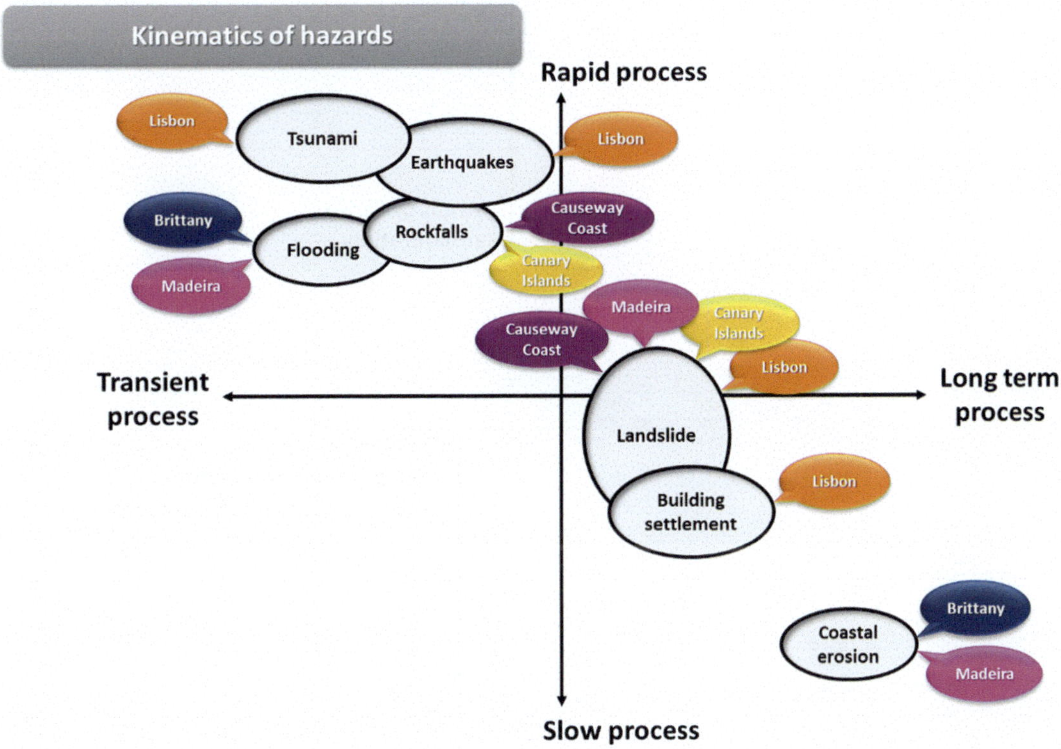

Fig. 8.6 Kinematics of the geohazards monitored by the AGEO pilots

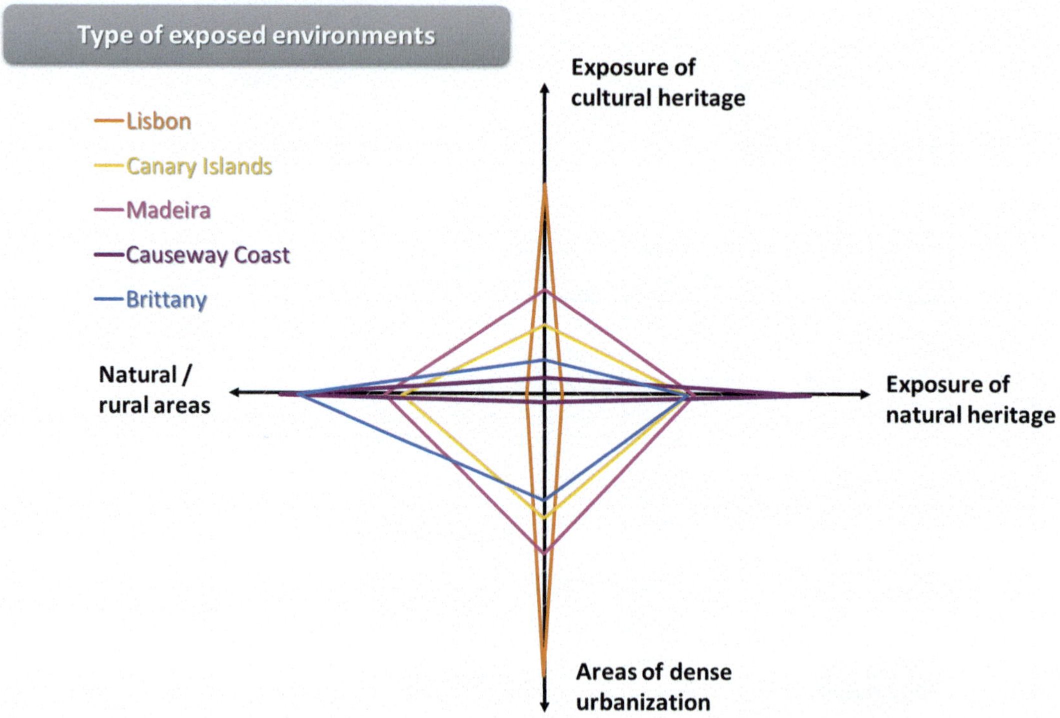

Fig. 8.7 Types of exposed environments in the AGEO pilots

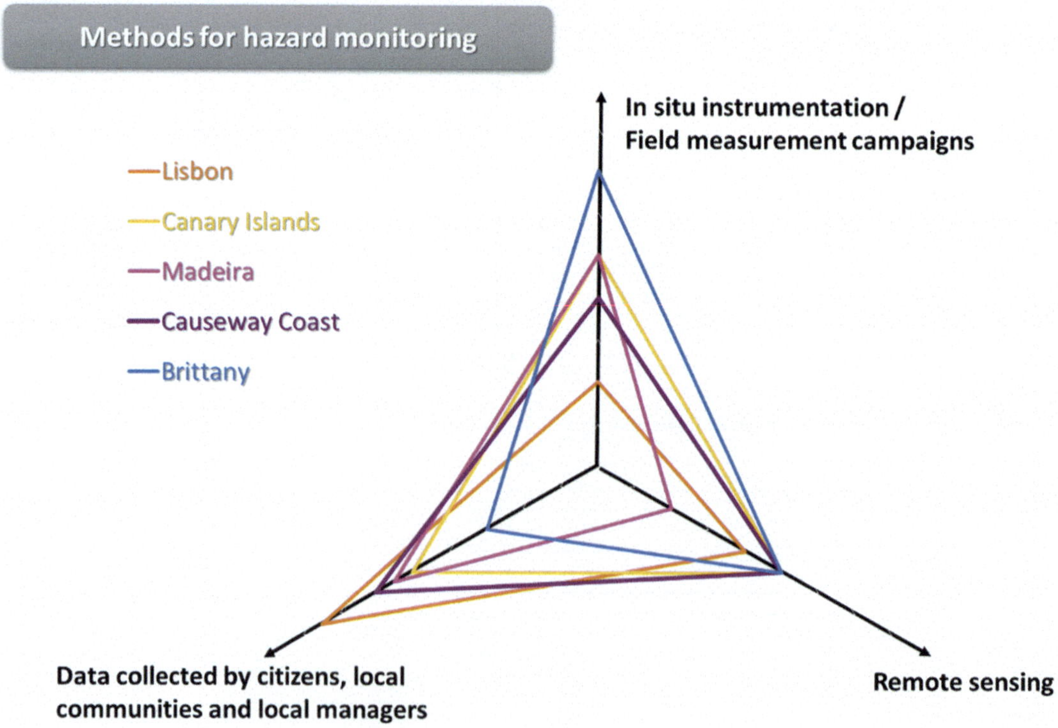

Fig. 8.8 Methods for hazard monitoring used by the AGEO pilots

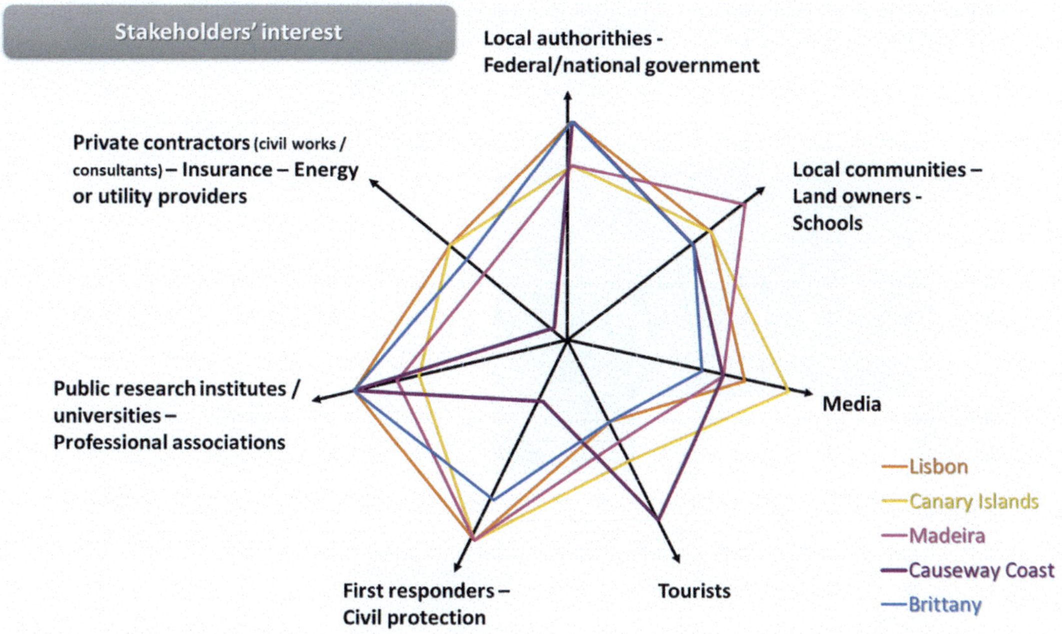

Fig. 8.9 Stakeholders' interest on the outcome of AGEO pilots

These classification matrices and the resulting QR codes confirm a variety of goals and overall strategic positions but also highlight similarities between the AGEO pilot observatories. Brittany and Lisbon pilots have similar QR-codes pilots, and so do Madeira and the Causeway Coast. It can be seen that all the observatories carry out activities related to scientific data collection and information/awareness-raising for citizens. The profiles reveal different types of missions ranging from scientific observation to operational decision support - sometimes in a crisis management context. As can be expected, observatories where populations are highly exposed to natural risks consider the relationship with emergency services among their priorities.

8.4 Conclusion

The set of AGEO pilot observatories covers a wide range of configurations in terms of the profile, goals and constraints of the pilots. In this context, AGEO's achievement is precisely to have developed common tools suiting the needs of such a diversity of pilots. These tools mainly aim at reinforcing the commitment of

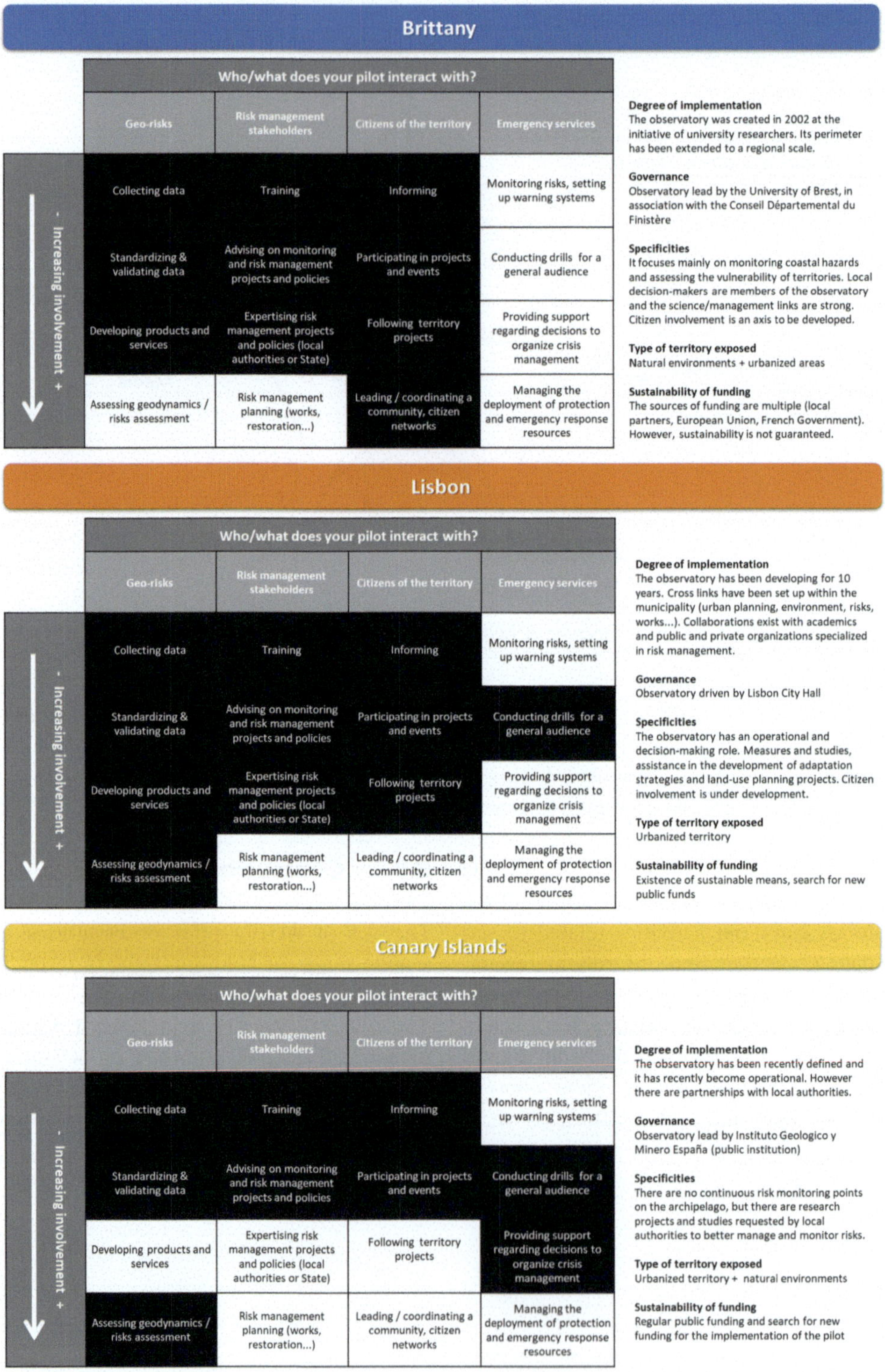

Fig. 8.10 Comparison of the "QR code" matrices for classification of the pilot observatories

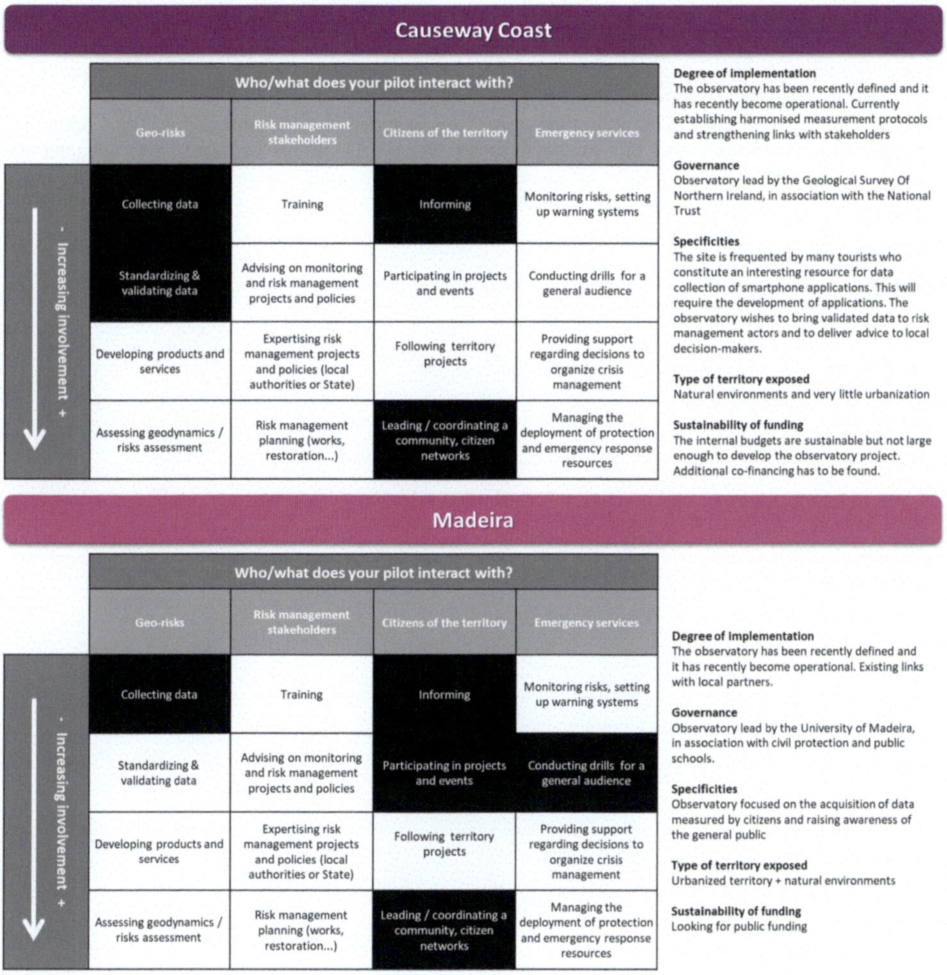

Fig. 8.10 (continued)

citizens so that they actively participate in the creation and dissemination of new knowledge on geohazards. Thanks to citizen involvement, AGEO pilots will be able to collect and exploit geohazards observation data to guide public action on the one hand, and to train exposed populations to increase their awareness and preparedness on the other hand. Given their applicability to the diversity of pilots AGEO, these tools are likely to be robust for potential transfer to new citizen geohazards observatories.

Acknowledgements This work has been undertaken in the framework of the INTERREG Atlantic Area AGEO Project (EAPA_884/2018 – AGEO).

Reference

Philippe, M. et Hénaff A., coordinateurs. (2021). Expérimentation pour un observatoire régional de la vulnérabilité aux risques côtiers en Bretagne. Projets OSIRISC et OSIRISC+. UBO, CNRS, Cerema. Soutenus financièrement par la Fondation de France et la DREAL Bretagne. 84 p + annexes. https://doi.org/10.48386/POPS-1519.

9 Citizens' Observatory on Rockfalls in the Canary Islands, Spain

Isabel Montoya-Montes, Inés Galindo, Nieves Sánchez, Juan Carlos García López-Davalillo, Inmaculada García, Noelia Cruz Pérez, Ariadna Ortega, Juan Carlos Santamarta, Joselin S. Rodríguez Alcántara, Mario Hernández Ruiz, Margarita Sanabria Pabón, Carlos Lorenzo Carnicero, Roberto Sarro and Rosa M. Mateos

9.1 Introduction

After floods, landslides are the most damaging geohazard in Spain regarding economic losses (Mateos, 2017). Rockfalls are described as rapid detachments of blocks, which are conditioned by morphological, geological and environmental characteristics.

Usually, they occur on steep slopes and cliffs in mountainous regions with planar and wedge-type fractures. Rock avalanches refer to very quick (>100 km/h) rock or debris falls resulting from large-scale slides or rockfalls on very steep slopes (Budetta, 2004).

I. Montoya-Montes (✉) · I. Galindo · N. Sánchez · J. C. G. López-Davalillo · I. García · M. Hernández Ruiz · M. Sanabria Pabón · C. Lorenzo Carnicero · R. Sarro · R. M. Mateos
Department of GeoHazards and Climate Change, IGME-CSIC, Madrid, Spain
e-mail: isabel.montoya@ulpgc.es

I. Galindo
e-mail: ines.galindo@csic.es

N. Sánchez
e-mail: n.sanchez@igme.es

J. C. G. López-Davalillo
e-mail: jc.garcia@igme.es

I. García
e-mail: i.garcia@igme.es

M. Hernández Ruiz
e-mail: m.hernandez@igme.es

M. Sanabria Pabón
e-mail: m.sanabria@igme.es

C. Lorenzo Carnicero
e-mail: c.lorenzo@igme.es

R. Sarro
e-mail: r.sarro@igme.es

R. M. Mateos
e-mail: rm.mateos@igme.es

I. Montoya-Montes
Instituto de Oceanografía y Cambio Global (IOCAG), Universidad de Las Palmas de Gran Canaria, Las Palmas de Gran Canaria, Spain

N. C. Pérez · J. C. Santamarta · J. S. Rodríguez Alcántara
Department of Agricultural and Environmental Engineering, University of La Laguna (ULL), La Laguna (Tenerife), Spain
e-mail: ncruzper@ull.edu.es

J. C. Santamarta
e-mail: jcsanta@ull.es

J. S. Rodríguez Alcántara
e-mail: jrodralc@ull.edu.es

A. Ortega
La Palma Research Centre, La Palma (Tenerife), Spain
e-mail: ariadna.ortega@lapalmacentre.eu

R. C. Gomes et al. (eds.), *Citizens' Observatories on Geohazards*, Geoenvironmental Disaster Reduction,
https://doi.org/10.1007/978-3-031-53371-6_9

In the Canary Islands, rockfalls and landslides are mainly triggered by meteorological phenomena, such as severe precipitations and wind, as well as dilation resulting from solar radiation. Seismicity associated to volcanic eruptions is also one of the main trigger in the Canaries. The volcanic submarine eruption in El Hierro (2011–2012) triggered numerous rockfalls near the Los Roquillos tunnel, a strategic infrastructure in the island. Anthropic activities can also contribute to slope instability (Cabildo de Gran Canaria, 2005).

Understanding information such as location, frequency, event type, rock type, block size, rainfall or wind intensity contributes to a better understanding of the process, which is critical for performing hazard analyses to improve territory management and risk reduction. Due to the high frequency of rockfalls recorded in the Canaries, the resultant damage can be significant (Cabildo de Tenerife, 2012). Damages depend on the velocity and magnitude of the processes, as well as the vulnerability of exposed elements. In the Canary Islands, small damages and some fatalities are associated with rockfalls, while severe material damages and numerous victims are usually due to rock avalanches and slides (Cabildo de Gran Canaria, 2005). In terms of economic impact, rockfalls result in the most significant losses, mainly affecting urban areas and road network. (Cabildo de Tenerife, 2020).

Geohazards are contemplated in the Spanish legislation in two well-defined areas (Mateos, 2017): (1) Land Urban Rehabilitation Bill (2015), which requires the preparation of natural risk maps in the context of land-use planning. Geohazards maps are incorporated in the environmental studies of the territory and must be finally approved by the regional government; and (2) the Basic Civil Protection Regulation (1992), which establishes the possibility of producing Special Plans focusing on particularly significant hazards in each region. In the Canary Islands, although rockfalls occur frequently and can cause significant damage, they are not included into any Special Plan for landslides, but are considered as secondary hazard associated with meteorological, volcanic processes, floods and earthquakes. For Tenerife, a landslide susceptibility map is included in the Territorial Special Management Plan for Risk Prevention (not a Special Plan) (Cabildo de Tenerife, 2012), but updates based on the frequently events occurring are needed.

It is noteworthy note that some insular and local civil protection plans have included information regarding landslides, underscoring the importance of reliable and updated data for efficacious management of mass movements (Insular emergency Plans in Gran Canaria and Tenerife, Local Emergency Plan in San Bartolome de Tirajana Municipality). However, despite improvements in identification, prediction, preventive actions and emergency systems, damages caused by mass movements are increasing exponentially. Uncontrolled urban expansion emphasizes social, economic and environmental impacts and it is leading to a large increase in people living in locations prone to mass movements (Cabildo de Tenerife, 2020). Consequently, it is still necessary to work on increasing public awareness of the rockfalls hazard and its prevention, as well as to further improve the hazard management plans.

There are not many experiences of risk prevention related to education and participation of the population in the Canary Islands. However, citizen science is increasingly being used as a tool capable of providing valuable information in real time about geohazards. Several Spanish initiatives have been established outside the Canary Islands, such as BAPA Project (mass movements register in Asturias region), FLOODUP Project (impact of floods register in Catalonia region), or ALQUIVIR Project (natural hazard Observatory in the upper Guadalquivir valley basin region).

The Citizens' Observatory on Rockfalls in the Canary Islands (Ortega Rodríguez et al., 2020; Cruz-Pérez et al., 2021), currently in the pilot stage, is intended as an instrument of active participation for society and citizen responsibility in the management of geohazards. It constitutes a tool that promotes citizen participation and seeks to empower society through citizens' training, oriented towards reporting mass movements that have occurred in the regions where they live. The Observatory encourages citizens to take an active part in risk prevention particularly when they face recurring hazards. The intention is to contribute to the well-being of the society as a whole by participating as a community in the management of rockfalls hazard, as well as its assessment and monitoring.

Citizens' engagement is one of the pillars of the Observatory, and numerous dissemination activities have been carried out to raise awareness of the Observatory, such as project dissemination days, workshops in schools, quarterly workshops with citizens participating actively, training activities, and social media campaigns. Moreover, an app, a communication exchange tool, was designed, implemented and tested. Therefore, any individual, whether a citizen or an expert collaborator, has access to communication channels that enable them to report any instances of rockfall events.

Not only data collecting from the app has been handled but also other information sources, such as historical data, site data and earth observation services data. The latter has aroused particular interest given their important role in the AGEO project. The use of the products and emergency services of the European Copernicus program as a key point in the implementation of the observatory, receives special attention in later sections as well as in Chap. 4 of this book.

Thus, the Citizen Observatory on Rockfalls in the Canary Islands is presented below as a pilot experience in the assessment of geohazards. The initiative involves the collaborative efforts of citizens, technicians, managers, and scientists, all working towards a common goal of disaster prevention.

9.2 Site Description

The Canary Islands is an archipelago of eight islands, belonging to Spain and located approximately 100 km from the Sahara Desert (Africa). It is also an outermost European region due to the fact it is 950 km from the European mainland coast. The most populated islands are Tenerife and Gran Canaria, which, in turn, are the capitals of the two provinces that make up the Autonomous Community of the Canary Islands (Fig. 9.1a).

These islands are of active volcanic origin, with the island of El Hierro being the youngest of the Canary archipelago, at approximately one million years old. 18 volcanic eruptions (prehispanic and historical) have occurred in the Canary Islands (Romero et al., 2020), being the most recent ones those of El Hierro (submarine) and La Palma, in 2011 and 2021 respectively. The rugged orography is a constant feature of the Canary Islands, especially in the westernmost islands of the archipelago. The highest peak in Spain, Teide (3,715 m high), stands in Tenerife. Therefore, erosion processes are also very active and are conditioned not only by orography but also by climatology. The latter is characterised by trade winds, which are the dominant winds (occurring more frequently in summer), while the main wind storms occur with the arrival of Atlantic storms that give rise to strong winds from the NW. However, those from S are particularly dangerous, since most of the infrastructures are not prepared to withstand unusual strong winds from this direction. The highest intensity of precipitation occurs mainly on the eastern slopes at medium altitudes.

The Observatory on Rockfalls in the Canary Islands aims to deepen the knowledge of these processes through citizens and risk managers participation by encouraging their collaboration and reporting of rockfalls observed in their region. Therefore, since the success of the Observatory depends on the degree of citizen involvement, three pilot areas were chosen: the island of El Hierro, Anaga Rural Natural Park (San Cristobal de La Laguna Municipality) in the island of Tenerife, and San Bartolomé de Tirajana Municipality in Gran Canaria to work hand in hand with the population (Fig. 9.1), disseminating the initiative and promoting awareness about the importance of rockfall prevention.

The selection of each one of the rockfall study areas in the Canary Islands within the

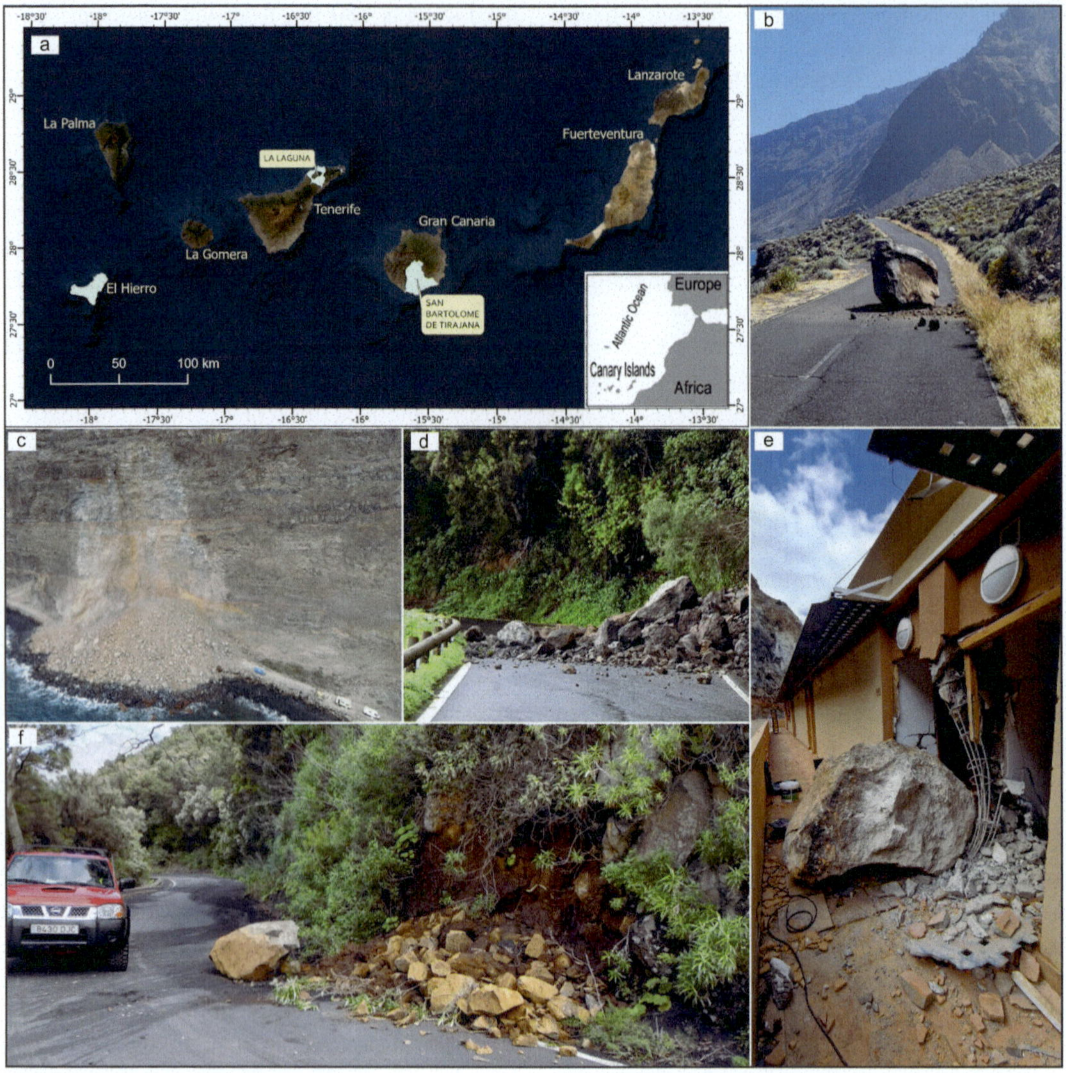

Fig. 9.1 **a** Map showing pilot area location and subpilots areas; **b** to **f** Rockfall examples in the Canary Islands

AGEO Project was based on the following criteria:

(i) El Hierro, with steep escarpments, mainly related to coastal cliffs and mega-slide scars, mean that most of the island's population is exposed to rockfall hazards (Rossi et al. 2020) (Fig. 9.1b). Rockfalls occurring during periods of heavy rainfall and/or strong winds have been observed to be frequent. During the volcanic-seismic crisis on since 2011 (Jurado et al., 2020), there were numerous rockfalls and analysis determining relationships between earthquakes and rockfalls were carried out (Llorente et al., 2015). The emergency situation caused the population throughout the island to organize themselves to be prepared in case of an earthquake or associated rockfalls occur. (Galindo et al., *in press*). Therefore, the entire citizen infrastructure for internal communication was fully established, i.e. the entire island became a Citizen Observatory, and this structure, which is still in place today, has been used by the Project for the development of the pilot on the island. As it will be presented in the following sections, that previous experience has resulted in the highest participation rates in the rockfall observatory in the Canary Islands archipelago. Thus, the island was selected in its full extent since natural and social characteristics turns El Hierro into an excellent pilot subzone for the implementation of the Observatory.

(ii) The island of Tenerife, being the largest and highest island in the Canary archipelago, is home to the Rural Natural Park called Anaga, which is characterized by high gradient slopes and the highest rainfall rates in the island (Díez-Sierra & Del Jesus, 2020). Moreover, Anaga has scattered rural settlements that are notably affected by rockfalls (Leyva et al., 2022) (Fig. 9.1d). These settlements can be cut off from the rest of the island because of the lack of alternative roads, making the impact of rockfalls on daily life during emergencies is a matter of significant concern.

(iii) On the island of Gran Canaria, the municipality of San Bartolomé de Tirajana is located in an erosive basin which contains at least 28 major landslides, some of which have reactivated over time (Lomoschitz & Corominas, 1997). The population is scattered across the municipality but it is particularly concentrated around some villages, so due to its scattered nature, population here is particular vulnerable to roads cut off due to rockfalls occurrence (Fig. 9.1) (Sarro, et al., 2020). Solana and Kilburn (2003) worked with communities in the area and suggested that population are generally aware about mass movements occur there, however, they rarely identify mass movements as hazardous events for themselves. Weak hazard perception, steep slopes, torrential rainfall regime, and a high level of willingness on the part of emergency managers to collaborate, were selection criteria in this case.

9.3 Objectives of the Pilot

The main objective of the Canary Islands Observatory on rockfalls, has been to achieve a community of citizens involved in what is happening around them, actively collaborating in the reporting of observed hazards. The idea is to prevent citizens from normalising their coexistence with this type of hazard and, instead, to take an active role that allows them to prevent and take part in the process of generating science, since the report of observed risks is analysed and filtered by the Spanish Geological and Mining Institute (IGME-CSIC), which uses this valuable source of data to increase its database and also to undertake studies aimed at improving knowledge about rockfalls in the Canary Islands. Thus, specific objectives are detailed below:

- citizens' sustainable engagement to raise geohazards awareness in general and rockfall hazard awareness in particular, and send information about rockfall events;
- rockfalls data compilation in order to get an updated inventory of events and developed products and also to contribute to complete the Spanish Geological and Mining Institute mass movements database, BD-MOVES (https://info.igme.es/BD2DMoves/), and therefore be able to initiate rockfalls hazard assessment for the region.
- support rockfall management in the prevention, preparedness and response stages helping to improve the current management plans.

9.4 Data Handled by the Pilot

Several data sources were handled in order to feed the Observatory with rockfall data and update the available mass movements inventories and contributing to complete the existing national database BDMOVES with mass movements data at regional scale. The data structure of BDMOVES is organized in two parts: one including intrinsic and relatively invariable characteristics of the

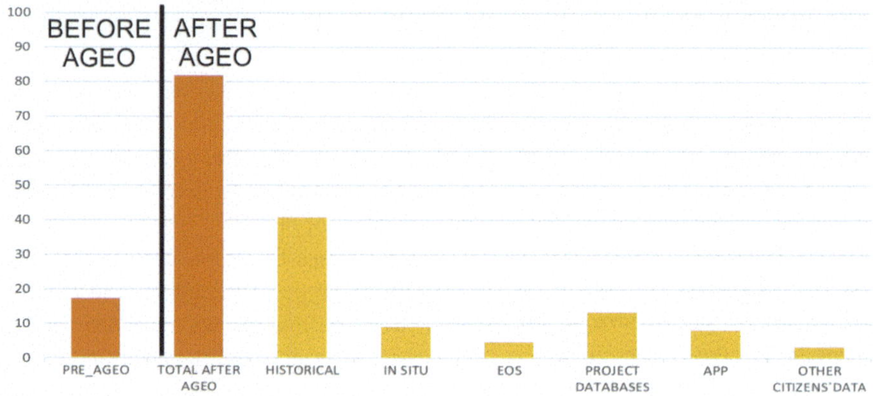

Fig. 9.2 Data source (%) handled by the pilot

movement, and another part including morphometry, triggering factors, damages, etc. Basic data to include are name, location, movement type, activity, methodology used to assess the movement type and activity status, date of most recent observation, and damage. Therefore, the data compilation in each source will be oriented to obtain such type of data. Historical data, in situ data, earth observation services data, project databases and citizen data were handled (Fig. 9.2).

- **Historical data**

Initial phases of the Observatory implementation consisted in a review of existing press reports on rockfalls in the Canary Islands archipelago, collecting for each event information on date, location, description, damage and also photographs or videos. Not only local press was reviewed but also regional and national, and it was not always possible to gather all the necessary information with a single news item, so sometimes additional sources such as other newspaper or social network publications were consulted in order to compare and complete the information to fully characterize the event. A total of 350 rockfalls from historical data were gathered, homogenized, structured and included in BDMOVES.

- **In situ data**

This type of data is collected from direct observation in field campaigns. Field data were gathered and included in the database. Both, data collected in the frame of AGEO and previous data were considered. This type of data could sometimes be incomplete since they are obtained through source areas identification and date data could be missing or with certain degree of uncertainty. A total of 81 rockfalls from in situ data were gathered, homogenized, structured and included in BDMOVES.

- **Earth observation services data (EOS)**

Rockfall data identified from LIDAR data have been collected by comparing two topographies from different dates (2009 and 2015) and using thematic data. Thus, their precise location is obtained but the exact date is undetermined since the date range between the two topographies is wide. In addition, activation of the Copernicus Emergency Management Service was requested for a rockfall and rock avalanche event in La Gomera and for the eruption of Tajogaite volcano in La Palma (see details in Chap. 4). A total of 44 rockfall events from EOS data were gathered, homogenized, structured and included in BDMOVES.

- **Project databases**

Synergies were also sought with other projects. This made possible to collect useful information on rockfalls in the framework of the Safety Project (2015), where data on rockfalls were collected on some roads to model the trajectory of the fallen blocks and thus determine their extent. They were data provided by the road maintenance services so in this case, only the kilometre number was available to locate it, so in the BDMOVES database these data (118 entries) were entered with some uncertainty.

- **Citizen data**

A questionnaire (https://arcg.is/5OK1H), was designed to be filled through the app ArcGIS Survey123 (ESRI) by the citizens, and two participation modes were possible: (i) as a citizen, by sending the data of any event recent, old or even very old; (ii) as an expert collaborator, attending training activities, training other collaborators, monitoring particularly susceptible areas, and sending rockfalls information with an extended description. The records registered by citizens and collaborators in the Observatory are validated or not by an expert. Records that present some type of inconsistency are checked by contacting the person who reported the record. If the problem is solved, the record is validated, otherwise it is rejected. Finally, validated registers are incorporated into BDMOVES (Fig. 9.3a).

In addition to the app, other communication channels were opened: social media (Facebook, Twitter, Whatsapp), e-mail (moves@igme.es)

A total of 106 rockfalls data coming from citizens were gathered, homogenized, structured and included in BDMOVES (Fig. 9.3b)

9.5 Citizens' Involvement and Platform (Including Application of AGEO Apps and Use of Social Media)

Implementation of a new citizen science project requires a strong effort focused on citizens' involvement and therefore an intense dissemination strategy is needed. Actions planned on the Canary Observatory were aimed at the general public and also to regional and local stakeholders.

A wide range of communication actions were carried out: workshops, publications, press releases, dissemination actions, conferences and congress, and collaborations with other initiatives (Fig. 9.3c), being dissemination actions the group with the highest number of activity especially during 2021.

Dissemination initiatives were aimed at promoting the project and starting to build the Observatory from the ground up. Thus, quarterly workshops with citizens participating actively in the Canary Islands Observatory were organized so that they were repeated quarterly on each island engaging population. The sessions were focused on increasing the rockfalls hazard awareness of citizens, inviting them to collaborate and send information about rockfalls events occurring. Citizens were invited to propose a topic of their interest in the field of geological hazards to deal with in the coming sessions (rockfalls, description of the process, typologies, mechanisms, causes, damage to the population and infrastructure, risk perception, citizen science, mobile applications as a communication tool between scientists and the population, etc.).

Educational proposals were also carried out: (i) Dissemination videos were designed and distributed, as well a Story Map web page was created (https://storymaps.arcgis.com/stories/2c978ee2f0b6465eb1d350fb0e37ffb2). (ii) Training sessions addressed to general public and related to mass movements (general issues) and specific characteristics of these movements in the Canary Islands were organized. (iii) In addition, practical activities in Secondary schools were implemented. The objective of these workshops was to involve the youngest public in the observatory, making them to participate in the research of rockfalls. A 3D model of each pilot area was designed by the students and the main zones with rockfalls problems were identified on the model. Successful work with students was achieved. Both training initiatives included the presentation and uptake of the Survey123 questionnaire designed.

Since communication with citizens is critical in a citizen science project, a particular attention was paid to the use of diverse and accessible communication tools that broadens the spectrum of audiences that can participate

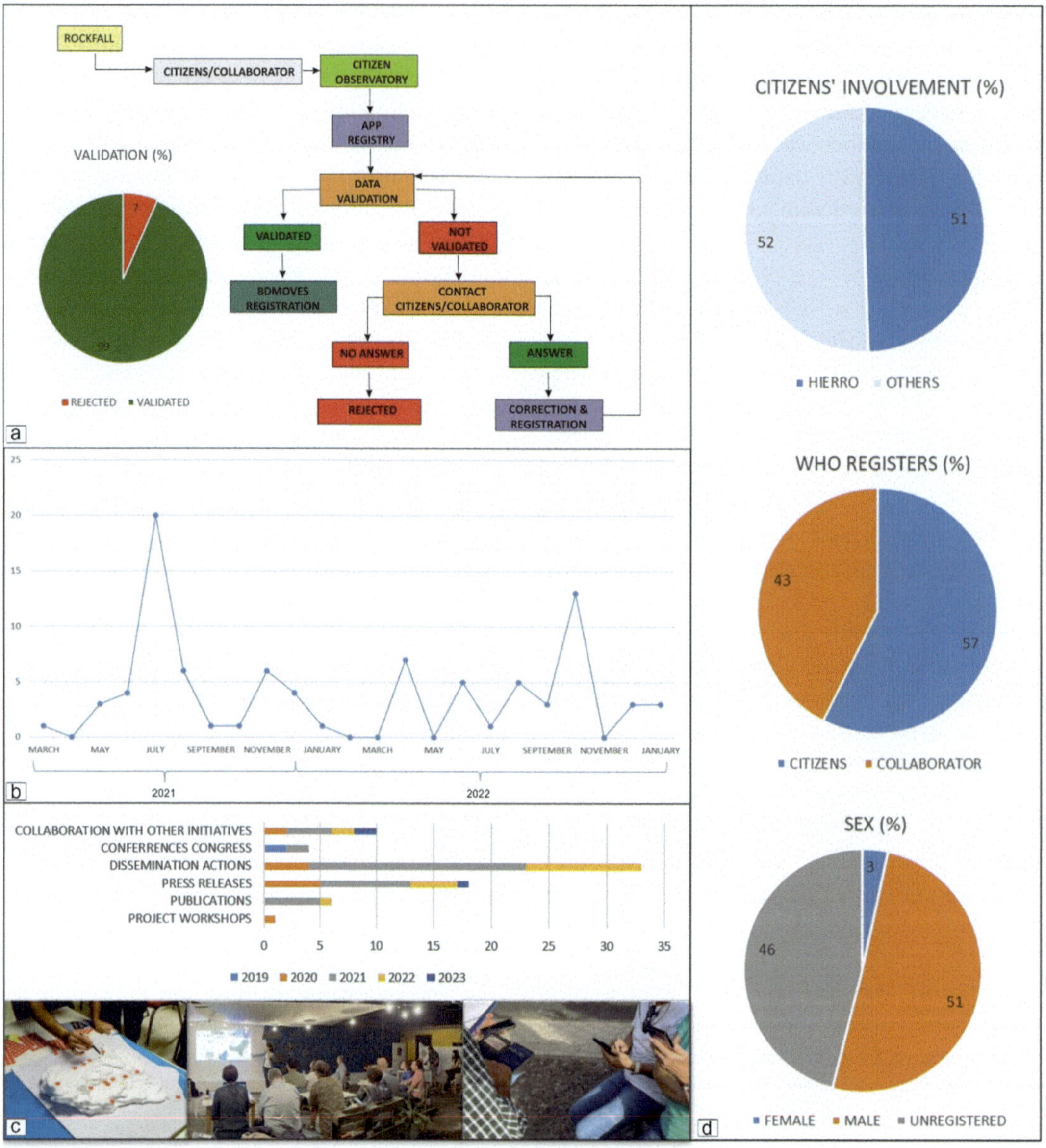

Fig. 9.3 **a** Proportion of app data validated and rejected and data collection and validation flow chart for app data. **b** Temporal distribution and number of each communication action; **c** Usage of the questionnaire designed in the Survey123 application: events submission over time; **d** Usage of the questionnaire designed in the Survey123 application: type of participants

in the observatory independently of their digital knowledge. Not only tools to send information to technicians and scientists are needed, but also communication channels must also be enabled to give feedback to the population involved in the initiative since they are essential. Social media (Facebook [https://www.facebook.com/Observatorio-Ciudadano-de-Desprendimientos-de-Canarias-103078888105514], Twitter [@ObservatorioCDC], Whatsapp), e-mail (moves@igme.es), phone and the Survey123 questionnaire were the communication tools used to exchange information with citizens: both to send information on rockfalls and to announce the activities of the observatory and communicate the milestones reached in the development of the AGEO project. Moreover, a preliminary version of a tailored platform for general public and stakeholders to visualize and query data from canary pilot was designed using Dashboard ESRI tool. First versions of the platform were presented to stakeholders in order to identify their necessities.

Throughout the organization and implementation of activities, it has been observed that the success in reaching the population is undoubtedly controlled by the possibility of having or not a person or organization closely involved in the project. The fact of having the support of a well-known or recognized citizen/stakeholder among the local population means that, on one hand, many more participants attend the events organized and, on the other hand, the message transmitted becomes much more credible and important and penetrates much deeper. Therefore, both the convening power and the degree of commitment are greater than without the support of these major actors.

In San Bartolomé de Tirajana, the municipal firefighters played a crucial role as the main key contact. Similarly, the Environmental Department of the City Council in La Laguna and the Geopark and Island Government in El Hierro were identified as the main key contacts. In the latter site, the high degree of involvement of the Island Operational Coordination Center (CECOPIN) was noteworthy. They provide a very relevant proportion of the data being data coming from El Hierro half of the data (51%) obtained in the whole Observatory (Fig. 9.3d).

The results on participation in the Observatory through Survey123 application show that users are diverse in terms of age and gender. Furthermore, approximately half of the participants were non-expert contributors (Fig. 9.3d). Since the application (Survey 123 questionnaire) was designed and launched, participation has been evenly distributed over time, with some increases coinciding with the face-to-face sessions with the population.

9.6 Other Achievements in the Timeframe of AGEO

The assessment of rockfalls hazard in the pilot area was addressed at an early stage through the characterisation of the region in terms of stability as well as directly involving the Observatory in the use of Copernicus products and services as well as other Earth observation systems. Additionally, we conducted a review and update of the inventory of historical rockfalls. This, join to the information obtained through citizen and stakeholder participation, allowed the update of BDMOVES database.

Source areas and susceptibility cartography

Detailed studies were carried out to characterize the pilot areas (San Bartolomé de Tirajana, La Laguna and Valle Gran Rey municipalities, in Gran Canaria, Tenerife an La Gomera respectively), in order to have the basic data necessary to carry out later analyses and rockfalls modelling; as well as to generate graphic products to be offered to citizens through the project website and the platform on rockfalls so that they become involved in the citizen observatory and maintain their interest and commitment to it. Four achievements were reached (Fig. 9.4):

- Compilation of historical events, and geological, geotechnical and geomorphological mapping.
- Mapping of source areas. Identification of the areas considering, on one hand, geology, orography and rockfalls marks identified in situ in the field; and, on the other hand comparing LIDAR data from 2009 and 2015.
- Rockfall susceptibility mapping.
- InSAR processing of Sentinel data provided by Copernicus Services in order to identify deformed areas after heavy rainfall events.

Copernicus Emergency Management Service (EMS) Activations

AGEO and the Citizens' Observatory on Rockfalls in the Canary Islands, encouraged the use of Copernicus services to help managers and interested entities to improve their efficiency

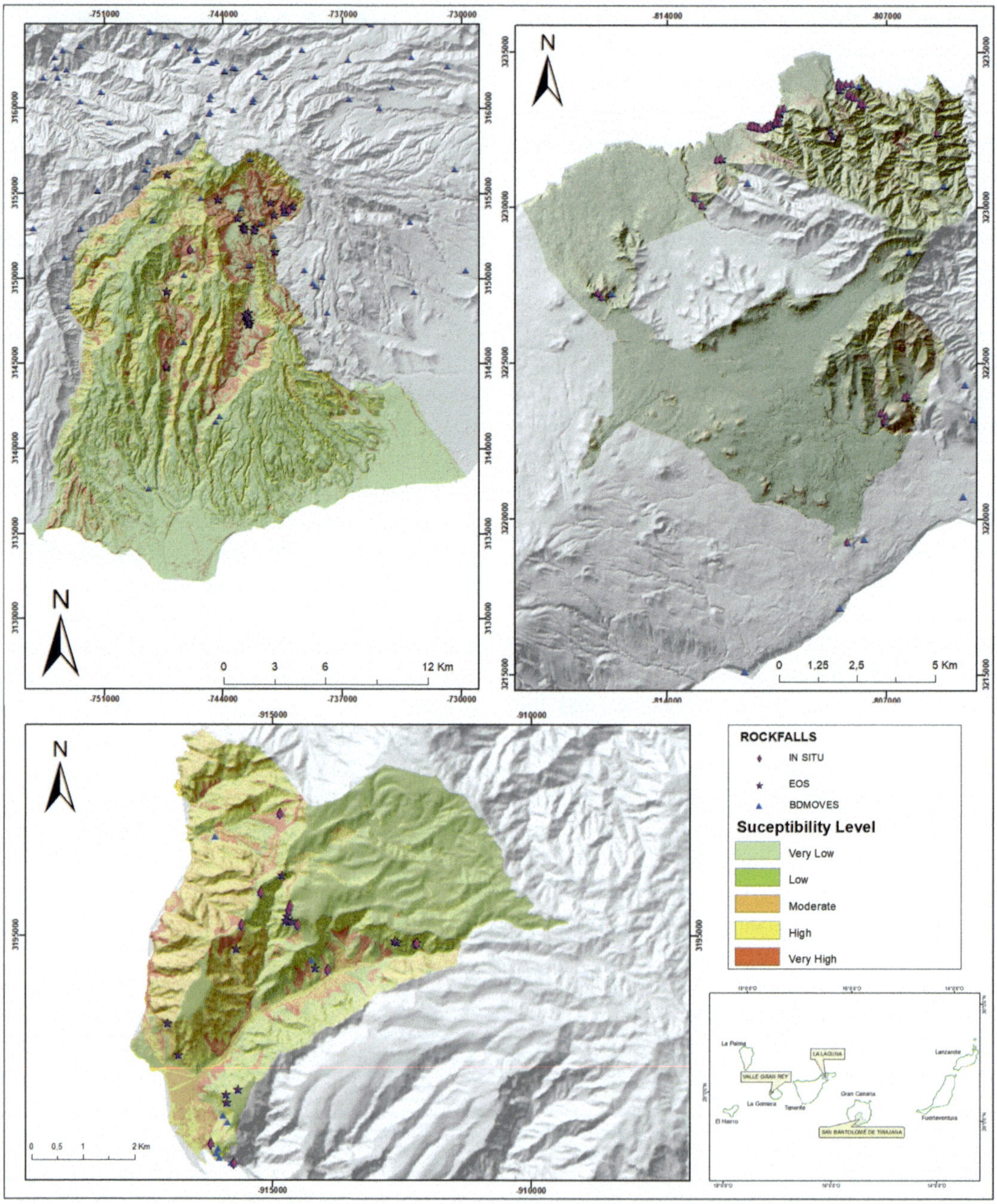

Fig. 9.4 Rockfall susceptibility maps and rockfall source areas identified in situ or by data from Earth Observation Services. **a** San Bartolomé de Tirajana (Gran Canaria), **b** La Laguna (Tenerife), Valle Gran Rey (La Gomera)

in emergency management. Thus, in order to implement the use of Copernicus Services on the Observatory, two activations of Copernicus EMS were submitted in the timeframe of AGEO (see details in Chap. 4):

- Activation EMSN093. Due to a rockfall emergency (Nov 14th, 2020) in Valle Gran Rey (La Gomera) Risk and Recovery Flex Mapping Service activation was requested to demonstrate that satellite-derived data could be used to analyse small to medium-sized rockfalls.

- Activations EMSN112/124, EMSR546 and EMSN119. Due to volcanic emergency in La Palma (Sep 19th, 2021), both Rapid Mapping and Risk and Recovery Mapping Services were firstly activated, to give support to emergency management through the obtaining of delimitation of the extent of lava flows, as well as the estimation of buildings and facilities affected. Moreover, products such as Landslide risk map or Soil risk erosion map were requested. Later on, Risk and Recovery Flex Mapping Service was also activated to get an updated post-eruption DEM.

Review of rockfalls inventory and update of BDMOVES

The continuous updating of BDMOVES with the records that come in from citizen participation, as well as its updating, including the records detailed in Sect. 9.4 (Fig. 9.2), allow us to confirm that the majority of the mass movements occurring in the Canary Islands are rockfalls, accounting for 89% of such events.. These are mainly recorded in Tenerife and Gran Canaria, although their amount is not negligible in the rest of the islands (Fig. 9.5). Located in steep areas, their occurrence is related to rainfall, wind, exposure to solar radiation, seismic activity and anthropic activity. The damage caused by them mainly affects roads, although cars and buildings have also get affected and injuries and fatalities have occurred (Fig. 9.5). Triggering factors are being analysed and statistics about the events collected up to date are being obtained.

9.7 Interaction with Stakeholders and Risk Management

The integration of the Observatory into risk management systems is another key point to consider in the implementation of a citizen observatory. The proposal from de Canary Observatory to this end had three focuses: (i) to involve the citizens in rockfalls emergency management; (ii) to contribute to risk management with technical support during rockfalls emergencies; (iii) to involve various stakeholders who can contribute not only to the stability and strength of the observatory, but also to its continuity over time by integrating it into their own structures.

Citizens involved in rockfalls emergency management

Previous experience of Civil Protection in dealing with citizen collaboration during the Tagoro eruption in 2011 was utilized for the establishing the Observatory in El Hierro. They created the figure of "groupers", citizens known by the community and responsible for helping and guiding people to meeting points during emergency. Thus, within the Observatory on Rockfalls, some proposals were presented to include these citizens in the current Civil Protection protocols in order to deal with a rockfalls risk emergency contributing to the implementation of the Observatory on Rockfalls in El Hierro and therefore to the improvement of regional and national management systems. Each proposed action was assigned to the corresponding phase within the risk management cycle: prevention and mitigation, preparedness, response and recovery. For example, aware citizens could be key in preventing, identifying and monitoring cracks or identifying and reporting rockfalls; preparedness, disseminating information provided by Civil Protection; response, gathering information about the event or observing and reporting subsequent rockfalls; recovery, assisting in reporting damages.

Technical support during and after rockfall emergencies

The Citizen's Observatory on Rockfalls in the Canary Islands played an important role in providing support to emergencies. Thus, scientific-technical team of IGME-CSIC, as part of the Observatory, contributed with technical support following the occurrence of rockfalls during the timeframe of the AGEO project on three

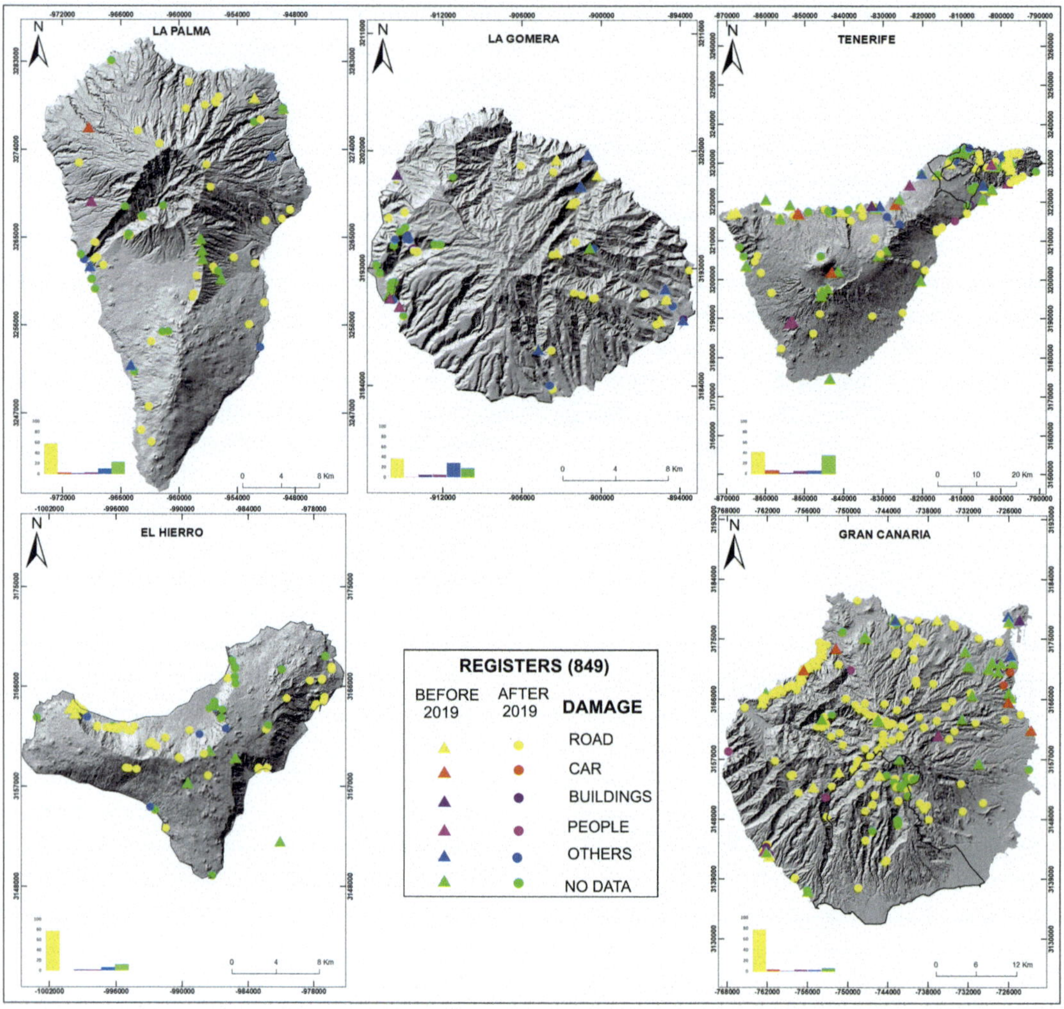

Fig. 9.5 Mass movements events registered currently in BDMOVES with different symbology if the register is before or after the starting of AGEO Project and different colour depending on the type of damage caused each time. Graphics below each island show proportion of damages represented by type (see legend)

occasions: La Gomera (Nov, 2020), La Palma (Sep, 2021), and Gran Canaria (Sep 2022).

On 14 November 2020, an important rockfall occurred on a cliff in the municipality of Valle Gran Rey (La Gomera), causing the total isolation of a neighbourhood as the cliff collapsed onto the road leading to the houses (Fig. 9.1c). Security and Emergencies Service of Canary Government requested technical support to study and advice on slope stability during and after the emergency. After emergency, an analysis of escarpment stability was carried out with drones in collaboration with the drone service of the Emergency and Rescue Group (GES) of the Government of the Canary Islands. Moreover, as mentioned before, the Observatory requested Copernicus EMS (Risk and Recovery Flex Mapping) activation because of this important rockfall. Likewise, during emergency, a new regional Civil Protection emergency plan for rockfalls was suggested to be designed since there was not any up to date. Regional General Emergency plan (PLATECA) had to be activated. The information provided by citizens through social media was crucial to characterize the rockfall event and manage the emergency.

On 19 September 2021, Tajogaite volcano on La Palma erupted after intense seismic and deformation activity, which caused the occurrence of rockfalls before and during the eruption. Security and Emergencies Service of La Palma Government requested IGME-CSIC technical service to monitor those rockfalls. Field visits were organized to give an assessment about the stability conditions of some slopes in the SE of the island. A visual inspection of the slopes was carried out, with photographs and videos taken at three locations, as well as a drone flight at one of the three sites. Moreover, field works were carried out in La Palma to identify rockfall source areas and rockfall deposits in Cumbre Vieja and littoral area with collaboration of the National Oceanographic Survey and Las Palmas de Gran Canaria University.

Recently, because of tropical storm Hermine (end of September 2022), flooding and several rockfalls occurred throughout the Canary archipelago, mainly in the western islands. In the south of Gran Canaria, the rockfalls in the municipality of Mogán were particularly damaging due to the size of the detached boulders (Fig. 9.1e). Fortunately, there were no casualties, but large boulders fell on tourist accommodation and cars. After the storm, an inspection of the escarpments and deposits as well as an overflight with a drone of the municipality was carried out.

Stakeholders involvement

Active collaboration of stakeholders is essential for the sustainability of the Observatory. Thus, agreements have been planned with the Department of Security and Emergencies of the Canary Islands Government to ensure the continuity of the Observatory on Rockfalls in the Canary Islands and to initiate first steps on developing a mass movement civil protection emergency plan. Likewise, collaboration has been initiated with the Operational Coordination Centers in La Palma, El Hierro, Gran Canaria and Tenerife, which inform the Observatory of all the rockfall events they register in order to contribute to keeping the inventory of events updated.

Interaction with authorities and stakeholders in the framework of AGEO has been based on presentations of the project (early stages), exploration of collaboration tools (intermediate stages), consolidation of cooperation agreements (final stages): project and app presentation to authorities, meetings with authorities to organize the activities in each municipality, meetings with teachers from pilot areas schools, dashboard presentation to local and regional stakeholders, meetings with Citizen Participation Service, meetings with stakeholders and authorities to discuss sustainability mechanisms of the Observatory further the end of the project, meetings with Operational Coordination Centers of each island to create protocols and communication channels to provide with data the Observatory.

9.8 Conclusion and Recommendations

Last update of the inventory of ground movements carried out in the framework of the Citizens' Observatory on Rockfalls in the Canary Islands reveals that at least 89% of the mass movements occurring in the archipelago are rockfalls. These, are rapid movements that many times occur in places that are difficult to access. Most often cause damage to roads, but sometimes resulting in injuries and fatalities. For this reason, citizen science is an excellent tool to obtain updated data to characterise the process in detail and thus estimate the hazard more accurately and improve the phases of prevention and mitigation.

Gaps have been detected in the current legislative frameworks as far as mass movements are concerned, therefore, initiatives such as the one proposed by the AGEO project count on particular importance and highlight the need for specific emergency plans for mass movements in the Canary Islands, which are currently non-existent. This type of hazard only appears as a secondary one associated with the occurrence of other hazards, but without a specific action protocol.

In citizen science, one of the greatest challenges is to engage population in such a way that their participation lasts over time. To this end, intensive hazard awareness campaigns must be planned and this work must be carried out in direct contact with citizens, involving them in the different phases of hazard assessment and management. It is also essential that local and regional stakeholders are also involved in order to make the observatory stronger and more sustainable. Likewise, it is also important to provide communication channels and tools that are accessible to the entire population. Thus, the Citizens' Observatory on Rockfalls in the Canary Islands designed and distributed a questionnaire to collect information on the events that occurred and also opened other channels such as social media and e-mail.

Despite the intense dissemination campaign, citizen participation was not as high as would have been desirable. Difficulties arisen clearly reveal that initiatives like this need more time to be implemented. However, it was always very well received and there has been a high level of involvement from local and regional authorities and stakeholders. Successful sessions with citizens were those including stakeholders as key contact with population. Protocols have been established with stakeholders participating actively in the Observatory in order to exchange information about the rockfalls occurred.

One aspect to be considered in the implementation of the observatory is to get a detailed characterisation of the environment, and to know both the conditioning characteristics and the triggering factors that can cause the rocks mass instabilities. Thus, the use of EOS, including Copernicus products, combined with detailed topography obtained with drones and field validation, will allow the identification of areas likely to be source areas of rockfalls and therefore update inventories of events.

Furthermore, the integration of the Observatory into risk management systems has been addressed in the Canary Islands Observatory through the involvement of citizens before, during and after the emergency by collaborating with emergency managers. Furthermore, the information provided by citizens together with the existing knowledge allows the Observatory's technicians to provide support in emergency management and issue assessments on the existing danger and recommendations aimed at prevention and minimising future risks.

The sustainability of the Observatory will be possible as long as it is integrated into the existing emergency management structure and complements and enhances the current organisation. For this to work, collaborative links must be created between institutions that will have to work in the same direction. The result of this integration will increase public awareness of these hazards, help to reduce them, increase the sense of community in society, foster learning and build bridges between scientists and society.

Acknowledgements The authors would like to acknowledge all citizens and stakeholders for their support and collaboration in this initiative. We are especially grateful for the great involvement of the Directorate of Security and Emergencies of the Government of the Canary Islands, the Government Delegation (Civil Protection), the San Bartolomé de Tirajana Municipality and the fire brigade of that municipality, the Government and the Geopark of El Hierro, the San Cristóbal de La Laguna Municipality, the Tenerife Mountaineering Group Service and the Nivaria mountaineering group. This work has been undertaken in the framework of the INTERREG Atlantic Area AGEO Project (EAPA_884/2018 – AGEO).

References

Ayuntamiento SBT (San Bartolomé de Tirajana). (2005). Plan de Emergencias Municipal de San Bartolomé de Tirajana. Ayuntamiento San Bartolomé de Tirajana, p. 315.

Budetta, P. (2004). Assessment of rockfall risk along roads. *Natural Hazards and Earth System Science, 4*(1), 71–81. https://doi.org/10.5194/nhess-4-71-2004

Díez-Sierra, J., & Del Jesus, M. (2020). Long-term rainfall prediction using atmospheric synoptic patterns in semi-arid climates with statistical and machine learning methods. *Journal of Hydrology, 586*, 124789. https://doi.org/10.1016/j.jhydrol.2020.124789

Galindo, I., Montoya-Montes, I., García López-Davalillo, J. C., Sarro, R., Llorente, M., Sánchez, N., Santamarta, J. C., Cruz-Pérez, N., Ortega, A., & Mateos R. M. (in press). Identification and management of indirect volcanic risks: The rockfalls citizens' observatory on the island of El Hierro. In: Gonzalez, PJ (eds) El Hierro. Active Volcanoes of the World. Springer, Berlin, Heidelberg.

Leyva, S., Cruz-Pérez, N., Rodríguez-Martín, J., Miklin, L., & Santamarta, J. C. (2022). Rockfall and rainfall correlation in the Anaga nature reserve in Tenerife (Canary Islands, Spain). *Rock Mechanics and Rock Engineering, 0123456789,*. https://doi.org/10.1007/s00603-021-02762-y

Lomoschitz, A., & Corominas, J. (1997). La depresión de Tirajana, Gran Canaria. Una macroforma erosiva producida por grandes deslizamientos. *Cuaternario y Geomorfología, 11*, 75–92.

Llorente, M., Galindo, I., & Paredes, C. (2015). Desprendimientos de rocas en la isla de El Hierro. In Hernández-Gutiérrez, L. E. & Santamarta-Cerezal, J. C. (Eds.). Ingeniería Geológica en Terrenos Volcánicos Métodos, Técnicas y Experiencias en las Islas Canarias. Ilustre Colegio Oficial de Geólogos, pp. 316–333.

Jurado, M. J., Ripepe, M., Lopez, C., Ricciardi, A., Blanco, M. J. Y., & Lacanna, G. (2020). Underwater monitoring of submarine volcanic activity: El Hierro (Canary Islands 2011–2012) eruption. *Journal of Volcanology and Geothermal Research, 107097.* https://doi.org/10.1016/j.jvolgeores.2020.107097.

Cabildo de Gran Canaria. (2005). Plan Territorial Insular de Emergencias de Protección Civil de Gran Canaria. Cabildo de Gran Canaria, Las Palmas de Gran Canaria.

Cabildo de Tenerife. (2012). Plan Territorial Especial de Ordenación para la Prevención de Riesgos. In: Boletín Oficial de Canarias, 94: 8690 - 8701. Cabildo de Tenerife. Santa Cruz de Tenerife.

Cabildo de Tenerife. (2020) Análisis de Riesgos Potenciales. In: Plan Territorial Insular de Emergencias de Protección Civil de Gran Canaria, 3. Cabildo de Tenerife, Santa Cruz de Tenerife.

Cruz-Pérez, N., Rodríguez-Martín, J., Galindo, I., Montoya, I., Sánchez, N., Hernández-Gutiérrez, L. E., Ortega Rodríguez, A., Leyva, S., & Santamarta, J. C. (2021). Citizen observatories applied to geological risk management, INTED2021 Proceedings, pp. 1022–1025.

Mateos, R. (2017). Geohazards into the land-use planning and urban development policies. Spanish experience and European overview. *Geonovas, 30*, 11–19.

Ortega Rodriguez, A., Carrilho Gomes, R., Telmo Jeremias, F., Santamarta Cerezal, J. C., Quental, L., Galindo Jiménez, I., Correia, V., Narciso Pinto, C., Le Dantec, N., Gouveia, F., Lemon, K., Hénaff, A., & O'Hare, G. (2020, May 4–8). AGEO – Natural hazard prevention and awareness raising through citizen observatories., EGU General Assembly 2020, Online, EGU2020–13519.

Romero, C. Sánchez, N., Galindo, I., Martín-González, E., & Vegas, J. (2020). Historical and prehistorical volcanic geoheritage in Canary Islands, *Authorea*. June 09, https://doi.org/10.1002/essoar.10503322.1.

Rossi, M., Sarro, R., Reichenbach, P., & Mateos, R. M. (2021). Probabilistic identification of rockfall source areas at regional scale in El Hierro (Canary Islands, Spain). *Geomorphology, 381*, 107661. https://doi.org/10.1016/j.geomorph.2021.107661

Sarro, R., Mateos, R. M., Reichenbach, P., Aguilera, H., Riquelme, A., Hernández-Gutiérrez, L. E., Martín, A., Barra, A., Solari, L., Monserrat, O., Alvioli, M., Fernández-Merodo, J. A., López-Vinielles, J., & Herrera, G. (2020). Geotechnics for rockfall assessment in the volcanic island of Gran Canaria (Canary Islands, Spain). *Journal of Maps, 16*(2), 605–613. https://doi.org/10.1080/17445647.2020.1806125

SAFETY Project - Sentinel-1 for geohazard prevention and forecasting. (2015). European Commission, Directorate-General Humanitarian Aid and Civil Protection (ECHO) (Ref. ECHO/SUB/2015/718679/Prev02).

Solana, M. C., & Kilburn, C. R. J. (2003). Public awareness of landslide hazards: The Barranco de Tirajana, Gran Canaria, Spain. *Geomorphology, 54*, 39–48.

10 OSIRISC: Co-Design of an Integrated Observatory to Monitor Vulnerability to Coastal Risks of Erosion and Marine Flooding in Brittany

Nicolas Le Dantec, Marion Jaud, Alain Henaff, Iwan Le Berre, Catherine Meur-Ferec, Quentin Ruaud, Mathias Rouan, Julie Pierson, Caroline Lummert, Manuelle Philippe, Thibaud Lami, Pauline Letortu, Elisabeth Guillou, Serge Suanez and Sylvain Lendre

10.1 Introduction

OSIRISC (French acronym for "*ObServatoire Intégré des RISques Côtiers en Bretagne*") is an integrated observatory addressing coastal risks of coastal erosion and marine flooding in Brittany.

The coastal zone has become very attractive over the last century, for the numerous amenities it offers. It now concentrates many infrastructures, economic activities, residential housing and a large population. At the same time, it is a dynamic zone, as the coastline is in constant

N. Le Dantec (✉) · M. Jaud · Q. Ruaud · C. Lummert
UMR 6538 GeoOcean, University of Brest-CNRS, Institut Universitaire Européen de La Mer, Rue Dumont d'Urville, 29280 Plouzané, France
e-mail: nicolas.ledantec@univ-brest.fr

M. Jaud
e-mail: marion.jaud@univ-brest.fr

Q. Ruaud
e-mail: quentin.ruaud@univ-brest.fr

C. Lummert
e-mail: caroline.lummert@univ-brest.fr

A. Henaff · I. Le Berre · C. Meur-Ferec · M. Rouan · J. Pierson · T. Lami · P. Letortu · S. Suanez
UMR 6554 Littoral-Environnement-Télédétection-Géomatique, University of Brest, CNRS, Institut Universitaire Européen de La Mer, Rue Dumont d'Urville, 29280 Plouzané, France
e-mail: alain.henaff@univ-brest.fr

I. Le Berre
e-mail: iwan.leberre@univ-brest.fr

C. Meur-Ferec
e-mail: catherine.meurferec@univ-brest.fr

M. Rouan
e-mail: mathias.rouan@univ-brest.fr

J. Pierson
e-mail: julie.pierson@univ-brest.fr

T. Lami
e-mail: thibaud.lami@univ-brest.fr

P. Letortu
e-mail: pauline.letortu@univ-brest.fr

S. Suanez
e-mail: serge.suanez@univ-brest.fr

M. Philippe
UMR 6308 Amure, University of Brest, CNRS, Institut Universitaire Européen de La Mer, Rue Dumont d'Urville, 29280 Plouzané, France
e-mail: manuelle.philippe@univ-brest.fr

E. Guillou
EA 1285 LP3C, University of Brest, 29238 Brest cedex 3, Brest, France
e-mail: Elisabeth.Guillou@univ-brest.fr

S. Lendre
CEREMA, Direction Eau Mer Et Fleuves, 134 Rue de Beauvais, 60280 Margny-Lès-Compiègne, France
e-mail: sylvain.lendre@cerema.fr

R. C. Gomes et al. (eds.), *Citizens' Observatories on Geohazards*, Geoenvironmental Disaster Reduction,
https://doi.org/10.1007/978-3-031-53371-6_10

evolution at different time scales, from storm events to seasonal and to multi-annual (including under the effect of climate change), in addition to longer time scales of coastline evolution, in relation to sea-level change. Coastal forms are continuously being reshaped according to the geomorphological response of coastal environments subject to varying marine and subaerial conditions, and to sediment erosion and deposition dynamics. Natural coastal environments are classified into different geomorphological types depending on whether their primary features were created by erosion of land (erosional coast) or deposition of eroded material (depositional coast). Erosional coasts include the wave-cut platforms as well as rocky and soft cliffs. Cliffs can only retreat under marine erosion, subaerial weathering and continental processes, while wave-cut platforms are shaped both through the evolution of their width and the downwearing of their surface. Depositional coasts include low-lying accumulations of shingle, gravel or sand forming barrier beaches and dunes on exposed environments. Phases of retreat of the coastline can alternate with beach and dune recovery as sediments are transported in cross-shore and alongshore directions due to the action of waves and currents, as well as aeolian processes on the exposed sandy beach and dune, giving rise to circulation patterns at the spatial scale of littoral cells. Finally, finer sediments are deposited in sheltered conditions in estuaries, tidal mud-flats or lagoons. Parts of the coastlines have been artificialized, either with coastal defences to protect against the sea or with harbour infrastructures. The coastline itself, meant to represent the limit of the marine and continental domains, is usually defined according to a morphological marker such as the top of a cliff or the foot of a dune, or to an elevation with respect to a vertical reference, such as the mean sea level or the maximum astronomical tide. The unstable nature of the coastal zone brings about 3 types of geohazards: coastal erosion, marine flooding and coastal dune migration. We will only consider the first 2, as the dune migration concerns nowadays only very few areas in Brittany and is often identified as coastline retreat.

- Coastal erosion is the loss of sediment, which eventually results in the recession of the coastline and therefore a loss of land to the sea (Fig. 10.1). Approximately 1/5 of the French coastline is affected, to varying degrees, by coastal erosion (Cerema, 2018). In the long term, it is a continuous process, with periods of more or less rapid retreat. Coastal erosion can be impacted by various factors including climate (sea-level change, frequency of occurrence of storms, wind regimes), topography, vegetation, tectonic activity, but also human activity (over-frequentation, sediment extraction, river dams, sea defence structures, urbanization close to the coastline leading to surface runoff, etc.).

Fig. 10.1 Coastline recession visible by comparing two aerial images from 1952 (coastline digitized in red) and 2021 (coastline digitized in green). Location: Ploudalmézeau and Lampaul-Ploudalmézeau, Finistère department, France (modified from: « Remonter le temps » website - © IGN)

- Marine flooding is the inundation of the continental part of the coastal zone by the sea during unfavourable meteorological and oceanic conditions (low atmospheric pressure, strong landward winds, storm surge and waves, and for macrotidal seas, during high tide) (Fig. 10.2). Three modes of marine flooding are generally distinguished:
- flooding by overflows (when the sea level is higher than the crest of the sea defence structures or the natural ground);
- flooding due to wave overtopping (when the instantaneous wave height at the coastline exceeds the crest of the sea defence structures or the natural ground);
- and flooding due to failure of a sea defence system.

Coastal erosion and marine flooding hazards can cause severe damage or losses to properties and infrastructures worldwide. Human lives may also be lost in dramatic flood events, such as the Xynthia storm during winter 2010 (Vinet et al., 2012) on the French Atlantic coast. In France, risk management policies consider coastal erosion as a slow and continuous process. However, during a single intense coastal event, the coastline can retreat several metres in a few hours and even up to tens of metres locally (Hénaff et al., 2013). Unlike other natural geohazards associated with sudden processes like earthquakes—where early-warning systems and crisis response plans under the responsibility of civil protection services are a large part of risk reduction policies—prevention of coastal erosion risks focuses mainly on land-use planning and mitigation measures to control or stop coastline retreat. These risk mitigation measures take many forms, ranging from active defence with hard engineering protection (seawalls, riprap, etc.) to other approaches such as beach replenishment or sediment bypass that attempt to prevent or slow down coastal retreat without shoreline stabilization. Alternative strategies of risk reduction through adaptation, consisting in relocating exposed elements when possible, are now actively being promoted, although their application requires a change in paradigm for all stakeholders until they become accepted. Risk prevention policies against marine flooding are also relying on protection with coastal defences and regulations on urban planning, as well as the development of a risk culture as marine flooding can occur in catastrophic ways, in case of breeching of protection systems. Again, relocation strategies are considered but their implementation remains limited. While coastal erosion and marine flooding hazards are treated differently in terms of regulations framework for risk management (mainly because financial compensation of losses due to marine flooding is on public funds, which is not the case for losses

Fig. 10.2 Marine flooding by wave overtopping in Penhors (Audierne Bay, Brittany, France) in 2016

due to coastal erosion), they are often interrelated and difficult to dissociate in practice.

Risk being the combination of hazards and exposure, the main driver in the increasing trend in coastal risks is in fact coastal urbanisation and anthropic actions causing perturbations in the sediment fluxes. This is measurable through the increase of damages reported due to coastal hazards in historical records and in particular over the last decades (Hénaff et al., 2018). The appeal for coastal environments has led to a major growth of exposed elements in areas subject to coastal hazards (Fig. 10.3). The upcoming sea-level rise and changes in wave climate induced by climate change will exacerbate coastal hazards of erosion and flooding. In this context, long-term, sustainable strategies of coastal risks management are needed because the addition or reinforcement of coastal defences not only fails to address the main cause of risk, the exposure, but also to restore equilibrium in the sediment budget and even prevents the natural resilience of the coastal system through coastline recession.

Results from interdisciplinary research studies on coastal risks have led us to progressively develop an integrated approach combining natural phenomena (hazard) and structural (socio-economic, cultural, functional, institutional) factors, within a conceptual framework of systemic vulnerability (*Hénaff and Philippe (Eds), 2014*). Following prior work such as D'Ercole (1994), this approach reflects the level of fragility of a socio-economic system as a whole in the face of risk, and the ability of society to respond to potential crises through adaptation and resilience. Instead of considering vulnerability and hazards separately, systemic vulnerability is the combined outcome of 4 inter-related components: the hazards, the significance and nature of the exposed elements, the management responses to address risks and the social representations of risks (Fig. 10.4). In this approach, all 4 components are assessed independently to generate indicators. Interrelations between the components are taken into account when combining indicators into indices. Exposed elements or exposure refers to assets (human, economic, structural, etc.), measured independently of their level of exposure to the hazards. Risk management comprises the public policies of risk protection, prevention and adaptation and their application at local scale by authorities in charge of coastal risks. Social representations address the perception people have on risks (awareness, relationship to

Fig. 10.3 Evolution of the coastal urbanisation in Île-Tudy (Brittany, France) images from 1950 and 2021. (GeoBretagne 1950' website)

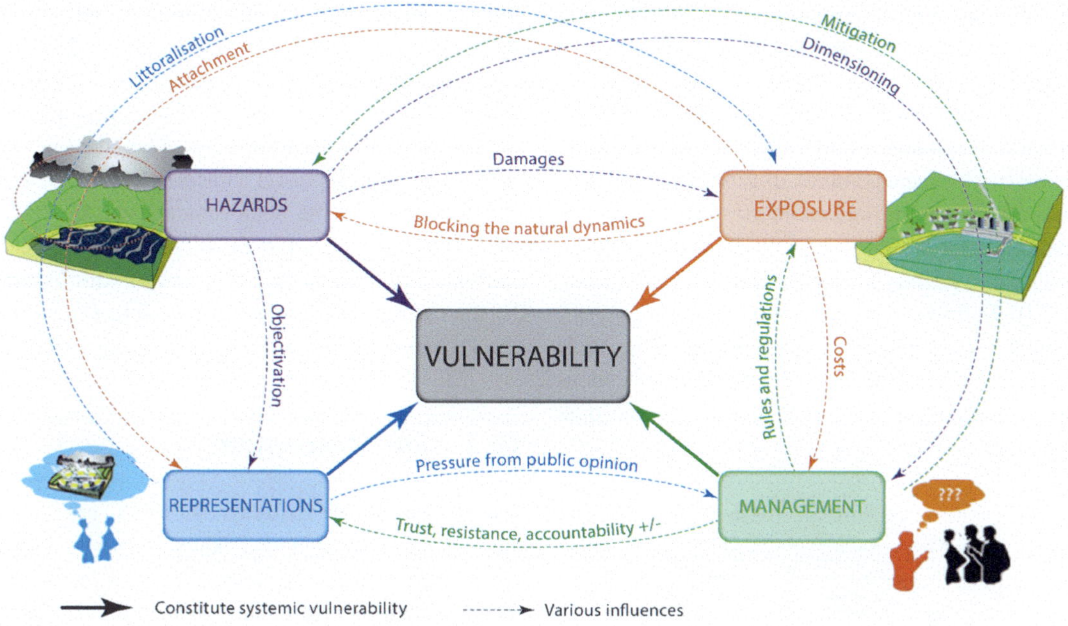

Fig. 10.4 The four components of systemic vulnerability, their interrelations and the nature of their contribution to systemic vulnerability (Meur-Ferec et al., 2020; modified)

the place they live, preferences for adaptation to risks, understanding and acceptability of management policies, etc.). These latter two components are the resources and levers available to cope with risks and its resulting impacts (Meur-Ferec et al., 2020). While increasing hazards and exposure will increase vulnerability, management is meant to reduce vulnerability. The influence of social representation on vulnerability is more difficult to characterize, as risk perception does not necessarily lead to behavioral changes acting towards reducing vulnerability. Rather, better understanding of how people who occupy a territory (as inhabitants or having activities in the territory, or being in charge of it) think about risks will bring insight on behaviors, which can help towards defining risk management strategies that may be better accepted.

As these four components evolve over time with their own temporalities, systemic vulnerability is dynamic as well. Beyond establishing systemic vulnerability diagnosis, monitoring trajectories of vulnerability over time, as a result of the evolution of the different contributing factors, through regular assessments, will improve our knowledge on coastal risks and help defining management strategies. The constitutive variables of each of the four components of vulnerability also vary over space, which raises the questions of the relevant size of territory, scale and granularity for the assessment of integrated (or systemic) vulnerability. A methodology based on indicators has been developed to evaluate systemic vulnerability (*Philippe and Hénaff, 2021*). Each of the four vulnerability components is approached using methods and tools specific to the corresponding scientific disciplines (geology, geography, law, environmental psychology…). The results can be presented in quantitative or qualitative form depending on the components. Thus, hazards and most of the elements-at-risk are measured via numerical data, whereas data on management and risk perception is collected via surveys using questionnaires and face-to-face interviews, and producing text information. Hazard, exposure and management data are converted into calibrated indicators. Results derived from social representation data are transcribed into textual form, providing contextual elements on the studied sites (*Philippe and Hénaff, 2021*). The production of indicators from the collected data is further described in Sect. 4.2.

This conceptual framework for an interdisciplinary monitoring system to promote integrated coastal risks management, in close relation with risks prevention stakeholders and decision-makers, was first implemented on test territories (*Philippe and Hénaff, 2021*; Le Berre et al., 2022). Following this development and experimental phase, OSIRISC observatory (Fig. 10.5) was created in Brittany in 2019, with this focus on "vulnerability" to coastal risks, and the dual goal to respond to scientific issues and societal demands through co-design and collaborative observation. In this chapter, we will first describe the territory where OSIRISC observatory is deployed, the objective of this collaborative observatory and the data handled in OSIRISC. We will then present the achievements of OSIRISC regarding the development of applications and platforms of citizen science, before discussing the interaction with stakeholders within OSIRISC observatory and the links with risk management.

Fig. 10.5 OSIRISC observatory logo

10.2 Site Description

The coastline of Brittany is 4,324 km long when including islands, islets and rias, and very convoluted. Brittany exhibits a variety of coastal types: low sandy and pebble beaches, rocky coasts, soft cliffs and deep rias (Fig. 10.6). Although 3 main provinces can be distinguished, with a low-lying Southern coast, mainly soft cliffs on the Northern coast and

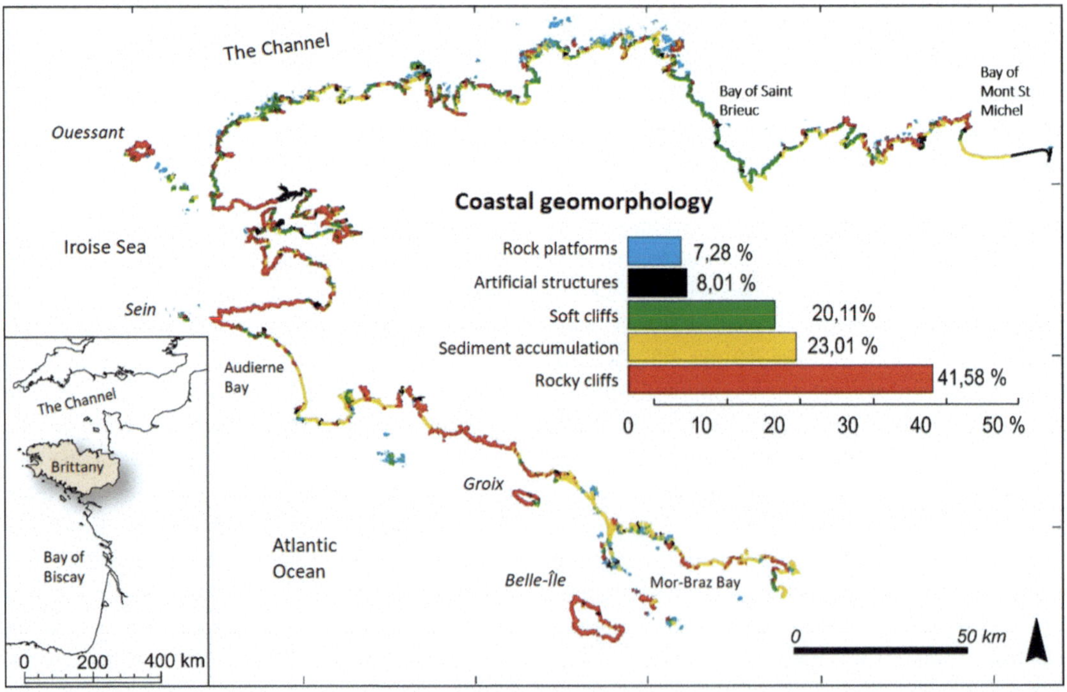

Fig. 10.6 The various coastal types in Brittany (from the Programme de recherche d'intérêt Régional «EROCOVUL»—2003)

rocky cliffs on the Western coast, the spatial heterogeneity of the coastal geomorphology is very high, with a succession of small stretches of coastline of different types. There is also a wide variety in the configurations owing to the spatial variations in coastal topography, tidal range, exposure to swells, fetch and sedimentary cover.

Overall, erosion concerns 6% of Brittany's coastline, according to a study based on diachronic analyses of ortho-photographs between 1950 and 2011 (Cerema, 2021a, b). This is lower than the national average, nevertheless, locally, some sites show retreat rates over 0.5m/year (Fig. 10.7). Accumulations of sand and pebbles represent about 24% of the region's shoreline. These are certainly the most monitored shorelines because of their relatively rapid dynamics and the many elements exposed behind. A specific analysis of changes in the shoreline of sand and pebble accumulations between 1950 and 2011 reveals that a third of them are in progradation, another third is in erosion and the remaining third is stable during the period (Fig. 10.8).

The damage produced during the historical period by coastal erosion and marine flooding (and also dune migration) on exposed territories is an indirect indicator of the growth of coastal risks in Brittany. The inventory of the damage observed since 1750 until nowadays counts some 8000 facts spread over the whole region with some coastal areas that are more frequently

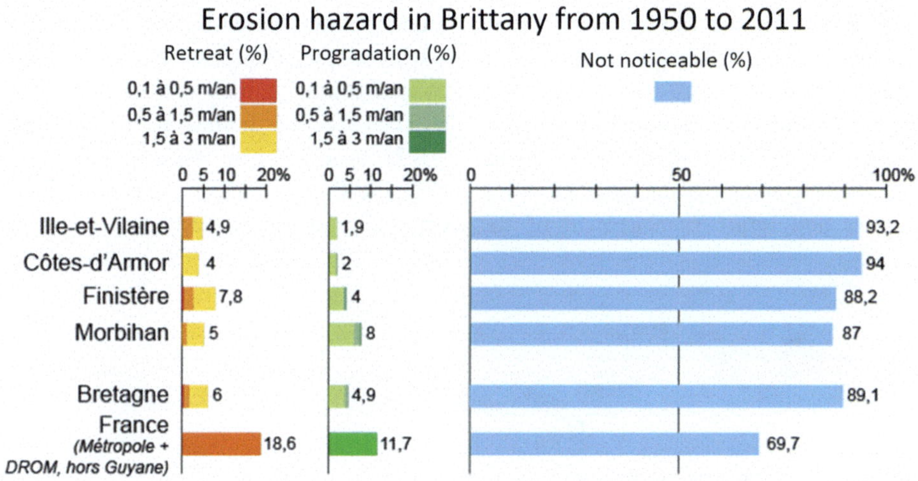

Fig. 10.7 Assessment of coastline trends in Brittany after Cerema (2021a, 2021b) and Hedou (2018)

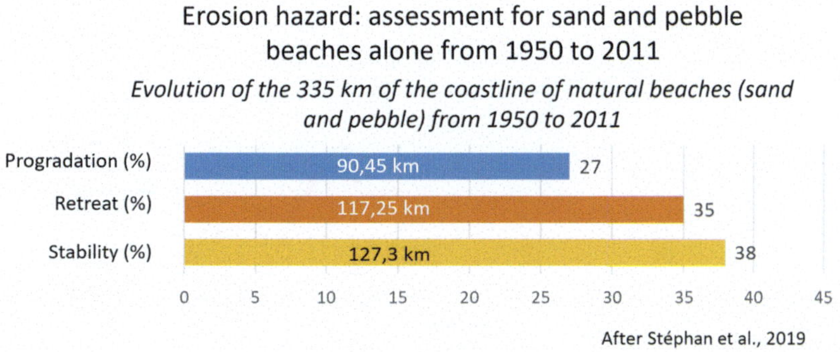

Fig. 10.8 Results of the diachronic analysis of the coastline of beaches from 1950 to 2011 in Brittany (after Stéphan et al., 2019)

affected. Diachronic analysis of hazards does not show any real growth in magnitude or frequency of occurrence of hazards during the historical period. This inventory shows periods of high occurrence of natural or anthropogenic hazards followed by quieter periods. If interannual fluctuations do exist in magnitude or frequency of hazard occurrence according to these periods, it appears that the observed increase in damage results mainly from the exposed elements that have continued to be developed in the coastal territories (Hénaff et al., 2018). The impacts of the most devastating hazards on the coasts are thus identified and they also allow to identify locally, when observed, the values of maximum shoreline retreat (Fig. 9a). Combined with the average rate of shoreline retreat, this value is crucial for predicting future coastal positions. It integrates the impacts of extreme events and allows stakeholders to obtain more reliable assessments of future positions of the shoreline than the only average rate of evolution. This is necessary in order to define a coastal risk management strategy that includes planning of prospective and future management of the coastal territory. Marine flooding mainly affects low-lying areas. Historically, an average of 8.7 incidents of marine flooding per year have been recorded from 1790 to 2021 (22% due to overtopping, 15% due to overtopping and destruction of coastal defences, 5% due to breach of the barrier beach, 45% unspecified (Chaumillon et al., 2017) (Fig. 9b).

The Breton coastline counts 257 municipalities, with a combined surface area of 4,874 km^2, or 18% of the region's surface area. According to the 2017 census by the French National Institute of Statistics and Economic Studies, the total population of coastal municipalities is 1.16 million inhabitants, representing 36% of the Breton total population. The density of the Breton coastline is 240 inhabitants per km^2, with the population of the coastal municipalities increasing by 13% from 1975 to 2012 censuses (Poupard, 2017). In addition, the coastal municipalities concentrate 74% of the 235,644 second homes in Brittany (Poupard, 2017). In addition to this population concentration, the Breton coastal municipalities also host economic (fisheries, harbours, tourist facilities,

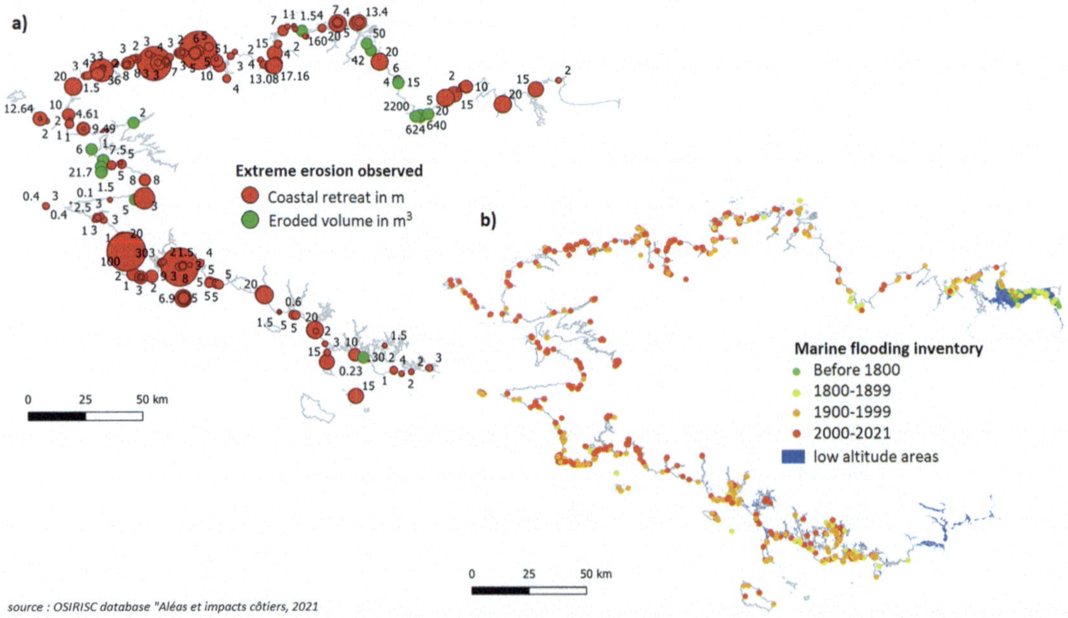

Fig. 10.9 Extreme coastal erosion observed (**a**) and recorded marine flooding events (**b**) in Brittany (source of the data: OSIRISC database "Aléas et impacts côtiers", 2021 (Chaumillon et al., 2017))

etc.) and structural assets (roads, public establishments, sea defences, etc.) as well as natural (protected areas) or patrimonial heritage (historic or prehistoric sites).

The OSIRISC observatory covers the 257 Breton coastal municipalities. The map in Fig. 10.10 shows the sites for which hazard measurements are (or have been) performed. These measurement sites were chosen either because they are particularly concerned by coastal risks or because they are representative of the evolution of a type of coastal environment. Given the diversity of coastal environments and evolution dynamics, measurement methods must be adapted to each site.

The anticipation of the long-term evolution of the shoreline is mainly based on the analysis of its present and past positions, going back to different periods quite far in time. Hence the importance of having long-term series of coastline measurements.

10.3 Objectives of OSIRISC Pilot

As previously mentioned, OSIRISC observatory aims to integrate all the dimensions of systemic vulnerability (combination of the four interdependent components: hazards, exposure, management and representations), in close relation with risks prevention stakeholders and decision-makers (*Philippe and Hénaff, 2021*). To understand the evolution of vulnerability, it is necessary to know the evolution of each of the components individually. This means "measuring" each of the components in a territory and monitoring them over time.

To achieve this, OSIRISC has carried out:

- The development of an interdisciplinary observatory methodology for coastal erosion and marine flooding risks by selecting or creating indicators and indices adapted to both research and management;

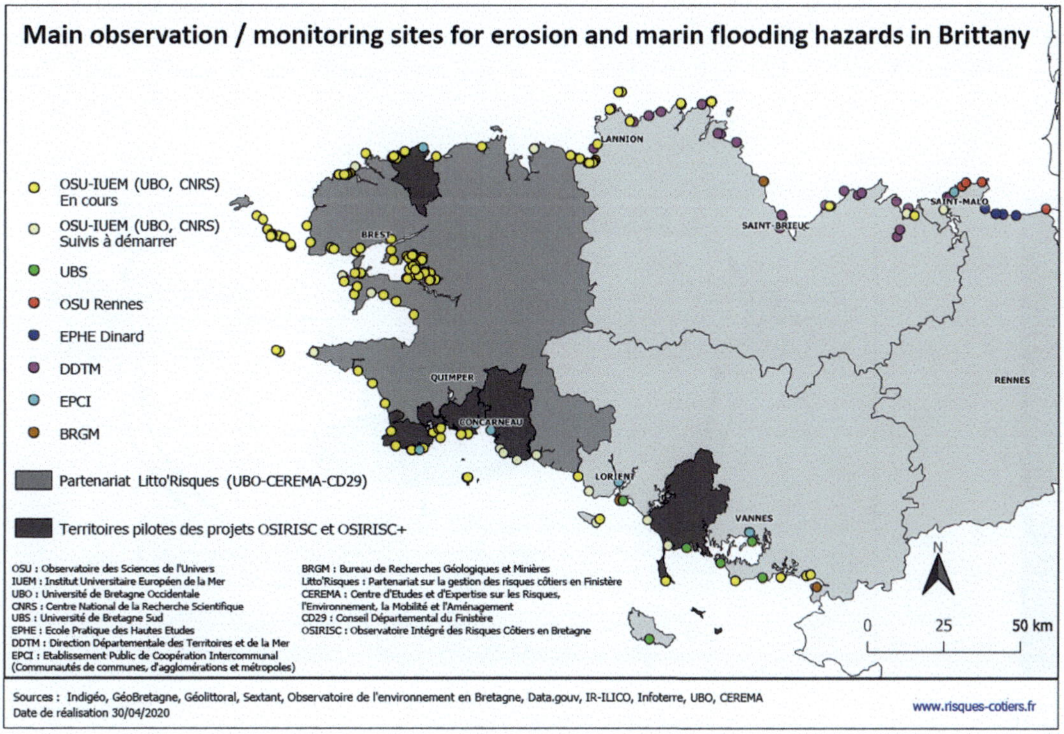

Fig. 10.10 Sites where morphologic surveys of coastal hazards are realized

- Evaluation and consolidation of this indicators approach and methodology through experimentation with local stakeholders on test sites;
- The implementation, within a Spatial Data Infrastructure (SDI), of specific tools for geographic information dissemination and sharing, as well as the study of its contributions toward scientific (improving our understanding of vulnerability to coastal risks) and applied (coastal risks management) objectives.

This approach (Fig. 10.11) is called "integrated" because it allows us to work on the different levers of vulnerability and analyse the determining factors. This will facilitate the identification of the most vulnerable areas where to focus actions, as well as the development of management strategies. This approach also allows for the monitoring of long-term trajectories in order to implement a long-term management strategy (not limited to sensitive areas).

Co-construction with local authorities implies providing technical and methodological support for coastal erosion and marine flooding risk management issues (through workshops, expertise, advanced students projects) and carrying out actions to raise awareness of coastal risks and bring together stakeholders. During the AGEO project, the OSIRISC team not only carried out in situ surveys, but also provided training and developed tools (using simple and low-cost measurement protocols, web platforms for data dissemination, etc.) to promote involvement of local authorities in coastal vulnerability observation.

Defining indicators is an essential step to make data of different types and formats compatible and accessible to a wide audience. Further details on this step are given in Sect. 4.2.

The collaboration within OSIRISC observatory between academics and local authorities is effective and was ratified by the Litto'Risques partnership between the University of Brest, the CEREMA (Centre for studies and expertise on risks, the environment, mobility and development) and the Finistère County council. Figure 10.12 shows the map of the territories involved in Litto'Risques. The OSIRISC team hopes to extend this collaboration to other territories in Brittany. Interactions with stakeholders, and specifically local authorities in charge of risk management, are presented in Sect. 10.6.

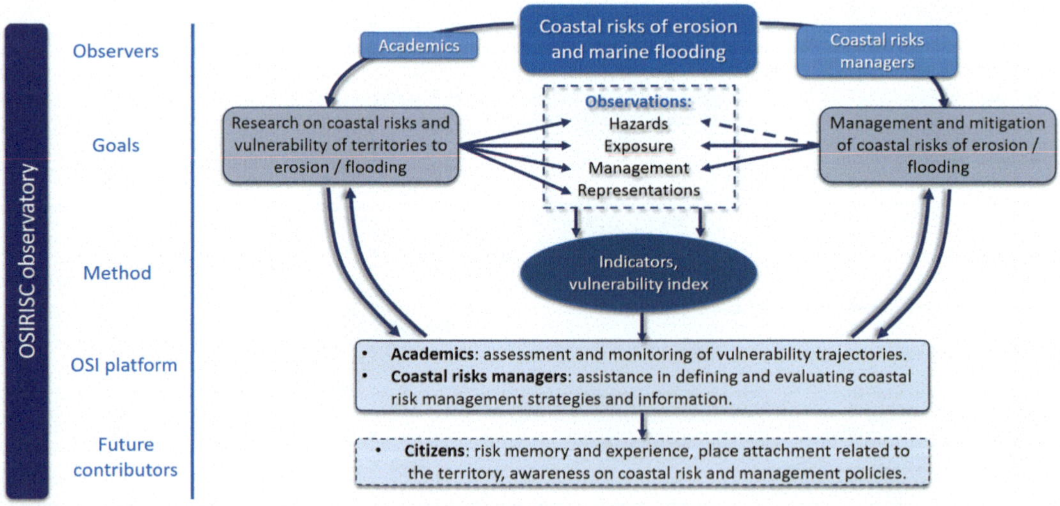

Fig. 10.11 Methodological approach developed in OSIRISC observatory

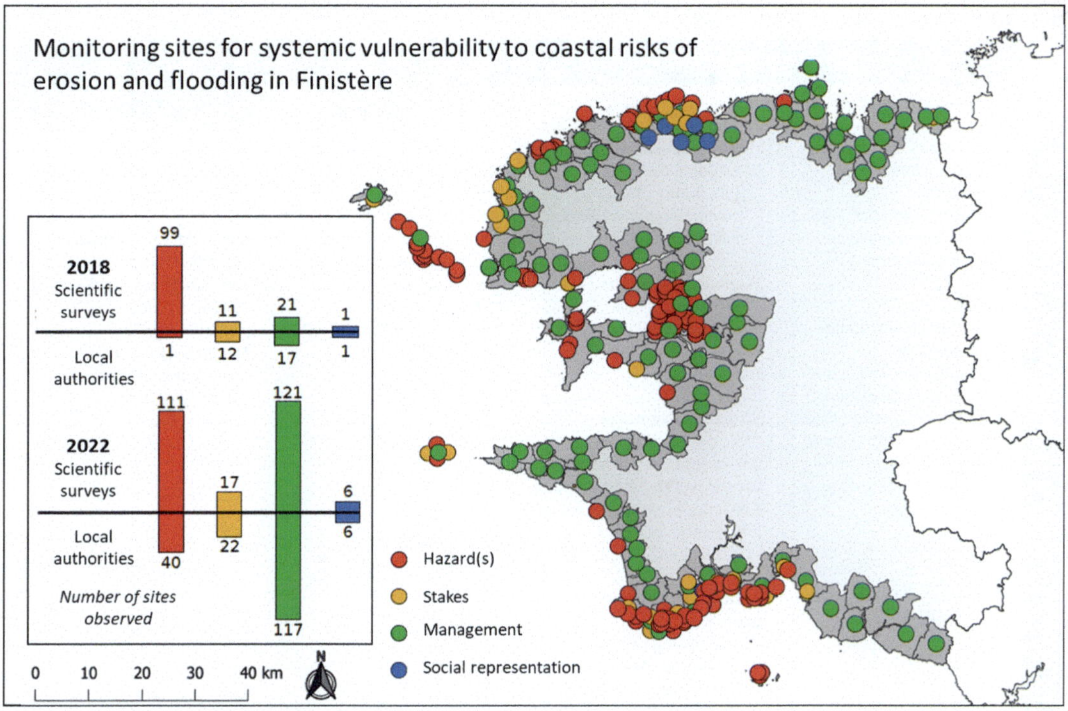

Fig. 10.12 Sites in Finistère where observations are collected on components of the systemic vulnerability to coastal risks of erosion and flooding

10.4 Data Handled by the Pilot

10.4.1 Combining Multi-Source Data

The OSIRISC observatory is based on multi-source and multi-scale data. Given the surface area of the territory concerned, data at regional level come from existing national or regional databases. Other data are collected in situ, at the local level, by academics or stakeholders. Table 10.1 summarises, for each vulnerability component, the data on which OSIRISC observatory is based.

From an academic point of view, since 2002, the marine sciences observatory of the Institut Universitaire Européen de la Mer (a Department of the University of Brest) has set up field observation data series on geomorphological evolution of the coastline. The initial objective of this monitoring is directly linked to fundamental research. The aim is to detect a signal of climatic and/or meteorological variations from the long-term morphological and sedimentary changes of the coastal fringe.

The use of modern field measurement techniques (tacheometer, Global Navigation Satellite System—GNSS, total station, Terrestrial Laser Scanner, photogrammetry, etc.) and image processing (aerial or satellite orthophotographs) has facilitated the acquisition of topo-morphological data of the coastal domain with a better accuracy, on more sites and at a higher frequency. Some of these technologies are now more widely available thanks to the emergence of low-cost instrumentation devices (Centipede collaborative GNSS network, distance meters), allowing data acquisition by managers-practitioners or student trainees (Fig. 10.13).

Observations are now collected at around 50 sites throughout Brittany, with several research laboratories contributing to the data acquisition efforts. Since 2014, some of the monitored

Table 10.1 Data handled by OSIRISC pilot

	Components of vunerability to coastal risks of erosion and marine flooding			
	Hazards	Exposures	Management	Representions
OSIRISC data	**Submersion:** - Extreme water levels probabilities - Inventory from municipal archives and newspapers - Flooding benchmarks **Erosion:** - Historical aerial/satellite photo database - Inventory from municipal archives and newspapers - Topographic measurements - Photogrammetric surveys - Terrestrial Lidar surveys - Collaborative data collected by local managers	**Human exposures:** - National databases (population, housing, vulnerable populations, infrastructure...) **Economic exposures:** - National databases (jobs, economic buildings, diversity of economic activities, tourism...) **Structural exposures:** - National and regional databases (cultural heritage, coastal works, harbours, road networks, reception capacity, emergency system...) - Local studies (inventory) **Agricultural and natural exposures:** - National and regional databases (protection of natural areas, agricultural areas...)	- Information provided by the land managers on the type and progress of the management strategy implemented by the territories. (control of urbanization, crisis management, awareness rising, knowledge/expertise, local initiatives...)	**Awareness of risk:** - Interviews with citizens - Online surveys **Evaluation of institutions and collective practices:** - Interviews with citizens - Online surveys **Meaning of the place:** - Interviews with citizens - Online surveys

Data at regional scale
Data at local scale

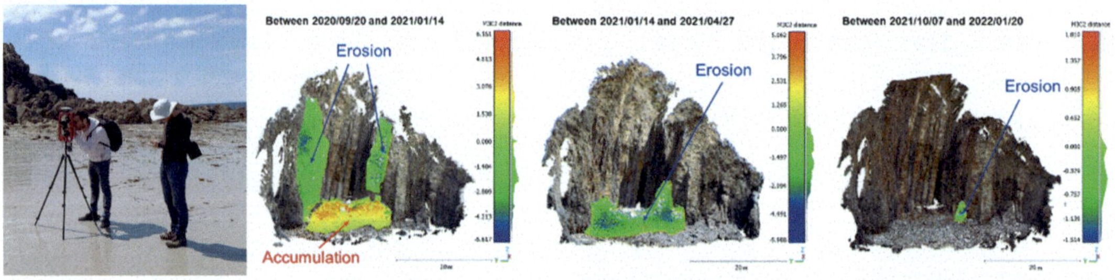

Fig. 10.13 Example of cliff erosion (between September 2020 and January 2022) measured with a photogrammetric low-cost system (Tromel cliff, Brittany, France)

sites have been integrated into the DYNALIT National Observation Service (SNO) labelled by the INSU CNRS. Similarly, several partners (local authorities, other stakeholders) produce data, either on their own or via contractors, particularly within the framework of the OSIRISC-Litto'Risques Observatory in Finistère.

Today, the aim of this monitoring program is also to meet a societal demand for applied research, especially since more than 70% of these survey operations are financed by local authorities. The aim is to compile a series of data on the evolution and morphology of the coastline to support local stakeholders in their coastal management and development policies. A final objective is to produce sufficiently long and relevant data series for the calibration of existing models in the field of engineering (1D and 2D models of beach evolution).

In France, data on exposed elements are increasingly rich and accessible via national reference databases. This makes it possible to have homogeneous information throughout the Breton territory and to reduce the very time-consuming recourse to field surveys.

Data for indicators on the management implemented in each territory are collected partly via information available on the Internet, and partly by individual surveys of technical staff in charge of risk management among local authorities. This implies close collaboration with local authorities.

Risks representation is the component for which it is most difficult to formalise information, as it involves psychosociological concepts. No pre-existing data is available for this component. Data is collected through individual surveys carried out among people who live (and work, play, ...) in the studied municipalities. Several means of communication were used to encourage people to answer an online questionnaire (town hall websites, tourist offices, social networks, associations, schools, posters in the municipality bulletin boards, etc.).

10.4.2 Creating Indicators and Indices

The indicators produced within OSIRISC observatory are a harmonised, numerically rated transcription of quantitative and qualitative data, allowing for the combination of multi-source data on the different components of vulnerability (Le Berre et al., 2022). Thus, for each component of vulnerability (hazards, elements-at-risk, management and representations) and for each type of observation, the indicators were defined with a calibration to provide a rating on a scale from 1 to 5.

These indicators are calculated spatially, at the scale of the territory where the data is available (see Table 10.1), on a grid defined to be compatible with the national databases. While the initial grid cell size was 200 m, which guarantees the anonymity of personal data, many indicators are now produced on grid sizes from 50 to 800 m. Each indicator is accompanied by a descriptive datasheet specifying its role (what the indicator means), the data sources and possibly their quality, as well as the protocol for filling it in, calibrating it into five categories and the time step for updating the indicator (Le Berre et al., 2022; Meur-Ferec et al., 2020).

A high value of hazard or exposure indicator contributes to increased vulnerability. Conversely, a high value of the management indicator (corresponding to a relevant management strategy) contributes to a decrease in vulnerability. As all indicators are homogeneous since they use the same scale and same grid, aggregated "indices" (either within a component or inter-component) can also be generated by crossing several indicators in order to have a more integrated assessment of vulnerability (Fig. 10.14). Only the representation component is not integrated into indices. This component provides contextual information that will facilitate the implementation of risk management strategies.

10.5 Citizens' Involvement and Platforms

10.5.1 AGEO App

The AGEO application (see chapter 6) makes it possible to report coastal damage including identification information in order to create or enrich an inventory. The application consists of geolocated photographs, accompanied by qualitative data concerning the characterisation of the type of hazard, the estimation of the magnitude of the hazard, the elements exposed and the estimation of the date of the hazard according to the frequency of visit of the site by the participants (Fig. 10.15). This app can be used by citizens, typically residents who often visit coastal areas, or managers-practionners. Reports on coastal impacts collected with AGEO app will be added to the existing OSIRISC database "Aléas et impacts côtiers".

10.5.2 CoastAppli App

Local authorities show interest in contributing to hazards monitoring, but sometimes lack both the time and the tools and technical skills to do so.

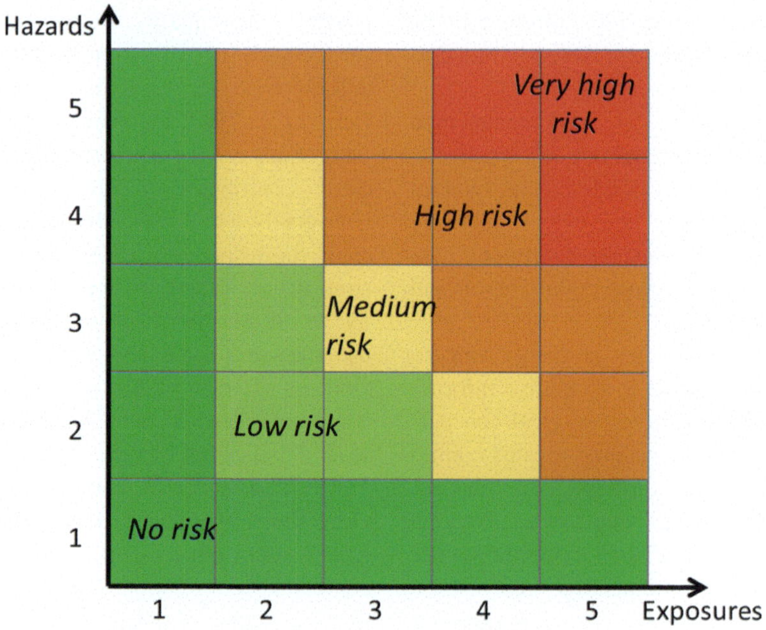

Fig. 10.14 Matrix for calculating the risk index by aggregating hazards and exposures

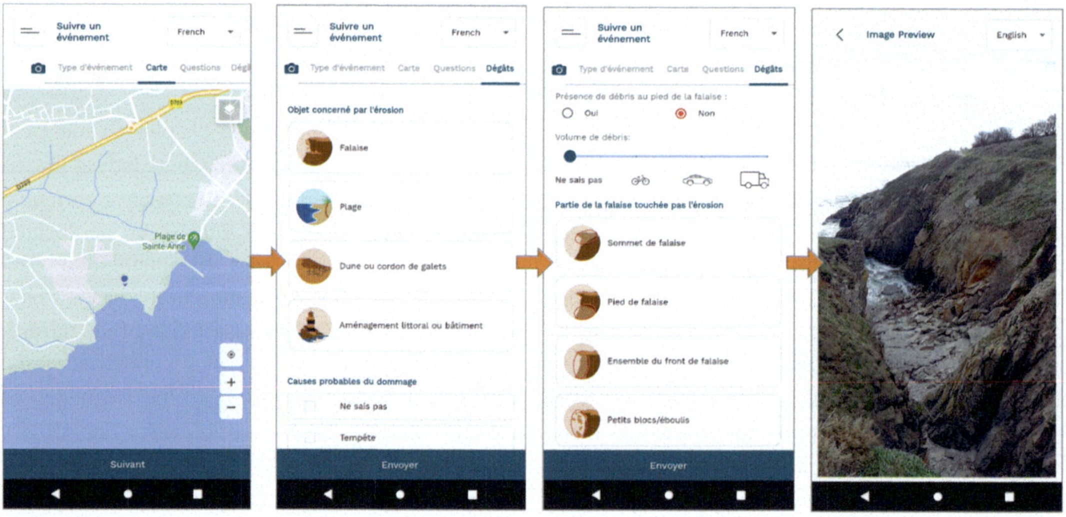

Fig. 10.15 Screenshots of the successive steps for damage reporting with AGEO app

Furthermore, the management strategies relying on adaptation measures, no longer attempting to maintain a fixed coastline position, are not always well received by the public, which make the dialogue about risk management more complicated.

The idea therefore emerged to set up an application called "CoastAppli" meant to facilitate the data acquisition for coastal surveys, to raise interest from the general public and enable its participation in monitoring programs on coastal dynamics.

This application (Fig. 10.16) was developed with the support of the EUR ISblue University Research School, the Sea-EU project and the INTERREG AGEO project. "CoastAppli" app is complementary to AGEO app. CoastAppli provides quantitative measures with resolution and accuracy close to typical positioning systems (up to ± 4 cm) for some of the protocols, using a simple approach such as measuring the size of a step and counting the number of steps, or simple tools like a measuring tape. As it relies both on quantitative and qualitative protocols, it is aimed at a targeted audience (technical staff from local authorities, trained citizens) as well as the general public, including children.

The first version of this application proposes a dozen of protocols for monitoring "erosion" and "flooding" hazards, using simple measurement tools. This version was tested and validated in the municipality of Guissény from December 2021 to October 2022, with technical staff from local authorities, schoolchildren from

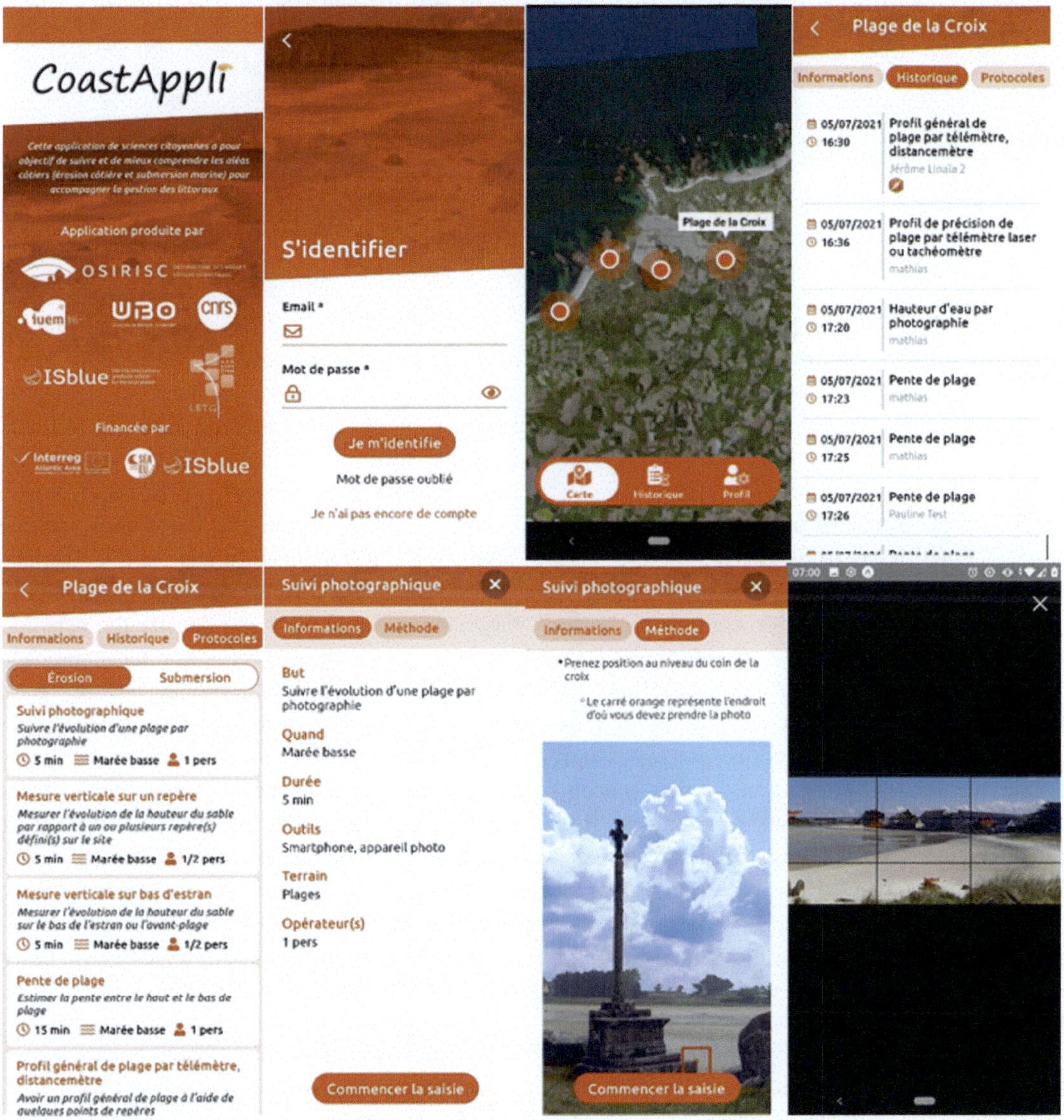

Fig. 10.16 Screenshots of the first version of CoastAppli app

Fig. 10.17 Test of data acquisition for coastal surveys using "CoastAppli" app with pupils in Guisséy (R. Ruault and Q. Millière)

the Diwan Grade School and the general public. We verified the reliability of the measurements, as well as the interest and the commitment of the users (Fig. 10.17).

10.5.3 Deployment of Centipède Geolocation System

Geolocation is one of the central tools for monitoring coastal hazards (positioning of the coastline, measurement of beach profiles, 3D modelling of cliff evolution, etc.). By default, the accuracy of satellite navigation systems is only a few metres. Differential correction systems can compensate for this lack of accuracy: Real Time Kinematic (RTK) positioning can approach centimetre accuracy. However, this approach requires a reference base set-up close to the data collection site, the accessibility of its differential correction signal and very expensive equipment to process it. For some years now, developers have been actively working with open-source/open-hardware software and components to create inexpensive, reliable, lightweight and easy-to-use differential GNSS solutions.

Centipede[1] RTK (https://docs.centipede.fr/) is a collaborative network of open GNSS bases available to anyone within the coverage area. Initiated in 2019 by INRAE (French National Research Institute for Agriculture, Food and the Environment), the network is being deployed thanks to public institutes, individuals, private actors such as farmers or other public partners who voluntarily contribute by deploying GNSS base antennas to extend and densify the network. The objective of the project is to offer a complete coverage of the French metropolitan territory. Base stations are also being set-up overseas and abroad.

Deploying this system appeared to be a way of making data acquisition financially and technically more accessible, and therefore usable for collaborative science as well as student training. Within the framework of AGEO project, UBO has contributed to the Centipède base network in Brittany by installing 9 Centipede bases (Fig. 10.18). In addition, UBO has equipped the local authorities of the Litto'Risques partnership with rover antennas to carry out their own coastal measurements.

UBO supported technical staff of local authorities across the Breton territory by training

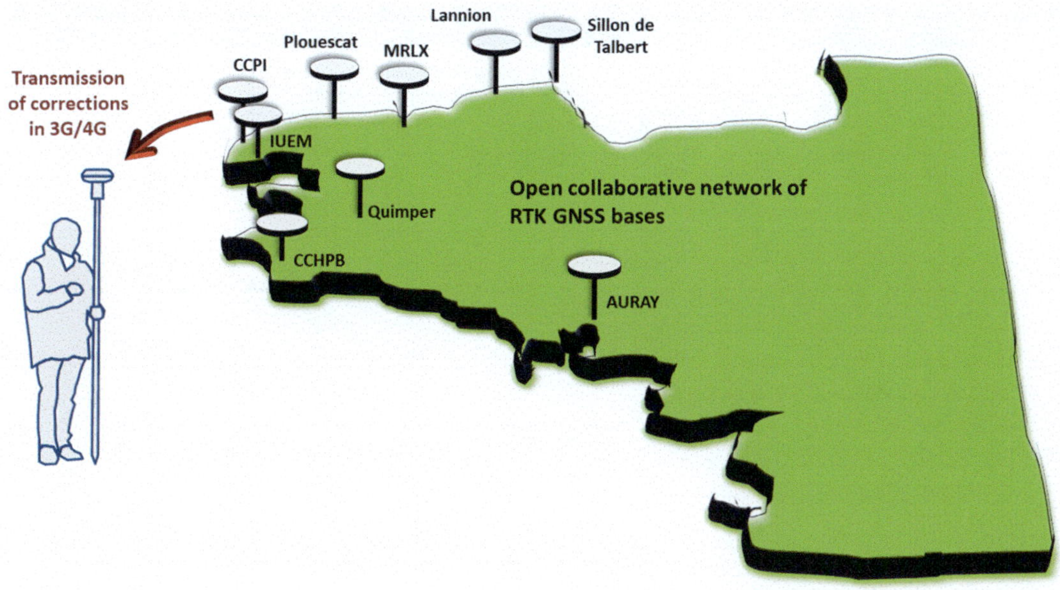

Fig. 10.18 The network of Centipède GNSS bases installed in Brittany by OSIRISC in the framework of AGEO

them in the use of this survey equipment, the measurement protocols, the choice of morphological features to survey and data processing. With involvement of local authorities, the number of measurement sites is rapidly increasing. About 40 survey sites have been added since Litto'Risques partnership was started. The data collected by local authorities using these methods and tools enrich the OSIRISC observatory databases.

10.5.4 Data cataloguing and diffusion: MADDOG and OSI Platforms

MADDOG[2], which stands for "Mise A Disposition des Données d'Observation Géomorphologique" is the OSIRISC coastline and coastal geomorphology observation data portal (https://portail.indigeo.fr/mviewer/?config=apps/maddog.xml#). It is a web platform for the visualisation of topo-morphological data series (coastline, cross-shore beach profile, Digital Terrain Model (DTM)) of the Brittany coastline: beaches, dune systems, sandbars, cliffs, river mouths.

MADDOG implements cataloguing and geographic information processing tools that allow the dissemination of these observation time series in the form of interactive databases. MADDOG also provides online interactive data processing services that allow a preliminary analysis of the morphological evolution of the coastline (Fig. 10.19). Three types of results are available on MADDOG:

- The planimetric evolution of the coastline allowing the analysis of the kinematics of the coastline by calculating a distance of shoreline retreat or advance between coastlines measured at two different dates.
- The altimetric evolution of beaches by comparing cross-sectional beach profiles measured at different dates. Morphosedimentary changes allow the calculation of the evolution of the sediment balance in m^3 per linear metre.
- The evolution of surface changes of the beaches in three dimensions (x, y and z) by comparison of digital terrain models. These data allow the calculation and mapping of the sediment budget (in m^3 per linear metre) in the form of accretion and erosion zones.

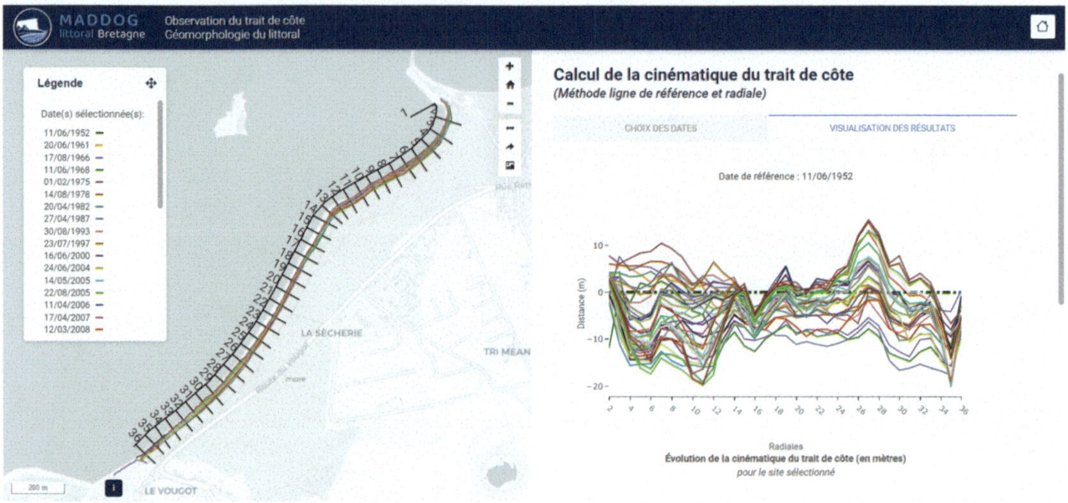

Fig. 10.19 Example of the altimetric evolution of beaches by comparing cross-sectional beach profiles on MADDOG interface (Guisseny beach, Brittany, France)

This platform is intended to be used by the scientific community as well as by managers-practitioners and citizens. It thus responds to the needs expressed by local authorities to have data readily accessible processing tools allowing preliminary exploitation of observations they carry out, and further to obtain interpreted results on geomorphological evolution of the coastline to support the definition of coastal risk management strategies (development, prevention, adaptation), as well as to raise awareness among all stakeholders.

The co-construction approach between researchers and territorial stakeholders (managers-practitioners, technicians and elected officials, and citizens) is a cornerstone of the OSIRISC regional observatory. Consistently, the data disseminated via the platform MADDOG comes from surveys carried out both by UBO research laboratories and by OSIRISC observatory partners: local authorities in charge of coastal management, State services and other public partners (the French coastal conservation agency, Natural Parks, etc.).

The European University Institute of the Sea (IUEM), which is the "marine and coastal sciences" component of UBO, runs the regional observatory OSIRISC and hosts the coastal observations database accessible via the MADDOG platform. IUEM manages (that is, hosting, archiving, maintenance, availability) the data collected by the many contributors from various marine sciences fields as part of its role as an "Observatory of the Sciences of the Universe" (OSU). The coastal monitoring data accessible on MADDOG are archived under the INDIGEO[3] Spatial Data Infrastructure (https://indigeo.fr/).

A common input data format has been defined for the series disseminated via MADDOG, in order to allow the visualisation of data collected by the various producers using different instruments.

The OSI[4] Web-GIS interface (https://osi.univ-brest.fr/) allows the visualisation of the indicators and indices developed to assess and monitor the vulnerability trajectories of coastal territories to coastal hazards. These indicators and indices are represented on an adaptive mesh ranging from 50 m to the scale of the municipality. The interoperability of data and geographic information processing tools allows for a back and forth navigation between the integrated analysis of vulnerability on the OSI interface and a more in-depth analysis of erosion and marine flooding hazard data on the MADDOG platform, where morpho-sedimentary evolutions are represented on a metric scale.

Fig. 10.20 Example of marine flooding indicator on the OSI web-platform interface (Île-Tudy, Brittany, France)

OSI displays each indicator filled in a territory for a given date (Fig. 10.20). The hot spots of high-risk territories are highlighted at both the local and regional levels. Decision-makers therefore have a general vision of the areas where risk reduction efforts should be focused. Based on these indicators, OSI calculates component indices (hazards, exposure, management and social representations). A vulnerability diagnosis can then be produced as the component(s) that most strongly influence vulnerability are highlighted. The action levers are therefore easily identifiable. Coastal risk management can then be based on this diagnosis. Once the management strategy is implemented, a new vulnerability diagnosis is obtained based on the regularly carried out measures of the indicators. This new diagnostic allows to evaluate the effectiveness of the means used to reduce the vulnerability of the coastal territory, while visualizing the evolutions of the other components. While risk management is too often aimed towards reducing hazards through the construction of sea-defences, the diagnosis can reveal the other possibilities such as reducing the exposed elements in the territories, educating citizens about risks, etc. Ultimately, these successive diagnostics define the territory's vulnerability trajectory. OSI is thus meant as a decision support tool for local authorities. OSI is also a scientific tool for geographic analysis and comparison to improve understanding of the changing vulnerability of territories and to improve coastal risks management.

10.6 Interaction with Stakeholders and Risk Management

In the framework of the OSIRISC observatory, the "Litto'Risques" partnership was signed in May 2019, aiming to strengthen the capacities of the local authorities in charge of coastal risks, in particular regarding observation of coastal vulnerability. To support coastal communities, UBO, Cerema and the County Council of Finistère have joined forces to provide methodological, scientific and technical support for the management of coastal erosion and marine flooding risks, through three main missions: coastal observation, expertise for coastline management and awareness-raising on coastal risks. Following the work carried out by the OSIRISC research teams with a co-construction approach between researchers and territorial local authorities, the scientific coordination of the OSIRISC observatory is carried out by UBO and its academic partners on coastal risks, in coordination with the stakeholders in charge of coastal risk management. This partnership aims to pool the skills of the partners in order to enable coastal communities to develop a long-term strategy for managing their coastline. The objectives of the partnership are to:

- contribute to the acquisition of data on coastal risks and to the dissemination of this data at the regional scale of the Finistère County;
- provide support to local authorities in the definition and implementation of their coastal development projects;
- support the coastal communities of Finistère in the definition of a prospective vision of their territory in order to anticipate the evolution of the coastline under the consequences of global change and coastal urbanisation;
- bring together the stakeholders in Finistère who are interested in monitoring coastal vulnerability;
- carry out dissemination and awareness-raising activities on the coastal risks.

Through this approach, there is a two-way sharing of expertise between academics and local authorities, increasing the skills of local stakeholders and disseminating the observation to decision-makers, as well as increasing the regional observation capacities and providing academics with feedback on risk management at an applied level. This is consistent with the twofold objective of addressing research questions and providing operational decision-aid tools. Coastal risk managers (from elected representatives and technical services of local authorities to government services) are both knowledge providers who contribute to monitoring the vulnerability components, and end-users of the results of the analysis of trajectories of systemic vulnerability.

Communities of municipalities can join the partnership with two levels of commitment. At a first level, they are informed of the activities and results of the observatory; they participate in the training sessions and seminars and can be assisted on coastal risk management issues. Membership is strengthened when they undertake the monitoring of vulnerability components in their territory, generally starting with hazards monitoring (Fig. 10.21).

Breton coastal territory managers can collect a certain amount of data that will improve the analysis of the systemic vulnerability of their territory. By providing them with measurement protocols, by helping them to choose appropriate methods and tools and by storing and disseminating information in the form of interactively mapped indicators, OSIRISC observatory is operated as a partnership with local authorities in charge of coastal risk management. A central requirement for the success of the partnership is to provide training on a regular basis, including:

- conceptual training to understand the observation approach, the processes involved or the consequences of a particular management strategy;
- technical training on the implementation of measurement methods and tools or the data processing.

As part of AGEO, technical staff and elected representatives from local authorities were offered several face-to-face training sessions (in classroom or on the field—Fig. 10.22). Online materials were also produced: methodological guides (Partenariat Litto'Risques, 2023a, 2023b, 2023c), video tutorials, Flot'Risco online course, etc. (Fig. 10.23). They are available through the website "Risques Côtiers"[5] (https://www.risques-cotiers.fr/).

Since OSIRISC is the result of co-design between scientists and stakeholders (at all stages, from the design of the methodology based on indicators to the implementation of collaborative monitoring), it enables capacity-building of technical staff and elected officials within local authorities in charge of coastal risks. As deployed within the Litto'Risques partnership, OSIRISC is more than an observatory.

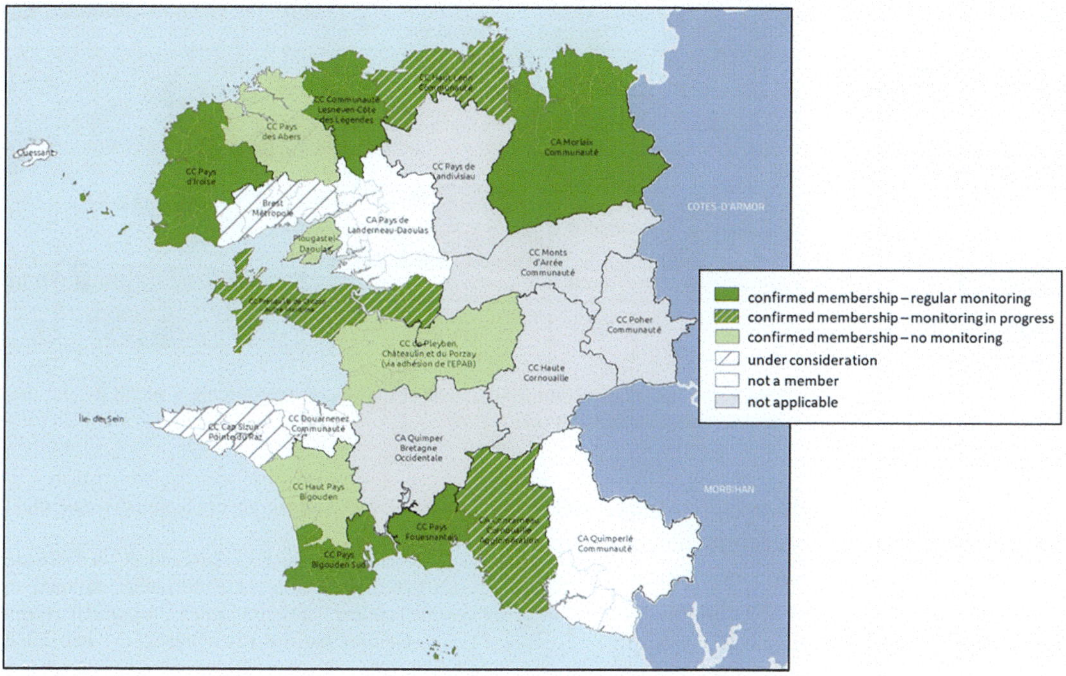

Fig. 10.21 Local authorities members of the Litto'Risques partnership

Fig. 10.22 Face-to-face training sessions with coastal risk managers

10.7 Conclusion and Recommendations

The success of OSIRISC observatory relies on the co-construction and the close interaction between academics and local authorities, which requires in-depth involvement of both parties in joint actions (data collection, sharing expertise, ...), as well as political support to achieve partnership agreements and engage in long-term observation of the vulnerability of coastal territories. The OSIRISC observatory already concerns many coastal territories in Brittany, mainly in the Finistère County through Litto'Risques partnership. OSIRISC observatory is currently expanding the Litto'Risques partnership model to other coastal territories in Brittany. This approach is intended to be generalised or

Fig. 10.23 Online training materials: methodological guide (**a**), video tutorials on measurement methods and data processing (**b**), online course on coastal risk management (**c**)

transposed to a larger territory, in Brittany and elsewhere. One of the short-term objectives is to further develop citizen observations, taking advantage of the apps developed thanks to AGEO project: AGEO App and CoastAppli.

Acknowledgements This work has been undertaken in the framework of the INTERREG Atlantic Area AGEO Project (EAPA_884/2018—AGEO).

References

Cerema, Ministère de la Transition écologique et solidaire, 2018. *Cartographie de l'indicateur national de l'érosion côtière*. http://www.geolittoral.developpement-durable.gouv.fr/indicateur-national-de-l-erosion-cotiere-r473.html.

Cerema. (2021). *Dynamiques et évolution du littoral.* Fascicule 4: Synthèse des connaissances du cap de la Hague à la pointe de Corsen. Collection: Connaissances, p. 272.

Cerema. (2021). *Dynamiques et évolution du littoral.* Fascicule 5: Synthèse des connaissances de la pointe de Corsen à la pointe de Chémoulin. Collection: Connaissances, p. 273.

Chaumillon, E., Bertin, X., Fortunato, A. B., Bajo, M., Schneider, J.-L., Dezileau, L., Walsh, J. P., Michelot, A., Chauveau, E., Créach, A., Hénaff, A., Sauzeau, T., Waeles, B., Gervais, B., Jan, G., Baumann, J., Breilh, J.-F., & Pedreros, R. (2017). Storm-induced marine flooding: Lessons from a multidisciplinary approach. *Earth-Science Reviews, 165*, 151–184.

D'Ercole, R. (1994). Les vulnérabilités des sociétés et des espaces urbanisés: concepts, typologie, modes d'analyse. *Revue de Géographie Alpine*, 4, LXXXII, 87–96.

Géoportail IGN. https://www.geoportail.gouv.fr/carte. GeoBretagne 1950'. https://geobretagne.fr/sviewer/dual.html.

Hédou, F., Roche, A., Trmal, C., Moraud, S., & Deniaud, Y. (2018). Elaboration de l'indicateur national de l'érosion côtière. *XVe Journées Nationales Génie Côtier—Génie Civil*, La Rochelle, 29–31 mai 2018, 648–654.

Hénaff, A., Lageat, Y., Bodéré, J. -C. (2003). *PRIR "EROCOVUL", Erosion et vulnérabilité du trait de côte en Bretagne*. Rapport annuel pour la Région Bretagne, année 3, Géomer LETG UMR 6554 CNRS, p. 43.

Hénaff, A., & Philippe, M. (Eds) (2014). *Gestion des risques d'érosion et de submersion marines, guide méthodologique*. Projet-ANR-10-CEPL-0001Cocorisco, p. 156.

Hénaff, A., Le Cornec, E., Jabbar, M., Pétré, A., Corfou, J., Le Drezen, Y., & van Vliët-Lanoë, B. (2018). Caractérisation des aléas littoraux d'érosion et de submersion en Bretagne par l'approche historique. *Cybergeo: European Journal of Geography* [En ligne], Environnement, Nature, Paysage, document 847. http://journals.openedition.org/cybergeo/29000; https://doi.org/10.4000/cybergeo.29000.

Indigeo: infrastructure de données et d'informations géospatialisées sur l'environnement, UMR LETG, OSU IUEM, ZABrI: https://indigeo.fr/.

Institut Géographique National "Remonter le temps": https://remonterletemps.ign.fr.

Institut National de la Statistique et des Etudes Economiques: https://www.insee.fr

Le Berre, I., Meur-Ferec, C., Cuq, V., Guillou, E., Lami, T., Dantec, N., Letortu, P., Lummert, C., Philippe, M., Rouan, M., Noûs, C., & Hénaff, A. (2022). Systemic vulnerability of coastal territories to erosion and marine flooding: A conceptual and methodological approach applied to Brittany (France*). International Journal of Disaster Risk Reduction, 78*, 103122. https://doi.org/10.1016/j.ijdrr.2022.103122

Meur-Ferec, C., Le Berre, I., Cocquempot, L., Guillou, E., Hénaff, A., Lami, T., Le Dantec, N., Letortu,

P., Philippe, M., & Noûs, C. (2020). Une méthode de suivi de la vulnérabilité systémique à l'érosion et la submersion marines. *Développement Durable Et Territoires [en Ligne], 11*, 1. https://doi.org/10.4000/developpementdurable.16731

Partenariat Litto'Risques. (2023). *Guide méthodologique de gestion des risques littoraux en Finistère—Volet 1: genèse, évolution et gestion du littoral, p.* 84.

Partenariat Litto'Risques. (2023). *Guide méthodologique de gestion des risques littoraux en Finistère—Volet 2: méthodologie de gestion des risques littoraux*, p. 196.

Partenariat Litto'Risques. (2023). *Guide méthodologique de gestion des risques littoraux en Finistère—Volet 3: recensement et retours d'expérience sur les solutions mises en oeuvre sur le territoire du Finistère dans la gestion du trait de côte, p.* 216.

Philippe, M., & et Hénaff A. (coord.) (2021). *Expérimentation pour un observatoire régional de la vulnérabilité aux risques côtiers en Bretagne. Projets OSIRISC et OSIRISC+*. UBO, CNRS, Cerema. Soutenus financièrement par la Fondation de France et la DREAL Bretagne. 84 p + annexes. https://doi.org/10.48386/POPS-1519.

Poupard, G. (2017). L'urbanisation du littoral : Un essouflement?—L'exemple de la Bretagne. *Population & Avenir, 733*, 4–8.

Hénaff A, Meur-Ferec C, Lageat Y (2013) Changement climatique et dynamique géomorphologique des côtes bretonnes. Leçons pour une gestion responsable de l'imbrication des échelles spatio-temporelles. Cybergeo: European Journal of Geography [En ligne], Environnement, Nature, Paysage, document 654, http://journals.openedition.org/cybergeo/26058; https://doi.org/10.4000/cybergeo.26058

Stéphan, P., Blaise, E., Suanez, S., Fichaut, B., Autret, R., Floc'h, F., Madec-Cuq, V., Le Dantec, N., Ammann, J., David, L., Jaud, M., & Delacourt, C. (2019). Long, medium, and short-term shoreline dynamics of the Brittany coast (Western France). *Journal of Coastal Research*, SI 88, 89–109. ⟨https://doi.org/10.2112/SI88-008.1⟩.

Vinet, F., Defossez, S., Rey, T., & Boissier, L. (2012). Le processus de production du risque "submersion marine" en zone littorale : l'exemple des territoires "Xynthia." *Norois, 222*, 11–26. https://doi.org/10.4000/norois.3834

Lisbon Multihazards Citizen's Observatory

11

Cláudia Pinto, Carlos Sousa Ferreira, Paula Pacheco, Carla Pousada, Mónica Ferreira, Pedro Leonardo, Alexandra Frias, Miguel Inácio, Ana Luísa Seabra, Maria Manuel Pinto, Inês Ascenso, Mário Fontes, Maria José Cerqueira, Carolina Almeida, Ana Runa, Rita Dórdio, Vanessa Fernandes, Joana Fernandes and Filipa Vilela

11.1 Introduction

Lisbon, the capital of Portugal is located at the confluence of the Tagus River with the Atlantic Ocean, a location that gave the city a prominent geostrategic position, but also a high exposure to natural and climatic hazards, such as earthquakes, tsunamis, coastal and urban floods, landslides, and an increasing occurrence of extreme weather events (Fig. 11.1).

Lisbon is a major economic centre and presents a high population density and urban occupation. The most part of the government and services buildings are located here, which, in case of a disruptive event could impact directly on the decision chain. Moreover, many of the city buildings are centuries old, and infrastructure is likewise aging so urban improvement will be required to reduce the impact of natural phenomena, which could affect structures and networks.

The city is preparing for other threats by integrating response services to increase efficiency and effectiveness, but Lisbon recognizes that it has more work to do in planning effectively for larger, more severe events brought on by geohazards and climate changes.

Considering the above, it is very important to prepare the society to better respond to disasters and demonstrate that citizens, properly oriented towards a problem, can provide value data about the evolution of a widespread phenomenon or a specific event in their area of residence, and play an active role in risk management systems, allowing a more informed decision making, and ultimately contributing to the implementation of proactive measures rather than reactive responses.

This was the proof of concept demonstrated in the Lisbon Pilot, managed by the Lisbon Municipality, which addresses seismic risk, urban floods, landslides, and geotechnical risk, and has the particularity of being developed in an urban context.

C. Pinto (✉) · C. S. Ferreira · P. Pacheco · C. Pousada · M. Ferreira · P. Leonardo · A. Frias · M. Inácio · A. L. Seabra · M. M. Pinto · I. Ascenso · M. Fontes · M. J. Cerqueira · C. Almeida · A. Runa
Lisbon Municipality, Campo Grande, 25, Lisbon, Portugal
e-mail: claudia.pinto@cm-lisboa.pt

M. Ferreira
e-mail: monica.ferreira@tecnico.ulisboa.pt

M. Ferreira
CERIS/Instituto Superior Técnico, University of Lisbon, Lisbon, Portugal

R. Dórdio · V. Fernandes · J. Fernandes · F. Vilela
Ursa Maior, Lisbon, Portugal

V. Fernandes
e-mail: vanessa.fernandes@ursamaior.pt

R. C. Gomes et al. (eds.), *Citizens' Observatories on Geohazards*, Geoenvironmental Disaster Reduction,
https://doi.org/10.1007/978-3-031-53371-6_11

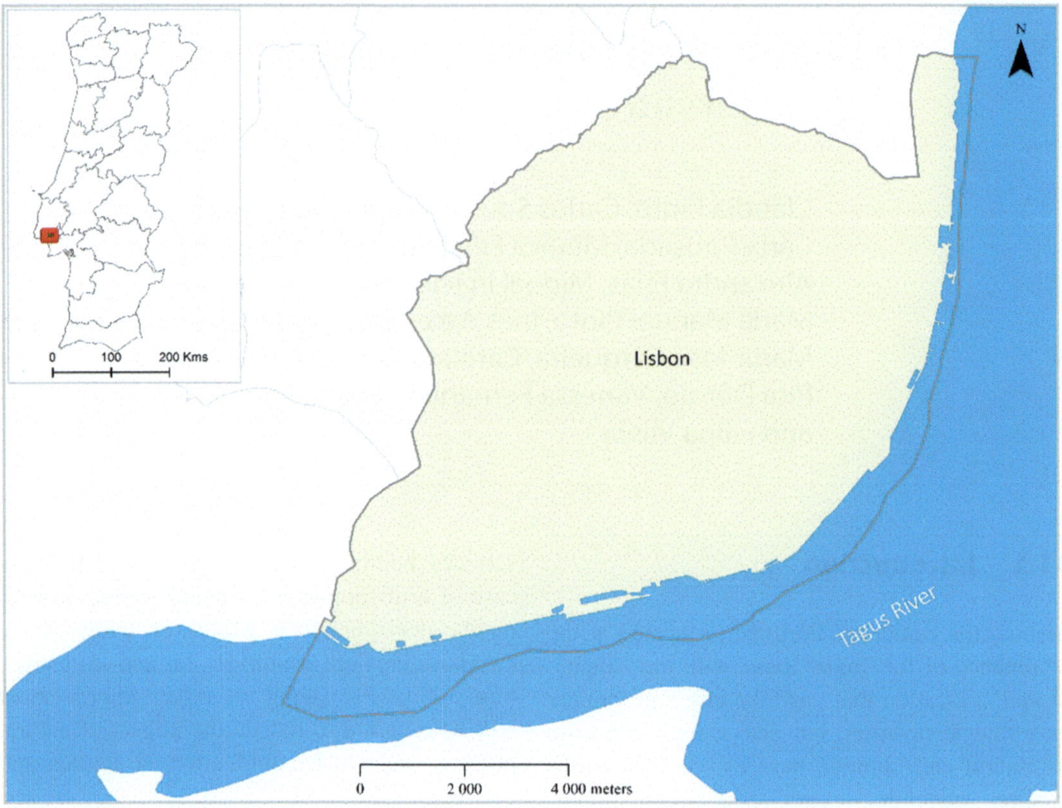

Fig. 11.1 Lisbon city location. *Source* Pinto (2020)

11.1.1 Society Engagement

In the framework of the Lisbon Pilot build awareness and involve specific groups of stakeholders, from the common citizen to specialized professionals, were the main goal. This requires providing the appropriate means to promote citizens participation, which includes the creation of a direct communication channel, considering the different levels of expertise.

The groups of stakeholders were selected considering their area of activity and their capability to promote an efficient monitorization of risk situations. In this case civil protection local groups, municipality services (fire department, sewage and pavements brigades), scientific community, schools, senior university students and groups of scouts were engaged.

11.2 Communication and Capitalization Strategy

Very few people are informed about geological hazards, and this takes on relevance when we talk about the impact of these phenomena on daily routine. This fact led to the need to develop a communication strategy. Everything starts with our mascot: RISKA.... first on drawings and then on digital, developed by Cacto Animation Studios (Figs. 11.2 and 11.3).

Using a simple chromatic universe, we intent to shape any situation in a clear way for a better understanding, using infographic and static animations. The representation of rocks and soils are drawn with organic lines, so we have a better understanding of texture and patterns involved in each one (Fig. 11.4).

Fig. 11.2 RISKA

Fig. 11.3 RISKA expressions

Fig. 11.4 Patterns involved in communication materials

Several communication products were developed to support the work with the stakeholders, particularly with educational community, such as flyers, posters, tray napkins with crosswords about AGEO and other games, bookmarks and rulers (Fig. 11.5). These materials were distributed through the several dissemination actions carried out in Lisbon schools, in which around 2500 students participated.

To involve the common citizen a partnership with a company called Delta Cafés was promoted, resulting on a campaign carried out to disseminate the AGEO Project through sugar packets that were distributed in the city of Lisbon. For this purpose, 10 layouts were developed regarding the risks managed by the Lisbon Pilot. An example is presented in Fig. 11.6.

Through seven animations formats RISKA explains, teaches, and alerts to some daily problems that occur in our city, how to prevent them and how to react in different situations. The videos are available on Youtube and at Lisbon web page (https://informacoeseservicos.lisboa.pt/prevencao/resiliencia-urbana/projetos/ageo).

Keeping in mind the audience diversity it was decided that the tone of the communication on social media should be informal, explicative, educational, and with low technical complexity. Facebook and Instagram were used as communication platforms and more then 50 posts noting specific days were published during the Project.

https://www.facebook.com/ageoatlantic
https://www.instagram.com/ageoatlantic/

RISKA became the communication identity of the Lisbon Pilot.

11.3 Risk Assessment and Copernicus Products

A risk assessment identifies and evaluates the possibility of natural or human induced phenomena occurring and their direct and indirect impact on elements at risk. The concept of risk is complex and depends on many variables comprising three main components: Hazard, Exposure, and Vulnerability (physical

Fig. 11.5 Educational contents

Fig. 11.6 Sugar packets

and social) of the elements exposed to hazards and their propensity to suffer adverse effects (Barbat et al., 2010; Ferri et al., 2020; Guillard-Gonçalves et al., 2015; Pereira et al., 2020).

Risk assessment directly support the local and national authorities in identifying vulnerable groups, infrastructure, and assets that are most vulnerable to hazards and to what extent injuries and damages may occur, and to develop evacuation plans and improve warning messages to include the potential impacts of hazardous events. Results from the risk assessment allow local governments to prioritize where and what disaster risk reduction interventions can be most effective to minimize future disruptions. To ensure appropriate measures are in place, disaster risk knowledge must be accessible to all (UNDRR & WMO, 2022).

11.3.1 Indicators to Assess Risk and Guide Decision-Making in an Urban Context

The Lisbon Pilot integrates different types of hazards, some more frequent than others, that

Table 11.1 Normalization matrix—Hazards

Hazards	Themes	Candidate set of Indicators (name and short definition & data source)	
	Earthquakes	Location of source/Focal Distance	*Lower Tagus Valley* *Distance Gorringe bank*
		Magnitude	*M6 and M8 (scenarios)*
		Soil Chart	*Soil Chart GeoSIG*
		Distance to faults	*IPMA*
	Geotechnical (Settlements, uplift, liquefaction)	Underground Water level	*PGIL (Lisbon Intelligent Management Platform)*
		Superficial water level	*PGIL,* *SENTINEL 1 interferometry*
		Standard penetration test (SPT)	*GEOSIG*
	Landslides	Slope	*Slope Chart*
		Geology/surficial fills	*Geological Chart/GEOSIG*
		Superficial water level	*GEOSIG*
		Precipitation	*IPMA*
		Thickness of coverage deposits	*GEOSIG* *PGA (peak ground acceleration)*
	Floods	Slope	*Slope Chart*
		Precipitation	*IPMA*
		Infiltration Rate	*Sentinel 1*
		Permeability/Hydrogeological Complexes	*GEOSIG*
		Natural water Lines	

pose serious consequences in everyday life. In the preliminary appraisal phase, many data types can be acquired and processed to model, simulate, and monitor risks. In the current study, the first step concerned on the selection of indicators that are suited to assessing risks (Table 11.1). These should be presented in a way that attracts organizations' and stakeholders' attention, with a thoughtful design that allows for easy understanding while also being robust and practical to guide the decision-making process.

A local strategy can be initiated to improve risk reduction management systems by identifying the spatial correlation between hazards and exposed elements. However, some questions need to be answered prior to the creation of this method:

i. What are the indicators that contribute to risk evaluation in the study area?
ii. What are their weightings?
iii. What are their sources and how reliable is the available data?

The methodology consists of a normalization matrix of hazard, vulnerability, and risk indicators towards the natural phenomena under study in the Lisbon Pilot, such as Earthquakes, Floods, Landslides and Geotechnical Risk.

For each topic (Hazard, Elements at risk, Impact and Disaster Management) the following tasks were took into consideration:

a. Definition;
b. Source of relevant data for risk evaluation;

c. Data quality: State the available format, coordinate system, precision, update frequency, date of the latest update, and the skills required for processing;
d. Calibration: Definition of the classes in which the results are presented;
e. References.

The choice of the variables was driven by several factors, including municipal coverage and direct relevance to the phenomenon that the indicators are intended to measure and qualify. A review of the technical features of hazards (such as their location, intensity, frequency, and probability) and the analysis of the dimensions of vulnerability and exposure (physical, social, economic) while considering the coping capabilities pertinent to the risk scenario. The Tables 11.1, 11.2, 11.3 and 11.4 shows a list of indicators, for each topic, as a starting point. The identification of indicators to populate the candidate set required the review of previous models available in the literature (Cadorna, 2006; Carreño, 2007; Ferreira, 2014).

The sequence of steps is as follows:

A. Definition of concepts inherent to the theme, such as susceptibility, hazard, exposure, vulnerability, resilience, and disaster risk management.
B. Definition of determinant indicators.
C. Definition of the weighting of each indicator based on expert opinion.
D. Definition of the combination of indicators that affect hazard, vulnerability, and risk.
E. Validation by comparison with past scenarios.

To compare different hazards in a single platform, i.e., Lisbon Multihazard Citizens' Observatory, each one has to be classified into standard classes based on the degree of its intensity and potential losses. The reclassified indicators are then combined based on individual weighting. Overlay analysis of files with a matrix format (raster files are made of cells represented by pixels, and each pixel has one or more numbers associated with it that specify the color in which it is displayed) generates a continuous representation of reality as a composite map. An example of the application of this method can be found in the flood risk assessment in Lisbon (Ferreira et al., 2022).

Therefore, with the latest up-to-date information available, multi-hazard risk analyses and maps can be produced to make risk-informed decisions in urban planning and land use development (with an application on the Lisbon Master Plan for example).

11.3.2 Use of the Copernicus Products

Copernicus Programme is served by a set of dedicated satellites (the Sentinel family). The information provided by the Copernicus services can be used by end users for a wide range of applications in a variety of areas, which includes risk management.

One of Lisbon Pilot goals was to use the Copernicus products to monitor critical areas in Lisbon where occurs landslides, flooding and settlements events and also to mapping update.

During the AGEO project we came across to several limitations on the use Copernicus in our Pilot. Earthquakes and tsunamis, fortunately, do not occur every day and cannot be predicted so we can only use the satellite data to evaluate damages after an event. Regarding settlements and landslides, in urban areas the movements are too slow and located so the Copernicus data scale is not compatible with the Lisbon Pilot needs. However an analysis to the application of Copernicus to manage landslides and settlement is ongoing by an IST team, so we hope that in the future we can use this data to manage this situations.

Lisbon Pilot also manage flood risk and Copernicus could give us an important improvement in flooding mapping update.

In the last years several precipitation events occurred in Lisbon city causing several million of euros in losses.

To implement efficient adaptation or mitigation measures we need to get a real picture of the areas flooded during the events, so we

Table 11.2 Normalization matrix—Elements at risk

	Themes:	**Candidate set of Indicators (name, short definition & data source/sub-indicator)**		
Elements at risk	Human	Population density	Residents	*Number of inhabitants (ex: pop. Density: % of population in high risk areas)*
			Workers	*Number of Workers (pop. Density)*
			Visitors	*Number of visitors/day (pop. Density)*
		Vulnerable population		*% of children (under 10 years old) and elderly people (over 65 years old) in high risk areas*
	Building and infrastructure	Building density		*Number of housing buildings (ex. % of building in high risk areas; Building age and corresponding building code)*
		Location of critical infrastructures		*Water, power, sanitary, telecommunication supply density (ex: % of Critical Infrastructures sited in/outside high risk areas)*
		Productive activities		*Surface area covered by businesses Productivity/benefit? (ex: % of commercial and manufacturing establishments sited in/outside high risk areas)*
		Essential services buildings		*Existence of hospitals (essential services buildings), schools, fire headquarters, emergency service or civil protection building (ex: % of essential services sited in/outside high risk areas; building age and correspondent building code; % of retrofitted hospitals, schools, fire stations, police stations,* etc.*)*
		Road network		*Road network density tunnels and overpass (ex: Level of system redundancy based on analysis of alternative routes and service lines)*
		Train railway		*Train railway network density tunnels and overpass*
		Harbours		*Presence of a harbour (trading, military port)*
		Natural and cultural heritage		*Existence of buildings or artefacts of heritage value (museums, theatres, archive, monuments,…) in high risk areas*
		Decision power		*Location of Government buildings in high risk areas*

Table 11.3 Normalization matrix—risk and disaster management

	Themes:	Candidate set of Indicators (name, short definition & data source/sub-indicator)	
Risk and disaster management	Governance, land use and urban planning	Hazard-specific resistant features and critical infrastructures	– *Housing improvement (retrofitting) and human settlement relocation from prone-areas (ex: % of retrofitted buildings)* – *% of Implementation of hazard-event control and protection techniques (PGDL)* – *Updating and enforcement of safety standards and construction codes* – *Reinforcement and retrofitting of public and private assets*
		Monitoring retaining walls collapse	*Use of satellite images (PSINsar)*
	Local strategy for urban planning	Municipality strategy for climate changes adaptation	*% of implementation*
		Involvement of stakeholders	*Number of stakeholders (national services, public institutions, associations…) contributing to local strategy for risk reduction*
		Actions with neighbour territories	*Degree of involvement in risk-related actions at larger territorial scale*
	Disaster management	Warning systems	*Implementation of warning systems*
		Resources/coherence of emergency response	– *Rescue and fireman manpower (ex: Nr of rescue and firemen manpower per 1,000 inhabitants)* – *Hospital beds (ex: Number of hospital beds per 1,000 inhabitants)*
			Endowment of equipment, tools and infrastructure
		National regulations	*National regulations—National Authority for Emergency and Civil Protection & National Commission for Civil Protection*
	Risk awareness	Schools, business and associations and parish	*Number of awareness-raising events (training) (ex: nº of students by school age per 1000 students; nº of employees per 1000)* *Education on risk (ex: nº of students by school age per 1000 students)*
	Knowledge	Scientific publications	*Number of scientific publications/population survey*
		News	*Number of news in newspapers/TV/social media*

Table 11.4 Normalization matrix—impact

	Themes:	Candidate set of Indicators (name, short definition & data source/sub-indicator)	
Impact	Costs		*Reparation costs*
			Evacuation costs
	Losses	*Population*	*Ex: number of dead people per 1,000 inhabitants; number of injured people per 1,000 inhabitants*
		Buildings	*(Ex: % damaged area—damaged area/build area)*
		Infrastructures	*(Ex: number of ruptures/km^2; % population affected; recovery time)*
		Cultural heritage	
	Economic	Employment	*Ex: % loss of jobs*
		Cultural/Leisure	*Ex: impact in number of tourists/day (%)*
		Tourist housing	*Capacity of tourist housing (number of beds)*

try to use information provided by Copernicus. In Lisbon the precipitation events are fast and intense, causing flash floods, that occur in specific parts of the city, usually in areas where the ancient water lines converge. One example of an area that is always affected by urban floods is Alcantara, closed to the Tagus River Bridge. Normally the water level reached more than 1.5 m high.

In this situation we try to use Copernicus products to update information but once again we concluded that the affected areas are small and isolated so several constraints regarding images scale were raised again. In addition, we notice that during the precipitation events the sky was cloudy, so it was impossible to use the optical images provided by Sentinel 2. After the rainfall episode and when the sky was clear, we try to use the optical images of Sentinel 2, but now, most of the water is no longer on the affected area.

The Sentinel 1 use interferometric radar imaging to monitor the Earth and could be used overcome questions regarding the cloudy weather. Although when we try to use post event information from Sentinel 1, the only flooded spots were at the airport, football fields and some others flat zones in the city. The affected areas do not appear in the data provided because they cannot be distinguished from paved roads.

11.4 Lisbon Ageo App

One of the key objectives of a Citizen Observatory is to assure citizen engagement. To accomplish that, it is necessary/fundamental to provide the means for active citizens to participate. One of the most direct ways to achieve this is to create an information channel for citizens allowing them to easily communicate events related with natural risks or geohazards, when they witness one. In the Lisbon pilot, this communication channel was materialized in a mobile application, accessible to everyone in every smartphone. Using the app citizens could report geohazards events in a straightforward and standardized manner.

Reporting geohazard events can be a tricky task. As the primary goal is to receive inputs from the biggest audience possible, the main target user is the regular citizen, that is, people with no technical expertise in the subjects at matter. On the other hand, even a non-specialized citizen, if effectively directed, could provide meaningful and crucial information about the witnessed events, which allows the first responders, or technical staff, to analyse the reported data and to take better and informed decisions, even before visiting the site.

Taking this into account, careful thought was put into the design of the geohazard reporting app, developing mock-ups of the app's user interface, and idealizing the user flow experience.

11.4.1 Designing the User Interface and Application Workflow

A set of mock-ups were developed to provide a visual overview of the expected user interface for the event report app. The prepared layouts were designed focusing on the expected user experience and with the main intent of providing an easy way to use and a visually attractive interface, with graphic buttons that facilitate and accelerate the user's perception of the required information.

The main steps of the application workflow while reporting an event are:

1. Choosing the type of event to report (Fig. 11.7).

In this step the user is presented with a grid of graphic buttons representing the different types of events.

An option of "Other events" was added, to facilitate the use by those citizens who don't know which event to choose or is not comfortable in identifying the specific event type, solely by the images.

2. Taking photos of the event and provide some additional descriptive info, if relevant.
3. Define the location and date of the event and indicate the major elements at risk (people, animals and/or assets).

For defining the location, the user can choose to use the device location info or, manually, pick the correct location in an interactive map available.

4. Reporting specific event information.

In this step, the user will be presented with different interfaces, according with the chosen event type, that will help the user provide meaningful information about the event.

Careful consideration was taken when defining the input interfaces for the user to provide information about the event, favouring self-explanatory graphic buttons that would allow the user to visually interpret what was expected of him and quickly report the information. When asking for measurement units, scales of measure with physical objects were provided so the citizen could easily visualize and identify the correct amount, even if he is not used to the standardized unit measuring systems. For instance, when asking the user to indicate the water level verified in a flooding event, a graphic scale using human figures was used, as shown in the image bellow (Fig. 11.8).

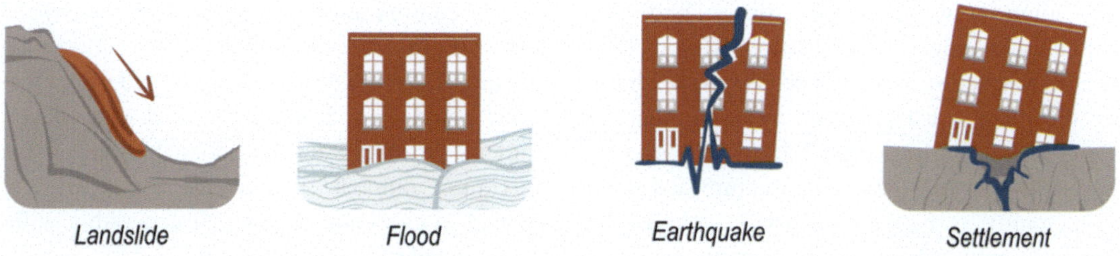

Fig. 11.7 Graphic buttons idealized in the designing phase to represent the different event types

Fig. 11.8 Examples of the interfaces designed for step 3 and step 4 for two event types: landslides and flood

In complement to reporting events, the citizen can check all the reported events through the app. Using a map-based interface, the user can download to its device the information related with the events reported in each location. Filtering or ordering the reported events by type of event or occurrence date is also available.

11.4.2 Implementing the Geohazards Report App

It was crucial, for the Lisbon pilot, to have an implementation of the idealized geohazards reporting app.

Since a quick, easy to implement and affordable solution was necessary, the technologies already available in the Lisbon Municipality data infrastructure were pondered, while choosing the most feasible option to implement the geohazards reporting app. Taking into account that all the spatial data infrastructure at Lisbon Municipality is based on ESRI's technologies and since enterprise licensing agreement (ELA) was already available for most of the ArcGIS platform solutions, it was a natural step to investigate, within the offered solutions in the ArcGIS platform, the one that best covered our requisites.

After careful analysis, we opted by the ArcGIS Survey123 solution, which, as stated in the product's webpage, "is a simple and intuitive form-centric data gathering solution" [1]. ArcGIS Survey123 provides easy and straightforward GUIs for creating survey forms associated with spatial data, as well as a simple way to make advanced configurations of the forms through the XLSForms [2] standard (based on an Excel datasheet template). It also provides a mobile app, accessible in Android, IOS and Windows versions, through which one can download and use the ArcGIS Survey123 forms.

The configuration of the ArcGIS Survey123 form for the intended geohazards report app thoroughly followed the designs and specifications established during the mock-ups' development. To facilitate the contribution of the maximum number of citizens, the form was opened to everyone, keeping it accessible without the need of any registration.

With the intent of disseminating the initiative and facilitating the implementation of similar

Fig. 11.9 Dashboard for reports visualization and global analysis

projects within other municipalities, regions or countries, the form template was published in an open code repository on GitHub (https://github.com/carlosmsf/geohazards_report), making it possible for other entities to use and/or further develop the designed form.

Detailed information on how to download and use the app can be found on: https://informacoeseservicos.lisboa.pt/prevencao/resiliencia-urbana/projetos/ageo#c24626.

11.4.3 Dealing with the Reported Data

While considering individually each one of the reported events, it can be hard to envision if the driving forces behind it are of geological nature or if they are related to some other site-specific circumstances. But when the reports are analysed altogether, evidence of more complex and systemic "problems" can emerge. For instance, if several houses in the same neighbourhood report the appearance of cracks in the building structure, subsidence or slope movement phenomena can be in place and thorough geological analysis of the area should be undertaken.

This ability to globally analyse the reports of several occurrences spatially related is one of great advantages behind multihazards citizens' observatories. Despite that, if a concerned citizen reports a hazard of any kind, local governments and authorities must assure that no immediate security concerns are in place and must take measures to correct any reported damages (Fig. 11.9).

To accomplish this, it is fundamental to guarantee that the reported information reaches the local authorities responsible for the first response operations, in case of an emergency and/or the services responsible for the maintenance of public space. With this objective in mind, Lisbon municipality is implementing an automatic workflow that forwards the information reported by the citizens to the adequate entities according with the circumstances reported through the geohazards report app.

The diagram in Fig. 11.10 summarizes the idealized data flow.

1. The Survey123 webhook service sends pushup notifications about the reported events whenever an event report is submitted.
2. An integration platform, according to the defined forwarding rules, sends the info about the reported event to the event management systems of the adequate local entities.

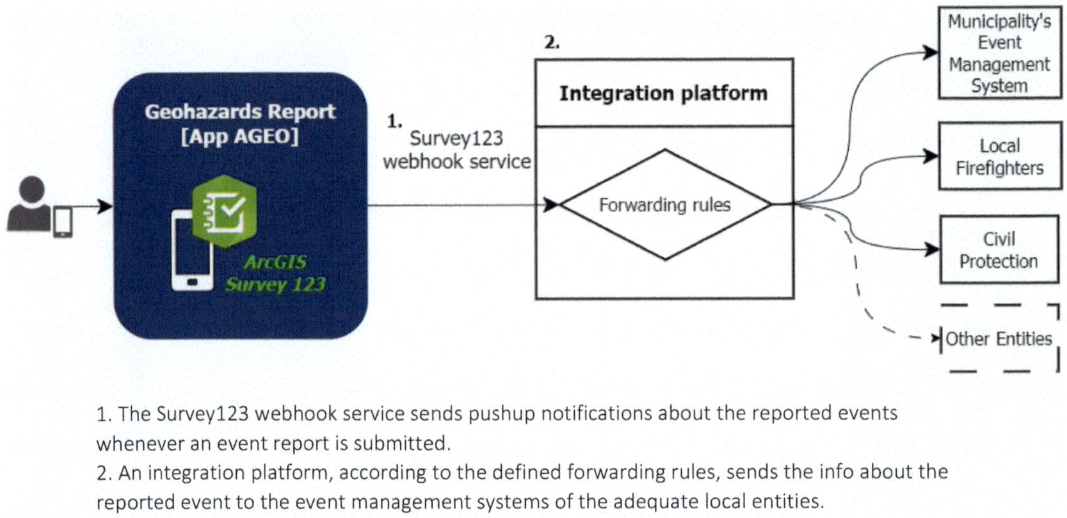

Fig. 11.10 Data flow of the reported event information

11.4.4 Final Considerations

As part of the Lisbon Multihazards Citizen's Observatory, which is integrated in the AGEO Project—Platform for Atlantic Geohazard Risk Management, Lisbon Municipality developed a communication strategy based on a mascot called RISKA. Educational contents were also developed aiming to raise the population's awareness to geological risks and allowing the implementation of preventive measures.

To receive information by citizens a platform was developed. The platform's interface aimed to simplify the report using simple language and based on graphic content, to enable ordinary citizens to give information about situations related with geohazards in Lisbon.

The simplicity of the app and the visual presentation of the screens had an excellent receptivity by the test public.

This solution will allow citizens to play an active role in the risk management system and will facilitate the monitoring of situations in which Lisbon Municipality technicians have increased difficulties due to lack of access. Ultimately, this tool will allow Lisbon Municipality to anticipate problems that may arise in some parts of the city, such as settlements in buildings, enabling a proactive response and not a reactive one, as it happens today.

The implemented proof of concept allowed to demonstrate that a citizen properly aware of a certain problem can contribute to a more efficient risk management in the city of Lisbon.

Acknowledgements The authors would like to thank to all the municipality technicians that supported the tasks developed, as well as the stakeholders who participated in the several dissemination actions and who introduced the AGEO app in their daily work. Mónica Amaral Ferreira is grateful for the Foundation for Science and Technology's support through funding UIDB/04625/2020 from the research unit CERIS.

This work has been undertaken in the framework of the INTERREG Atlantic Area AGEO Project (EAPA_884/2018—AGEO).

References

Barbat, A. H., Carreño, M. L., Pujades, L. G., Lantada, N., Cardona, O. D., & Marulanda, M. C. (2010). Seismic vulnerability and risk evaluation methods for urban areas. A review with application to a pilot area. *Structure and Infrastructure Engineering, 6*(1–2), 17–38. https://doi.org/10.1080/15732470802663763.

Cardona, O. D. (2006). A system of indicators for disaster risk management in the Americas. In Birkmann, J. Measuring Vulnerability to Natural Hazards: Towards

Disaster Resilient Societies (pp. 189–209). United Nations University, New York.

Carreño, M. L., & Barbat, A. H. (2007). Urban seismic risk evaluation: A holistic approach. *Natural Hazards: Journal of the International Society for the Prevention and Mitigation of Natural Hazards, Springer, 40*(1), 137–172.

Ferreira, M. A., Mota de Sá, F., & Oliveira, C. S. (2014). Disruption Index, DI: An approach for assessing seismic risk in urban systems (theoretical aspects). *Bulletin of Earthquake Engineering, 12*, 1431–1458. https://doi.org/10.1007/s10518-013-9578-5

Ferreira, M. A., Pacheco, P., Pinto, C., Ferreira, C. S., Oliveira, C. S., & Gomes, R. C. (2022). Flood risk assessment in Lisbon: for planning purposes and citizen involvement. Proceedings 2nd International Conference on Urban Risks (ICUR2022), pp. 23–25 June, Lisbon, Portugal.

Ferri, M., Wehn, U., See, L., Monego, M., & Fritz, S. (2020). The value of citizen science for flood risk reduction: Cost-benefit analysis of a citizen observatory in the Brenta-Bacchiglione catchment. *Hydrology and Earth System Sciences, 24*(12), 5781–5798. https://doi.org/10.5194/hess-24-5781-2020

Guillard-Gonçalves, C., Cutter, S. L., Emrich, C. T., & Zêzere, J. L. (2015). Application of Social Vulnerability Index (SoVI) and delineation of natural risk zones in Greater Lisbon. *Portugal. Journal of Risk Research, 18*(5), 651–674. https://doi.org/10.1080/13669877.2014.910689

Pereira, S., Santos, P. P., Zêzere, J. L., Tavares, A. O., Garcia, R. A. C., & Oliveira, S. C. (2020). A landslide risk index for municipal land use planning in Portugal. *Science of the Total Environment, 735*, 139463. https://doi.org/10.1016/j.scitotenv.2020.139463

Pinto, C. (2020). Modelação de dados geológicos e geotécnicos visando a produção de cartografia geotécnica em áreas urbanas—Aplicação a Lisboa. Dissertação apresentada à Faculdade de Ciências da Universidade de Lisboa para obtenção do grau de Doutor em Geologia, especialidade Geotecnia, p. 454.

UNDRR and WMO. (2022). Global status of multi-hazard early warning systems: Target G. United Nations Office for Disaster Risk Reduction.

[1]—ESRI—ArcGIS Survey123—https://survey123.arcgis.com/.

[2] XLSForm Docs—https://xlsform.org/.

12 Causeway Coast, Northern Ireland: Developing a Geohazard Citizen Observatory to Assist Risk Management

Kieran Parker, Kirstin Lemon and Rui Carrilho Gomes

Abstract

The Causeway Coast is located on the northern coast of Northern Ireland. A rugged coastline, it is a area of high tourism due to its natural beauty and an area of significance in Irish mythology. Two sites on the Causeway Coast were chosen to develop a geohazard citizen pilot observatory, the world-renowned Giants Causeway and Carrick-a-rede. Both sites attract a high number for visitors throughout the year and both are also managed by the National Trust. The Causeway Coast is geologically dominated by Paleogene mafic volcanics and is highly susceptible to frequent landslide occurrences. It is also an area with a high number of international, European and national designations due to its geological importance and biodiversity. The objective of the pilot was to (i) carry out routine monitoring of the sites, (ii) engage local communities and stakeholders to provide them with education awareness of geohazards, (iii) encourage their active participation in hazard monitoring through a mobile application developed by the AGEO project, (iv) evaluate satellite data from the Copernicus programme as a geohazard monitoring tool and (v) promote better understanding of geohazard risk to stakeholders to enhance risk management at the sites using evidence based information. To meet these objectives, we initiated research and monitoring to better understand and characterise the risk through carrying out geological site surveys, analysis of available datasets, completing Laser imaging, Detection And Ranging (LiDAR) and photogrammetry surveys, assessed satellite derived data and incorporation of citizen monitoring data. The pilot also initiated a series of outreach events within the local communities to raise awareness of geohazard and the participatory role citizens could take in monitoring their landscape. The work carried out resulted in the recording of numerous landslide events including images of each event leading to an increased understanding of the areas and exposure of risk across the sites. A common reference dataset of the geomorphology, to millimetre scale, was attained providing a future reference for change detection analysis. Citizen involvement was also achieved through local engagement and collection and evaluation of citizen data as a means of geohazard monitoring to enhance risk management of the sites.

K. Parker (✉) · K. Lemon
Geological Survey of Northern Ireland (GSNI), British Geological Survey (BGS), Belfast, UK
e-mail: Kiepar@bgs.ac.uk

R. C. Gomes
CERIS, Instituto Superior Técnico, Universidade de Lisboa, Lisboa, Portugal

R. C. Gomes et al. (eds.), *Citizens' Observatories on Geohazards*, Geoenvironmental Disaster Reduction,
https://doi.org/10.1007/978-3-031-53371-6_12

12.1 Introduction

The Giant's Causeway and Causeway Coast World Heritage Site (WHS) is located just outside Bushmills, on the north coast of County Antrim in Northern Ireland. Known locally as the Giant's Causeway, the site was inscribed on the World Heritage List in in 1986 due to its geological importance, but it is also a site of cultural, historical and touristic significance. Adjacent to the WHS, close to the village of Ballintoy, is Carrick-a-rede, famous for its rope bridge that connects the mainland to a small island, once used by fishermen but now used for tourists. Together, these sites make up the Causeway Coast citizen observatory with both being major tourist destinations which, prior to the Covid-19 pandemic, attracted a combined 1.5 million visitors each year (NISRA, 2020). The visitor numbers are derived from the paid attractions such as the visitor centre at the Giants Causeway and the Carrick-a-rede rope bridge. It does not include local and other recreational users to the Causeway Coast, which when taken into account, significantly increase the footfall at each location. Both sites are affected by slope instability that has resulted in site closures on a number of occasions due to the risk to the public. The Giant's Causeway and Carrick-a-rede sites were chosen as pilot citizen observatories due to their geological, biodiversity, cultural and economic significance to Northern Ireland as well as the geohazard risk to the site users. Both sites also have the additional benefit of attracting large number of visitors making it possible to communicate geohazard information and the potential for visitors to engage with monitoring of the landscape to assist in site management.

The landscape of the Causeway Coast citizen observatory is composed of rugged cliffs and secluded bays that display limestone overlain by a series of basalt lava flows. The cliffs are subject to regular mass movements with rockfalls very common across the Northern coastline. Such mass movements at the Giant's Causeway resulted in the permanent closure in 1994 of a section of the cliff path that runs mid-slope at Port Reostan. Subsequent monitoring and assessment of the site established that the ongoing risk of slope instability exists across the majority of the Giant's Causeway (McDonnell & Smith, 2000). It has also been identified that the frequency, distribution, and intensity of slope movements is increasing, perhaps as a response to climate change through increased rainfall and storm intensity (Smith et al., 2011). Following the closure of the public path, an exercise was undertaken to characterise the types of slope failures and map their occurrence across the site (www.qub.ac.uk/geomaterials). This detailed for the first time the scale of slope failures and complexity of the processes (Fig. 12.1).

The Causeway Coast citizen observatory has a dynamic landscape (Fig. 12.2) that has attracted visitors for centuries. Tourism at the Giant's Causeway dates back over 300 years with visitors attracted by the rugged cliffs and approximately 40,000 regular jointed basalt columns. The area was popularised by poets and artists who produced works inspired by its dramatic landscape. Tourism increasingly grew in the nineteenth century with the first tourist facilities established in the mid-1800's. Today, the Giant's Causeway is Northern Ireland's top tourist destination, attracting up to 1 million visitors each year.

Carrick-a-rede is a small rocky island located off the County Antrim coast, approximately 14 km east of the Giant's Causeway. The island is connected to the headland by a narrow-suspended rope bridge spanning 20 m across a chasm and 30 m above the Atlantic Ocean (Fig. 12.2). First erected in 1753 by salmon fishermen to access the island and its surrounding productive fishing grounds (National Trust, 2023), the rope bridge today is a major attraction for visiting tourists, with up to 500,000 visitors each year. Areas of both the headland and island have experienced slope failures with a small landslide in 2019 resulting in the closure of the rope bridge.

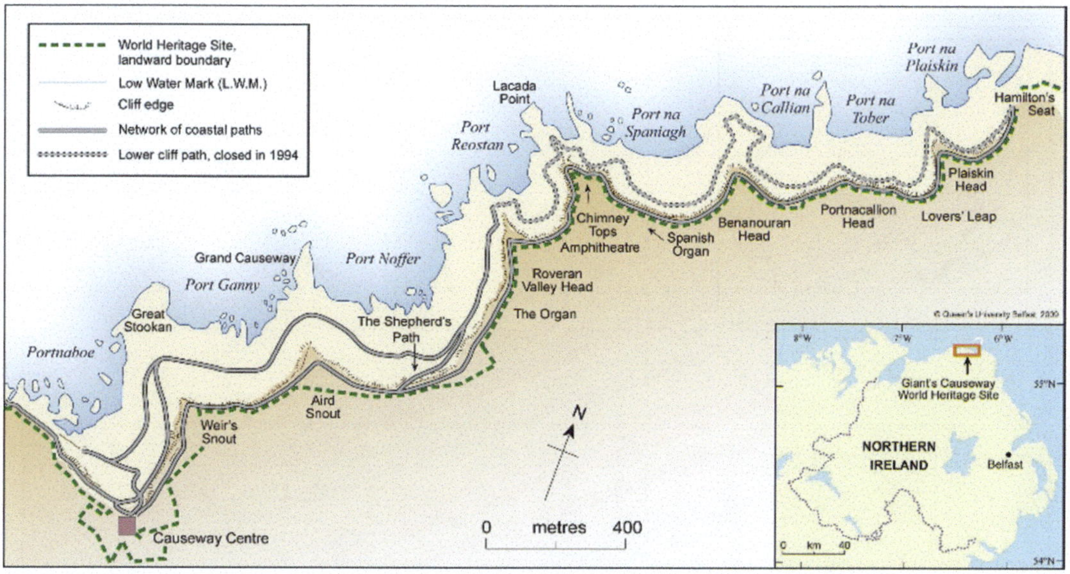

Fig. 12.1 Map showing the Giants Causeway bays and public paths (http://www.qub.ac.uk/geomaterials/)

Fig. 12.2 Aerial photograph of Giants Causeway (right) looking east across Port Granny and Port Noffer. (photo courtesy of SNAPMAP, Northern Ireland Environment Agency). (Left), Aerial view, looking north-west of Carrick-a-rede rope bridge spanning from the headland to the island *(Photo:Tourism NI Media Library,* 2023)

The geological hazards at each site are well known with signs of increasing frequency and impact. Due to the high number of visitors to the area, managing this risk is been becoming increasingly difficult. In 2019, the AGEO project identified the area as ideal to set up a citizen observatory for geohazard assessment, citizen participation and to assist in risk management.

12.2 Site Description

12.2.1 Giant's Causeway

The Giant's Causeway and Causeway Coast is one of only two WHSs in the UK and Ireland that have been inscribed purely because of their geological heritage (Fig. 12.3). The WHS was inscribed under Criterion (vii) due to the cliff

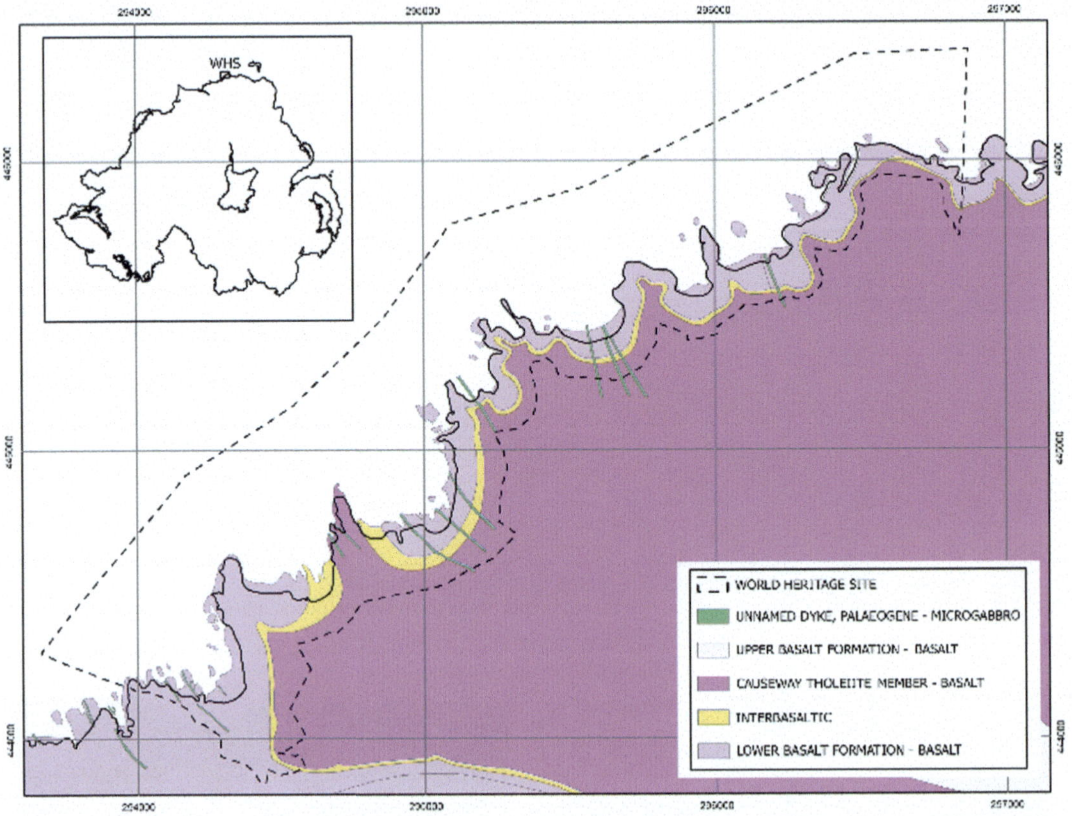

Fig. 12.3 Map showing the area of the Causeway Coast covered by the UNESCO World Heritage Site (WHS) and the geological map

exposures of columnar and massive basalt at the edge of the Antrim Plateau and under Criterion (viii) due to the evidence provided to demonstrate geological activity of the Palaeogene era that allows for a detailed analysis of the events associated with the opening of the North Atlantic (UNESCO, 2023). The Giant's Causeway has a huge cultural significance in relation to Irish mythology, as well as playing a key role in the development of geological science. The site is also home to a diverse range of plant and animal life, including many rare and endangered species. In total, it covers an area of 71 ha (0.71 km^2) along a coastal strip approximately 5 km in length from Causeway Head and Benbane Head (Crawford & Black, 2010).

12.2.2 Geology

The Giant's Causeway and Causeway Coast WHS is composed of layers of basalt that are part of the Antrim Lava Group (ALG) which overlie a compact limestone that forms the Ulster White Limestone Formation (UWLF). The basalt sequences formed as a result of extensive volcanism associated with the opening of the North Atlantic Ocean during the Paleogene period around 60 million years ago. Volcanic activity continued for c.4ma in three phases separated by two periods of inactivity (Mitchell, 2004). The initial lava flows created the Lower Basalt Formation (LBF) which was followed by a hiatus. During this time, the uppermost flows of the LBF underwent weathering in, what was then, a wet tropical environment, resulting in sequences of laterite, lithomarge and bauxite seams known as the Interbasaltic Formation (IBF). Following this, lava extrusion resumed with the flows draining into valleys up to 90 m deep, where they ponded and cooled slowly and evenly to develop the regular columnar-jointed patterns (Lemon, 2012) forming the Causeway Tholeiite Member (CTM). Recent research indicates that subsidence associated with magma draining led to the formation of a depression creating the deep valleys (Simms, 2021). Another lull in volcanic activity was followed by further lava extrusions forming the Upper Basalt Formation (UBF).

At the Giant's Causeway a total of approximately 40,000 basalt columns exists. The columns range in height from 1 to 12 m and have been eroded by sea, wind and rain over time, resulting in a variety of shapes and sizes (Lyle, 1996).

Today the Giant's Causeway is characterised by the LBF, weathered horizons of the IBF and CTM visible along with high angled scree slopes and a raised beach on its western side.

12.2.3 Carrick-A-Rede

The island of Carrick-a-Rede and the adjoining mainland represent a section through an explosive volcano forming during the earliest stages of Palaeogene volcanic activity, immediately prior to the formation of the rocks at the Giant's Causeway.

The narrow chasm over which the rope bridge spans is the remains of a volcanic vent, the composition of which can be seen on either side of the bridge. The vent is composed of agglomerate consisting of blocks of basalt within a matrix of volcanic ash and occasional fragments of limestone. This material is preferentially weathered, with the island being made up of dolerite, and the cliffs of the mainland composed of limestone and overlying basalt, all of which is more competent and resistant to weathering and erosion than the vent agglomerate.

The basaltic rocks at Carrick-a-rede and Giants Causeway are part of the Antrim Plateau, a geological formation that stretches across the north-eastern part of Northern Ireland. The rocks were formed by the solidification of lava flows that were extruded between 60-50ma (Mitchell, 2004) (Fig. 12.4).

12.2.4 Geohazards on Site

The geohazards that exist within the Causeway Coast citizen observatory are due to a long series of natural and human interactions and adjustments (Smith et al., 2011) all of which have created the spectacular landscape of coastal cliffs and relict mass movements. These include:

- Major episodes of early Paleogene lava flows separated by lulls in volcanism resulting in some layers of weathered weak lateritic palaeosoil. Thes lava flows are further cut with later dyke intrusions. The area is also impacted by tectonic events and later volcanic extrusions.
- Exposure to environmental conditions, including humid tropical-like conditions during the early stages of volcanism, exposure to high-energy coastal marine conditions leading to undercutting of the slopes leading to slope failure, a land carved by the advance and retreat of ice sheets and the readjustment of the land from post-glacial rebound after the last ice retreat 13 k years ago (Mitchell, 2004).
- Human intervention through the extraction of stone, recreational and commercial use, transport infrastructure and building and footpath construction.

This has led to the development of a steep coastline, relict deep-seated rotational block failures and ongoing mass movement across the site. The area undergoes a series of complex landslide events that occur across the site of the

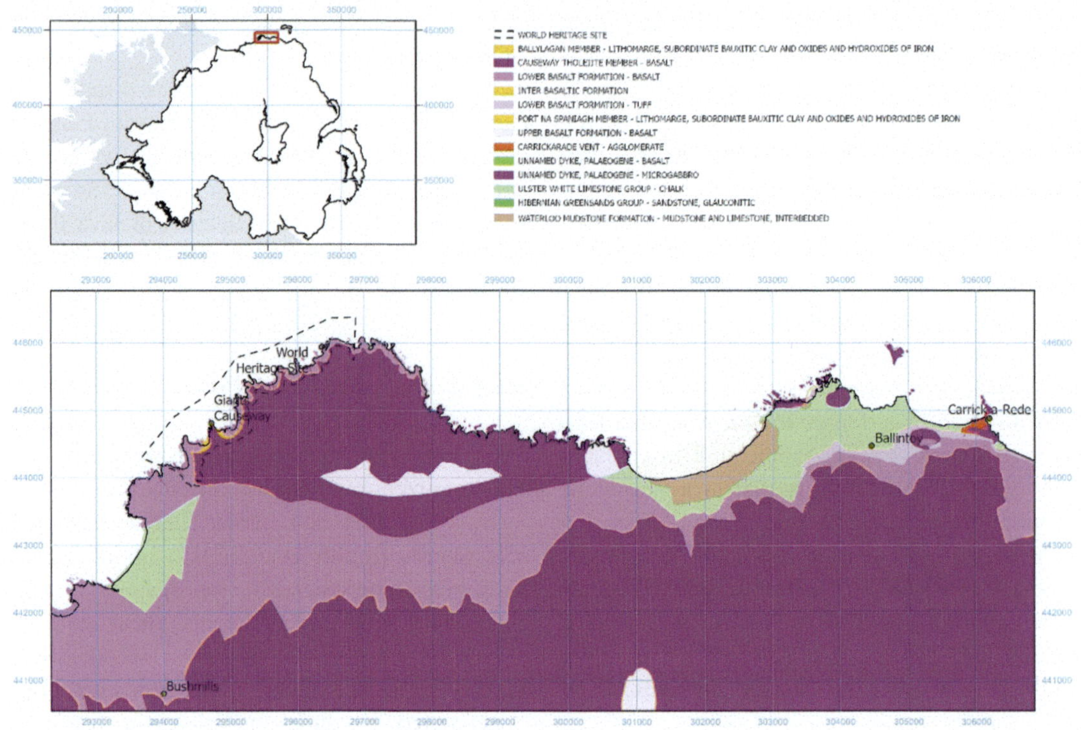

Fig. 12.4 Geological map of the causeway coast

Giants Causeway. Rainfall is the primary trigger for landslides at both the Giants Causeway and Carrick-a-rede sites.

12.2.5 Site Management and Designations

There are six bodies with legal rights to the Giant's Causeway and Causeway Coast WHS: The Crown Estate; the National Trust; Causeway Coast and Glens Borough Council and; three private landowners (CCGHT, 2021). Carrick-a-rede is under the sole ownership of the National Trust.

Despite the complicated legal rights to the WHS, the National Trust operates day- to -day management activities for both sites, and they are regarded as the Property Manager. As the largest conservation charity in Europe, the National Trust manages over 2,500 km^2 of farmland, over 1000 kms of coastline, and over 500 historic properties, gardens and nature reserves, all across England, Wales and Northern Ireland.

In addition, the WHS has a Steering Group, made up of a number of stakeholders with responsibility for the World Heritage Management and Action Plan. These stakeholders include: NIEA, National Trust, Causeway Coast and Glens Borough Council, the Crown Estate, CCGHT, Tourism Northern Ireland, Council for Nature Conservation and Countryside, the Geological Survey of Northern Ireland (GSNI), and Queen's University, Belfast.

The National Trust manages the outstanding universal value (OUV) of the WHS and the various designations and protections assigned by the relevant agencies for both sites (see below for further details). These responsibilities are mirrored by its own organisational strategy, 'Playing our Part' and its focus on landscape protection, welcoming everyone and providing opportunities for people to get involved. The National Trust usually operate a three-year

Property Business Plan which directs and details management activities at both sites. This allows for emerging challenges or opportunities to be addressed (CCGHT, 2021).

The geological and cultural significance of the Giants Causeway has resulted in a number of national and international designations based on its geology, landscape, geomorphology, habitats and biodiversity. Such designations bring with them additional responsibilities for site management.

12.2.5.1 International Designations

As mentioned previously, the Giant's Causeway and Causeway Coast was inscribed on the WHS in 1986. WHSs are designated for being of outstanding universal value (OUV) to humanity and have been placed on the World Heritage List to be protected for future generations to appreciate and enjoy (UNESCO, 2023). World Heritage status is not a statutory designation however, Planning Policy Statements (PPSs) for Northern Ireland set out policies on land-use and other planning matters, two of which specifically refer to World Heritage (Lemon, 2015).

12.2.5.2 European Designations

Almost all of the Giant's Causeway and Causeway Coast WHS has been designated as the North Antrim Coast Special Area of Conservation (SAC) and the marine portion of the site falls within the Skerries and Causeway marine SAC. These areas are protected under the Conservation (Natural Habitats, etc.) Regulations (Northern Ireland) 1995 (as amended) for their biodiversity value, having been transposed and implemented as part of the EU Habitats Directive. are designated to safeguard habitats and species which are rare or threatened in a European context (Lemon, 2015).

12.2.5.3 National Designations

i. Areas of Outstanding Natural Beauty

Both the Giant's Causeway and Causeway Coast WHS and Carrick-a-rede fall within the Causeway Coast Area of Outstanding Natural Beauty (AONB). This designation gives formal recognition to distinct landscapes of high scenic value in Northern Ireland. AONBs are a statutory designation that makes provision for the conservation and enhancement of the area's natural beauty together with increasing awareness and public enjoyment. This is usually achieved through the development and delivery of AONB management plans by non-governmental AONB Partnerships. In the case of the Causeway Coast citizen observatory, this is delivered by the Causeway Coast and Glens Heritage Trust (CCGHT).

ii. Areas of Special Scientific Interest (ASSIs)

ASSIs are designated through legislation contained within the Environment (Northern Ireland) Order 2002. ASSIs are areas of land that have been identified by scientific survey as being of the greatest importance for their conservation value for either biological or Earth science interest. The Causeway Coast citizen observatory falls within the Dunseverick and Causeway Coast ASSI and the Carrickarede ASSI, primarily for their geological value.

iii. National Nature Reserves (NNR)

NNRs are protected under the Nature Conservation and Amenity Lands (Northern Ireland) Order (1985) in recognition of their wildlife, habitats and geology. The Giant's Causeway is a NNR, and the primary aim of this designation is to manage the site is to conserve its features through the development of experience and techniques in managing land for conservation. However, they have a secondary aim of providing opportunities for the public to experience these natural habitats and landscapes (Lemon, 2015).

iv. Other Designations

Part of the marine element of the Giant's Causeway and Causeway Coast WHS is protected by the Protection of Wrecks Act (1973) due to the location of La Girona, a warship of the Spanish Armada which off the coast in 1588. This prevents any access to the site without a licence.

12.3 Objectives of the Pilot

The Causeway Coast pilot focused on two sites. The Carrick-a-rede area that has experienced recent hazards due to landslip and the much larger site of the Giant's Causeway which undergoes regular numerous complex landslides. The Causeway Coast pilot has 5 key objectives.

- Initiate routine monitoring to increase understanding of the geohazard processes at each site, record the dynamic changes at the sites to better characterise both natural and human induced geological hazards.
- Engage with local communities and key stakeholders to deliver educational resources on geological hazards that impact the region.
- Engage with local communities and stakeholders to actively participate in hazard monitoring and risk preparedness.
- Evaluate satellite data from the Copernicus programme for its usefulness as a site-specific monitoring tool. Encourage regional level uptake and use of Copernicus products and services.
- Promote better understanding of geohazard risk so that regional authorities and stakeholders are better prepared to manage the risk and adapt using evidence-based information gained through more focused monitoring.

12.4 Data

Throughout the project a number of activities were undertaken to acquire, review and analyse data to gain a better understanding of the processes on site, assess monitoring tools including citizen monitoring and analyse how this can benefit risk management. The methodology included:

- Literature review.
- Walk-over surveys of the sites.
- Analysis of current and historic datasets.
- Acquisition of data using Laser imaging, Detection And Ranging (LiDAR) scanning and photogrammetry.
- Assessment of satellite derived data from the Copernicus Programme.
- Incorporation of citizen monitoring data.

12.4.1 Historic Literature/Research

A review of previous research carried out at each site was carried out. The Giant's Causeway has been the focus of numerous research projects over the years from the early geological interpretation to more recent investigations on the landslide occurrences (Smith et al., 2011), risk management (Bratton et al., 2013) and recent evidence to suggest subsidence events, pre-lava flow, are the origin of valley formation leading to thick accumulations of basalt and columnar formation (Simms,2021). Research carried out by Queens University Belfast, as part of the Geomaterial project identified and mapped 11 different categories of slope failure on the site (Fig. 12.5). This included rockfalls, toppling failure, translational slides, rotational slides, complex landslides, mudflows, debris flows, slumping, creep, scree and marine undercutting. The information gained highlighted the complex nature of slope instability at the site and identified areas where slope movement could have catastrophic loss of access to the site. The main paths and road, which are located at the base of the slope, winds its way up upslope at Port Noffer are very susceptible to being impacted by instability therefore placing site users at great risk.

The Carrick-a-rede site is one that has been under-researched in terms of geological hazards,

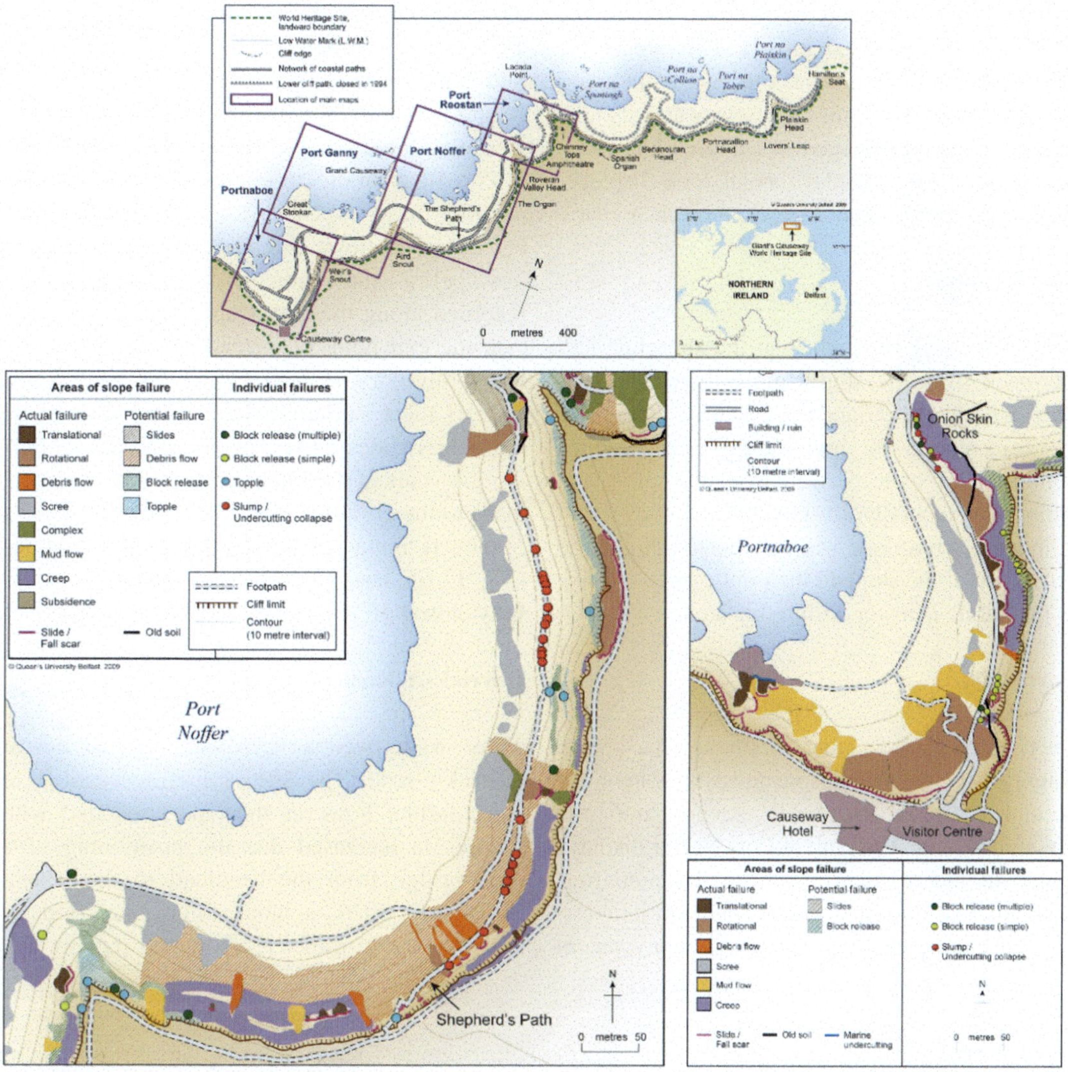

Fig. 12.5 Slope failure characterisation map of Port Noffer and Portnaboe spatially showing the various types of failure (http://www.qub.ac.uk/geomaterials/)

partly due to the lower occurrence of landslip, its smaller size and perceived lesser susceptibility of being impacted by slope failure.

12.4.2 Walk-Over Surveys

Six site walk-over surveys were carried out at the Giants Causeway to verify mapping, identify relict landslides and assess landslides that occurred during the project duration. Walking from the Giant's Causeway Centre the path goes through Portnaboe, Port Granny, Port Noffer and terminates at Port Reostan in the east. Three walk over surveys were carried out at Carrick-a-rede.

Portnaboe

The path runs from the visitor centre downhill at the base of the slope. The area sits on a dormant mudflow complex and raised beach. Descending the path, there is evidence of slumping occurring in the overlying slope as well as rockfalls. The location of the path, directly beneath large unstable cliff overhangs in places, exposes the path to being impacted greatly. This area is of strategic importance for the site as it is the main access point to the Giant's Causeway and also the route for emergency access. The area known as the 'Onion Skins' displays small rock displacements from its weathered structure on each visit.

Port Granny

At Port Granny the path is at the shoreline and sits back from the slopes more than any other area within the Giant's Causeway. The area is characterised by high angled steep slopes composed of scree derived from the cliff face. Numerous rockfalls are evident and movement within the scree slope was also noted. Due to the proximity of the slope to the public path, the risk to the public is lower within this area.

Port Noffer

This is the area that contains the Main Causeway and the most visited spot where the columnar basalt causeway extends into the Atlantic Ocean. The path at Port Noffer displays abundant indent markings from the numerous rockfalls that impact it from above. On its western side, the steep slope displays many recent debris slides, mudflows, slumps and rockfalls. The path extends upslope where it diverges mid-slope towards Port Reostan to the east and steeply towards the crest of the cliff in a south westerly direction. The eastern path terminates at Port Reostan due to a large landslide in 1994 resulting in its closure. Along this path, a number of debris flows and slumping was recorded just below the path. Six 'outbursts' were recorded along the path (Fig. 12.6). These indicate the exit point of water flow that use pathways through the basalt fractures and joints. Two of these outbursts were recorded as causing debris slides on two or more occasions during the visits at the same location. The path that leads to the cliff crest, known as the 'Shepherd's Steps', is impacted by numerous debris slides (Fig. 12.6). The path that runs upslope is regularly closed off following landslide events.

Carrick-A-Rede

Three walk over surveys were carried out at Carrick-a-rede. The site is less susceptible to slope failure however the risk associated with failures is increased due the positioning of a rope bridge from the headland to the island. In January 2020, a block failure occurred on the island side directly below the rope bridge

Fig. 12.6 Photogarphs taken of landslides during walkover surveys.(Left to right), An 'outburst' causing a small debris slide on the Path leading to Port Reostan, large debris slides on the slope of Port Noffer and a mud flow within Portnaboe

Fig. 12.7 Looking from the headland towards Carrick-a-rede island, a photograph taken of the block failure that occurred in January 2020 directly below the rope bridge—Failure plane marked in red

causing the site to close for a period of time (Fig. 12.7). On each visit, only minor erosion of the rockfaces was noted.

12.4.3 Current and Historic Datasets

Since 2009, the National Trust have been recording all landslide events that occur on site. This has resulted the recording of 402 events from April 2009-August 2020 excluding the 2010 calendar year where data is not available. On average, a landslide event was recorded every 9.5 days during this period. Throughout the duration of the AGEO project, the National Trust continued to record events on their own systems and through the AGEO mobile application when it became operational in September 2022. GSNI also hold landslide information collated from historic events from past research and media reports. Combined with the National Trust datasets, it is apparent that landslide occurrences are becoming more frequent, as predicted by Smith, 2011. This may in part due to climate factors including prolonged wet and dry periods, higher intensity rainfall and storms. Better awareness and recording of events may also be a factor.

12.4.4 LiDAR Scanning & Photogrammetry

A digital documentation of the slope and cliff features was undertaken to create a common reference system of the morphological features to assist change detection in future surveys. This was carried out using LiDAR and photogrammetry.

A LiDAR scan was carried out in February 2022 and covered the area from Portnaboe to Port Reostan. LiDAR measurements determine the range of objects with a laser and through measurements of the time of the reflected light to return to the receiver it calculates the distance of the objects to create a high resolution 3D map. This provides a high-resolution record

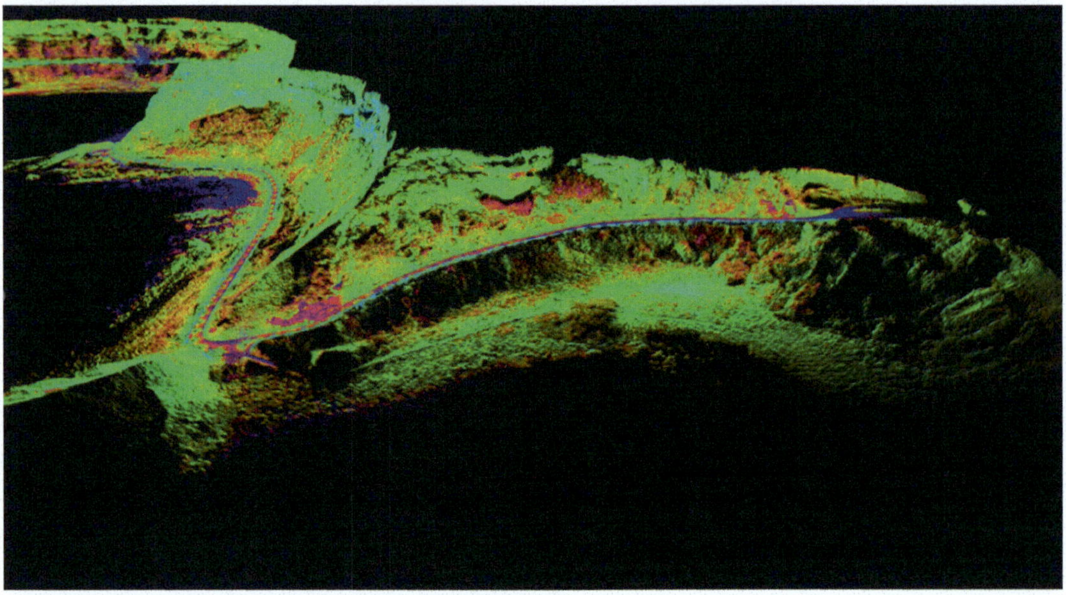

Fig. 12.8 A visual output of the LiDAR scan carried out in February 2022

of the morphological features at the time of the scan. The survey was carried out using a terrestrial Faro Focus S120 3D laser scanner. The instrument has a range of up to 120 m, ranging error ±2 mm and measuring rate of up to 976,000 points/second. The survey was carried out from the accessible public path that runs at the base of the cliffs and transverses upslope at Port Noffer. Each survey point was referenced using GNSS to create a 3D georeferenced point-cloud dataset (Fig. 12.8).

In November 2021, a study was carried out at the Giant's Causeway to assess photogrammetric reconstructions developed using participatory citizen science using Structure-from-Motion (StM) photogrammetry approach. The objective was to evaluate the quality of integrating citizen-sourced photos. As a baseline, an 'expert reference dataset' was developed using an RTK-GNSS assisted SfM photogrammetry using a Nikon D800 reflex camera. A total of 544 photos were collected including 204 georeferenced photos from eight GNSS-RTK positioned stations using a GNSS Leica 1200 + antenna (Fig. 12.9) (Jaud et al., 2021).

To assess citizen-sourced photography, seven different smartphones were used with variable focal length and image size. A total of 700 + photos were acquired of varying quality with 684 of sufficient quality to incorporate into the reconstruction. The photo dataset included georeferenced and non-georeferenced photographs. A series of six tests were carried out to reconstruct the slope using the 'expert reference dataset' and citizen-sourced photos alone, with other tests incorporating a mix of the two. The results indicated that citizen-sourced photos improved the quality of the reconstruction which in turn, provides greater confidence in using citizen-sourced photographic images as a beneficial tool to integrate as part of slope monitoring and assist in risk management. Further reconstructions will also enable the evaluation, over time, of erosion rates, indicative areas of hazards and record changes in the morphology (Fig. 12.10).

12.4.5 Copernicus Products and Services

A number of datasets acquired through the Copernicus Global Land Services was assessed to classify change within timeframes ranging

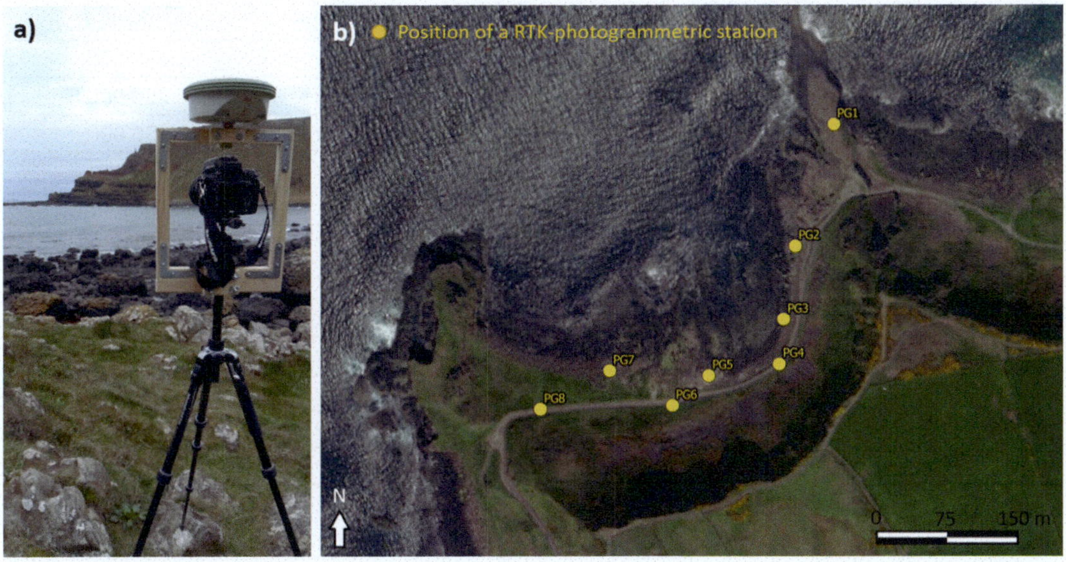

Fig. 12.9 **a** Measurement system developed for the Real Time Kinematic (RTK) Global Navigation Satellite System (GNSS)-assisted terrestrial Structure-from-Motion (SfM) photogrammetry method (equipped here with the Nikon D800 Reflex camera and the Leica 1200+GNSS antenna). **b** Positions of the measured camera stations for the RTK GNSS-assisted photogrammetric survey. *Source* Jaud et al. (2021)

Fig. 12.10 Acquiring the 'expert reference dataset and acquisitions from different types of mobile phones taken on 22nd November 2021. View from Main Causeway looking towards Port Noffer (left). Output of the photogrammetric model at a small section known as 'The Onion Skins' (right)

from 2006 to 2018 (Fig. 12.11). The data was mosaiced to a new raster and cropped by the pilot site boundary. The datasets provided a regional picture of land changes however at a resolution of 20 m, the data proved to be of limited benefit as a monitoring tool for the pilot sites. While satellite data derived from Copernicus programme may complement monitoring over large geographic areas, for more localised small-scale sites, terrestrial and aerial monitoring techniques remain the most beneficial monitoring tools.

Further to data from the Global Land Services, Interferometry of Synthetic Aperture Radar (InSAR) data was assessed to measure ground surface motion at the pilot sites. The data was based on multi-temporal processing of C-Band SAR images acquired between 2014–2022 by Sentinel-1 satellites from the Copernicus programme. Historic InSAR data

Name	Version	Data Type	Scale/Resolution
Corine Land Cover	2006	Vector	1:250 000
	2012	Vector	1:250 000
	2018	Vector	1:250 000
EU-DEM	V1.1	Raster	25m
Imperviousness Density	2006	Raster	20m
	2009	Raster	20m
	2012	Raster	20m
	2015	Raster	20m
	2018	Raster	20m
Imperviousness Classified Change	2006-2009	Raster	20m
	2006-2012	Raster	20m
	2009-2012	Raster	20m
	2012-2015	Raster	20m
	2015-2018	Raster	20m
Grassland	2015	Raster	20m
Grassland	2018	Raster	20m
Water & Wetness	2015	Raster	20m
	2018	Raster	20m

Fig. 12.11 Table showing the type of datasets acquired from Copernicus Global Land Services including the year of acquisition, classified change and resolution. (https://land.copernicus.eu/)

from 2002–2010 (Cigna et al., 2017), acquired from pre-Copernicus European Space Agency Envisat monitoring satellite was also reviewed to look at ground motion over the past twenty years. Envisat data comprised 40 Envisat scenes acquired from a 100 km swath between 2020–2010 in both ascending and descending modes (Cigna et al., 2017) with a 25 m ground resolution. Sentinal-1 SAR data was acquired from a 250 km wide image swath incorporating three sub-swaths at 5 x 20 m spatial resolution.

Multi-reference (MR) was used to estimate the temporal evolution of displacements as these are best suited for rural areas and take into account non-linear evolutions. The data showed that in build-up and non-vegetated areas, SAR imagery is an effective monitoring tool to measure ground surface motion. However, due to the slope aspect of the pilot sites relative to ascending and descending orbits, SAR was not able to estimate, with confidence, north–south displacements at the pilot sites.

12.5 Citizens Involvement and Platform

I. Impact of the Covid-19 pandemic—site closure

The onset of the global pandemic in 2020 impacted citizen involvement severely. While progressing the setting up of the citizen observatory, in March 2020, a number of long running 'lockdowns' and restrictions came into force. This resulted in the closure of the Giant's Causeway visitor centre & Carrick-a-rede rope bridge to the public. Travel was also severely restricted and the high number of international visitors to the sites reduced from 1.5 m annually to almost zero overnight. Throughout 2020 and 2021, visitors to the site were predominately local users. This impacted on AGEO project plans for citizen outreach events at each site and acquisition of citizen science data. With the

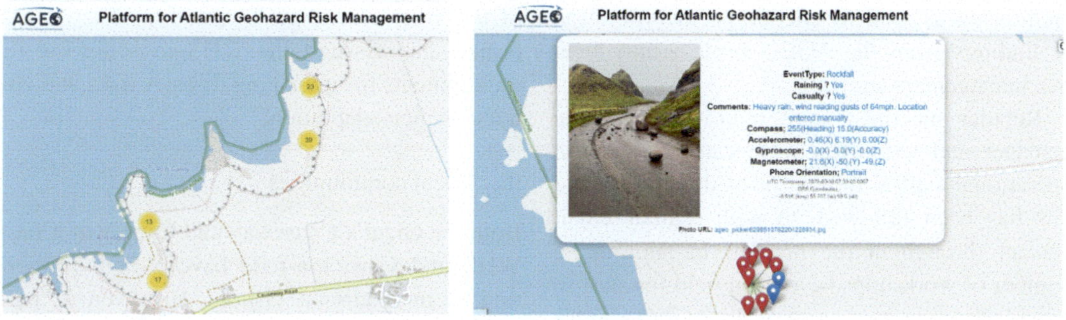

Fig. 12.12 Image taken from the AGEO online platform showing the number and location of citizen monitoring uploads taken using the AGEO application (left) and a zoomed in view of the portal showing the location of slope failures and the associated data captured through the application including event type, GPS, date, time and comments from the citizen (right)

easing of restrictions in Northern Ireland and globally, site users returned however this was a slow process and visitor numbers have yet to recover to the pre-pandemic levels.

II. Citizen Monitoring

The AGEO mobile application was released for testing in September 2022. At the Causeway Coast pilot site, the National Trust staff were provided with training on its use and a series of public workshops was delivered to promote citizen involvement through using the application. Between September 2022 and January 2023, a total of ninety-three citizen monitoring images, identifying recent slope failures, were uploaded to the AGEO online platform through the app (Fig. 12.12). All were recorded at the Giants Causeway with the exception of one image.

12.6 Other Achievements

a. Educational resources

Formal education on geohazards was identified as a key area of work within the the Causeway Coast citizen observatory. Having developed a number of geoscience resources in association with the CCGHT in the past it was decided to build on this success and develop a suite of educational resources for high school (15–18 year old) students. Resources were developed to highlight all the geohazards experienced by the AGEO partners and were designed to meet the requirements of the national curricula requirements not only in Northern Ireland, but across the entire geographic area of the AGEO project. These were accompanied by a fact sheet on the AGEO project as a whole and one on each of the five pilot sites.

Presentation of these resources has begun in a number of high schools in Northern Ireland and they have been accompanied by a training webinar for teachers. The overall aim being to embed the AGEO project in each high school and allow for a ‘real’ example of geohazard risk management at a local level.

12.7 Stakeholder Engagement

The development of a successful citizen observatory in Northern Ireland required direct engagement with a number of key stakeholders who were identified at an early stage within the AGEO project.

i. The National Trust

As the Property Managers, the National Trust were identified as the most important stakeholder for engagement. This was essential to allow site access, to implement on-site monitoring as well as encourage incorporation of findings from the AGEO project into local risk management strategies.

Regular meetings were held with the site manager and site visits were carried out with operational staff at the outset of the project and this has been followed up with regular email contact throughout the project. In addition, a number of workshops have been held for staff to provide training on the use of the AGEO app for on-site monitoring. This is now part of the daily monitoring across the Giant's Causeway allowing for detailed information to be gathered.

This has been advantageous in the delivery of the objectives of the AGEO project but has helped to strengthen the relationship between the GSNI and the National Trust allowing for better collaborative work in the future.

ii. Giant's Causeway and Causeway Coast WHS Steering Group

With oversight of the WHS Management Plan, the Giant's Causeway and Causeway Coast WHS Steering Group is made up of all of the relevant organisation that have an interest in the management of both the WHS and Carrick-a-rede. Engagement with the WHS Steering Group has taken place with assistance from the CCGHT who provide the secretariat for the Group.

The WHS Steering Group meets four times a year, and updates on the AGEO project have been provided twice yearly since the project began. This has provided vital support for the project from a number of key organisations including the Causeway Coast and Glens Borough Council and the Northern Ireland Environment Agency (NIEA). Both of these organisations have championed the AGEO project to local and national decision-makers and presented on the project's behalf at World Heritage UK.

Engagement with the WHS Steering Group has also allowed for the AGEO project to be included within the Periodic Reporting, one of the core conservation monitoring mechanisms of the World Heritage Convention. The AGEO project has assisted the WHS in achieving its requirements for managing disaster risks and climate change adaptation.

iii. Local communities

Both the Giant's Causeway and Causeway Coast WHS and Carrick-a-rede have a high proportion of international visitors. However, as part of the AGEO project it was identified that it was important to engage with local communities to ensure that there was an increased understanding of local geohazard risk, and a sense of ownership developed to allow for better management of those risks.

Several locations were identified within the wider area of County Antrim to allow for engagement with local communities on geohazards. A series of geohazard workshops were delivered with the assistance of the CCGHT and local community representatives in November and December 2022 (Fig. 12.13). These workshops provided information on the AGEO project, training on geohazards with a specific focus on the local area, as well as information on how to use the AGEO app for geohazard monitoring.

A total of 50 community members were engaged and included local business representatives, officers from the local council, mountain rescue members, wider National Trust representation and individual community members.

The workshops made it clear that further engagement was needed in the area of geohazards and identified areas of focus to be developed beyond the AGEO project, many of which have now been incorporated into the GSNI work programme.

12.8 Conclusion and Recommendations

The Causeway Coast is an area affected frequently by landslides resulting in an economic, social and environmental impact. An issue that could be exacerbated with climate change

Fig. 12.13 (left) Photograph of AGEO geohazard public workshop taken on 9th November 2022 at Carnlough Heritage Hub, Co. Antrim and (right) one of the AGEO information points at the Causeway Visitor Centre used to inform the public of the project and how to become involved

and will have a direct impact to locals and the wider region, given its significance to Northern Ireland. With the high number of tourists that visit the sites, it is important to understand processes, morphological changes and manage the sites in a sustainable manner. The AGEO project enabled a close working relationship with site managers including the National Trust and Causeway Coast WHS Steering Group who were very engaged with the project from the outset. The proposal of including the public in geohazard monitoring is an attractive asset and one that can potentially greatly assist both understanding and managing the risk at the sites. Through citizen engagement, it has also provided the public with increased knowledge of geohazards within the area including their occurrence, impact and importance of managing the risk. Since the release of the AGEO application, there has been good take-up enabling the recording of landslide events in a quick and secure way. The occurrence of landslides at the sites are unlikely to decrease in the future. Options for engineered mitigation are also limited given the sites designations and importance of natural conservation carried out the National Trust. Therefore, it is critical that the risk is managed in a way to minimise public risk and environmental damage. The project has created a common reference datasets of the geomorphological features through the capture of data using LiDAR and photogrammetry techniques. The incorporation of citizen photos into future photogrammetry models is an important milestone and one that will enable repeated assessments and change detection through the use of the common reference dataset. Satellite derived data is increasingly becoming a standard monitoring tool globally however the data reviewed within the AGEO project indicated that it has limited usefulness for smaller sites at current resolutions. Higher resolution data available from commercial satellite data providers may have the potential to enhance satellite monitoring for such sites.

Recommendations

Early planned engagement with citizens to educate the public on geohazard processes, their impact and how they can assist in monitoring their local area. This should include clear information on how they, as contributors, will benefit from the collective actions of participants.

Citizen Monitoring tools, such as the AGEO application, needs to be easy to use for users as well as attractive by returning information of interest to them and also clearly demonstrating how their actions are assisting. This is crucial in a world where he public are exposed to a very high quantity of mobile applications in a highly competitive market.

Geohazard monitoring benefits from citizen involvement. It enables data to be collected at numerous times and locations. It is a cost-efficient approach that has the added benefits of providing educational information to the user.

References

Bratton, A., Smith, B., McKinley, J., & Lilley, K. (2013). Expanding the geoconservation toolbox: Integrated hazard management at dynamic geoheritage sites. *Geoheritage, 2013*(5), 173–183. https://doi.org/10.1007/s12371-013-0082-8

CCGHT. (2021). Giant's Causeway and Causeway Coast World Heritage Site Management Plan 2021–2027. Retrieved from https://ccght.org/wp-content/uploads/2021/07/WHS-Management-Plan-FINAL-compressed.pdf.

Cigna, F., Banks, V., Donald, A., Donohue, S., Graham, C., Hughes, D., McKinley, J., & Parker, K. (2017). Mapping ground instability in areas of geotechnical infrastructure using satellite InSAR and small UAV surveying: A case study in Northern Ireland. *Geosciences,7*(3). https://doi.org/10.3390/geosciences7030051.

Copernicus Global Land Services. https://land.copernicus.eu/.

Crawford, K., & Black, R. (2011). Visitor understanding of the geodiversity and the geoconservation value of the Giants Causeway World Heritage Site, Northern Ireland. *Geoheritage, 2012*(4), 115–126. https://doi.org/10.1007/s12371-011-0049-6

Geomaterials. http://www.qub.ac.uk/geomaterials/.

Jaud, M., Le Dantec, N., Parker, K., Lemon, K., Lendre, S., Delacourt, C., & Gomes, R. C. (2022). How to Include Crowd-Sourced Photogrammetry in a Geohazard Observatory—Case Study of the Giant's Causeway Coastal Cliffs. *Remote Sensing*, 14(14), 3243https://doi.org/10.3390/rs14143243

Lemon, K. (2012). The geological society of London, 100 Geosites, Giants Causeway, Northern Ireland. Retrieved from https://www.geolsoc.org.uk/GeositesGiantsCauseway.

Lemon, K. (2015). Giant's Causeway and Causeway Coast World Heritage Site: Geoconservation. Retrieved from https://ccght.org/wp-content/uploads/2021/02/Giants-Causeway-and-Causeway-Coast-World-Heritage-Site-Geoconservation.pdf.

Lyle, P. (1996). A geological excursion guide to the Causeway Coast, Environment and Heritage Service, (Department of the Environment, NI), Belfast.

McDonnell, B, A, M., & Smith, B. J. (2000). Slope failures at the Giants Causeway 1995–1998. Report for Environment and Heritage Service.

Mitchell, W. I. (ed) (2004). The geology of Northern Ireland—Our natural foundation, Geological Survey of Northern Ireland, Belfast. Retrieved from https://earthwise.bgs.ac.uk/index.php/Geology_of_Northern_Ireland:_our_natural_foundation.

National Trust. Retrieved from https://www.nationaltrust.org.uk/visit/northern-ireland/carrick-a-rede/history-at-carrick-a-rede.

National Trust. (2023). Carrick-a-rede. Retrieved from https://www.nationaltrust.org.uk/visit/northern-ireland/carrick-a-rede.

NISRA. (2020). Northern Ireland Visitor Attraction 2019 Survey Report. Retrieved from https://www.nisra.gov.uk/sites/nisra.gov.uk/files/publications/NI-tourism-publication-VisitorAttraction-Survey-2019-Report.pdf.

Simms, M. J. (2021). Subsidence, not erosion: Revisiting the emplacement environment of the Giant's Causeway, Northern Ireland. *Proceedings of the Geologists' Association*, 132(5), 537–548. https://doi.org/10.1016/j.pgeola.2021.07.001

Smith, B., Ondicol, R., & Alexander, G. (2011). Mapping slope instability at the Giants Causeway and Causeway Coast World Heritage Site: Implications for site management. *Geoheritage, 3*, 253–266.

Tourism NI Media Library. (2023). Retrieved 26 March, 2023, from https://northernirelandscontentpool.com/.

UNESCO. (2023). https://whc.unesco.org/en/list/369/

Citizens' Observatory on Rockfalls in the Madeira Island, Portugal

13

Domingos Rodrigues, Fátima Gouveia, Paulo Lobo, Ana Peixoto, Nuno Caldeira and Rui Figueira

13.1 Introduction

Natural hazards occur frequently on Madeira Island and their impact can be devastating. Precipitation-triggered hazards including debris flows, landslides and floods, are the main causes of damage and death. Coastal slides and erosion are also important in the region, and earthquakes have also been recorded.

On February 20, 2010, debris flows and landslides resulted in the loss of 47 lives and caused great material losses. On that day, there were periods of very intense rainfall, especially in the morning, reaching 58 mm/h in Funchal. In fact, the amount of precipitation measured during that month was extremely high, approximately 7 times higher than the average, and the third highest since 1865, as reported in the Monthly Climatological Bulletin of Madeira for February 2010 by the Portuguese Institute for the Ocean and Atmosphere (IPMA).

On March 7, 2020, at 20:58, an earthquake of magnitude 5.2 occurred with an epicentral location about 35 km south of the city of Funchal. The earthquake was felt by the majority of the population across the island, raising awareness about seismic activity in the Region. Improving the resilience of the community regarding geohazards by engaging citizens in hazard preparedness prevention, mitigation, emergency response, and recovery is crucial. Participatory methods and tools like Citizen Observatories should be applied to improve risk management. The AGEO Platform and Citizens Observatory are important tools to engage citizens and improve risk management in Madeira, contributing to mitigate the impact of geohazards in the future.

D. Rodrigues (✉) · F. Gouveia · P. Lobo
Department of Civil Engineering and Geology, Madeira University, Campus da Penteada, 9020-195 Funchal, Portugal
e-mail: domingos@uma.pt

F. Gouveia
e-mail: fatima.gouveia@staff.uma.pt

P. Lobo
e-mail: plobo@staff.uma.pt

A. Peixoto · N. Caldeira · R. Figueira
Municipal Service of Civil Protection, Municipality of Funchal, Estrada da Vitória 46 E 48, 9000-784 Funchal, Portugal
e-mail: ana.peixoto@funchal.pt

N. Caldeira
e-mail: nuno.caldeira@funchal.pt

R. Figueira
e-mail: rui.figueira@funchal.pt

D. Rodrigues
Institute of Earth Sciences, Braga, Portugal

F. Gouveia · P. Lobo
CERIS, Instituto Superior Técnico, Universidade de Lisboa, Lisboa, Portugal

R. C. Gomes et al. (eds.), *Citizens' Observatories on Geohazards*, Geoenvironmental Disaster Reduction,
https://doi.org/10.1007/978-3-031-53371-6_13

13.2 Site Description

The Madeira Archipelago is in the North Atlantic, 900 km southwest of Lisbon. The islands, Madeira, Porto Santo, Desertas, and Selvagens, are of volcanic origin, dating from the Miocene, and are interpreted as "hot spots" that formed in an initial stage by submarine volcanic activity and later subaerial eruptions. Madeira Island is part of the emerging shield volcano Madeira-Desertas, whose lava is mainly composed of alkaline basalts ranging from basanite to mugearite and pyroclastic rocks.

Madeira is the biggest island of the archipelago, with 728 km^2 and a population of about 250 000 inhabitants. Madeira is highly urbanized on the south coast, especially in the capital, Funchal, with 106 000 inhabitants.

The orography of Madeira Island is characterized by a strong relief with altitudes that reach 1862 m at Pico Ruivo, high mountain ridges with ravines and steep deep valleys. Its topography and geomorphology strongly conditions land occupation (Figs. 13.2 and 13.3) leading to construction next to riverbanks, occupation of flood plains, modification of drainage systems, execution of excavation slopes for the construction of structures and roads, amongst others. In fact, some urban areas, as Funchal, Camara de Lobos, Santa Cruz, Ribeira Brava and Machico, are located next to important river mouths, increasing their vulnerability to flooding and debris flow.

Most of the occurrences are triggered by extreme weather events. According to Madeira Regional Water Plan (PRAM—Plano Regional da Água da Madeira), the average annual precipitation is around 1628 mm, varying between 600 mm in the south coast and 1000 mm in the north coast. In higher altitude areas, in the central region, cumulative annual precipitation is higher than 2800 mm. Rainfall data collected at different stations has shown a high spatial variability and the intensity measured at specific sites is significant.

Floods and debris flow

Extreme precipitation events occurring over a very short period, such as those observed on the 20th of February 2010, can have catastrophic consequences. This was the most recent event with greatest consequences. The volume of precipitation during 24 h period, in higher mountain

Fig. 13.1 Location of Madeira Archipelago. *Source* Esri—World Imagery (WGS84)

Fig. 13.2 Relief on the south coast (Funchal)

Fig. 13.3 Relief on the south coast (Funchal)

elevations, was nearly two-thirds of the annual amount of precipitation for the island (Nguyen et al., 2012), triggering landslides, intense flooding and debris flow.

In the past, other events had high impact, such as the flash flood of 1920, which caused two deaths and 500 homeless people, and the debris flow of 1929, which caused 34 deaths (Rodrigues & Ayala-Carcedo, 2000).

Rockfalls and Landslides

Mass movements are quite frequent along the island and tend to be more frequent between October and March, a period that tends to exhibit higher precipitation. Besides natural factors (precipitation, elevation, slope gradient, shape, lithology and soil properties), other factors associated with human activity may also increase the risk of mass movements and impact of floods (Nguyen et al. 2013). Rockfalls and landslides occur across much of the island and are more frequently reported when happen along roads (Figs. 13.4 and 13.5).

Earthquakes

Madeira is considered, according to Eurocode 8, a very low seismicity region, with a reference peak ground acceleration of less than 0.05 g. However, there are records of an earthquake that happened on 1748, with a maximum intensity of VIII Mercalli; also in 1816, 1866, 1918 and 1923. On May 26, 1975, an earthquake left around 50 people homeless in Funchal (Madeira et al., 2007). Several earthquakes have occurred in recent years. The most recent one was felt by most of the population on March 7, 2020. Although most of these events have relatively low intensity, they may trigger mass movements in areas with steep slopes.

Coastal Erosion

Coastal erosion is also one of the most active geomorphic processes and a major concern, threatening buildings and roads that are close to the coastline. The cliff coast retreat depends on mass movements in the coastal slope and the

Fig. 13.4 Rockfall in Ribeira dos Socorridos

Fig. 13.5 Landslide in Curral das Freiras (Pico Furão)

direct action of the sea, coastal orientation, and geological structures. In Madeira, coastal retreat rates can reach, in some areas, several m/yr. Ponta Delgada, Ponta do Sol, S. Vicente are areas with significant coastal retreat risk (Fig. 13.6).

13.3 Regional Civil Protection

The Portuguese Civil Protection Structure is organized at different levels: national, regional, and municipal (Fig. 13.7). The activities of the different players and their responsibilities and competencies are defined by law (Civil Protection Basic Law).

Civil protection agents include firefighters, security, the armed forces, maritime and aviation authorities, and medical emergency services, among others. Other entities, such as volunteer organizations, health services, and scientific and technological institutions, have the duty of cooperation.

The National Emergency and Civil Protection Authority (ANEPC) is the national authority and a central service with direct administration from the state. It is responsible for planning, coordinating, and defining the National Civil Protection policy regarding the prevention and reaction to severe accidents and catastrophes. It is also responsible for the coordination of the national risk assessment.

The Civil Protection system in Madeira is under the administration of the Regional Government. There are Civil Protection commissions, coordination centers and relief operation command, with tasks similar to the ones performed by the National Command for relief operations, but adapted to the scale and specificities of the Region.

Emergency response is regulated by the Regional Emergency Plan, which aligns with the National Commission's guidelines and articulates with the National and Municipal emergency plans. These emergency plans, along with land use planning and the Flood Risk Management Plan, are tools currently used for regional risk management.

The Municipal Civil Protection services are responsible for risk prevention and assessment at the local scale, the development of municipal emergency plans, the planning and coordination of civil protection operations, providing support to victims of catastrophes and raising public awareness.

To promote behavioural changes, social responsibility and increasing resilience to natural hazards, the Funchal Civil Protection Service established Civil Protection Units. These voluntary community organizations that aim to enable school communities and residents to build a collective response in terms of risk mitigation and prevention (Figueira et al., 2022). These units have been active since 2016 (after an important forest fire) and their activities are adapted to each local reality.

The Funchal Civil Protection Service has mantained regular contact with citizens and conducted several activities, including training sessions and emergency situation simulations. Their approach has proven to be effective in terms of risk preparedness and mitigation, and the impact of their action is especially important in areas of difficult access and high vulnerability. In recognition of their efforts, these Civil Protection Units were awarded a national prize for exemplary participatory democracy practices in Portugal, an initiative promoted by ANEPC.

13.4 Madeira Pilot

The AGEO project created a resource platform on Atlantic geohazards risk assessment and confirmed the capacities of Citizen Observatories in contributing to the improvement of risk management systems.

The AGEO Madeira Pilot established a Citizens' Observatory with a suite of tools and methodologies to foster citizen engagement and active participation amongst local communities and schools in the prevention, mitigation and management of geohazards. Focusing this goal, the pilot implemented a School Station Network that collects and analyses data from weather stations and seismic stations made available in a group of schools, providing real-time data for risk management, which is especially relevant at the local level.

The aim is to engage citizens, local communities, and schools to actively participate in risk preparedness and monitoring of geohazards for integration into local risk management systems.

The AGEO Madeira Pilot demonstrated a new form of engagement between civil society, communities, schools, and the local authorities for geohazards-related capacity building.

13.4.1 Methodology

The citizen engagement strategy was designed to operate at three distinct levels: citizen participation level, community level and school level (Fig. 13.8).

1. Citizen participation level

During the initial stage of the project, a new layer was designed and added to the mobile application "Funchal Alerta". This mobile application has been used by the citizens of the Funchal municipality in recent years to report incidents related to public space and mobility, and issues. The new feature, developed as part of the AGEO project, introduced the capability to report occurrences related to geohazards. Reports include written information (via a questionnaire), photos and georeferenced locations. This layer was tested with students and citizens who primarily reported mass movements such as rockfalls and landslides. Following this, the mobile application developed within the AGEO Project, which is exclusively dedicated to reporting situations of risk related to natural geohazards, was also made available and tested.

2. Community level

The Funchal Civil Protection Service has regularly been conducting several activities with the community, including areas such as Curral dos

Fig. 13.6 Coastal erosion study area in Ponta Delgada

	POLITICAL DIRECTION	POLITICAL COORDINATION	INSTITUTIONAL COORDINATION	OPERATIONAL COMMAND
NATIONAL LEVEL	Prime Minister	National Commission for Civil Protection (CNPC)	National Operational Coordination Centre (CCON)	National Command for Relief Operations (CNOS)
REGIONAL LEVEL	President of the Regional Government	Regional Commission for Civil Protection (CRPC)	Regional Operational Coordination Centre (CCOR)	Regional Command for Relief Operations (CROS)
MUNICIPAL LEVEL	Mayor	Municipal Commission for Civil Protection (CMPC)		Command for Relief Operations (COS)

Fig. 13.7 Civil Protection structure at National, Regional and Municipal levels

Romeiros and Barreirinha, where Civil Protection Units were stablished. These activities are tailored for each community considering their needs and concerns. These include lectures, practical activities, collaborative mapping, safety at home related tasks, drills, Basic Life Support (BLS) and the implementation of preventive practical measures against natural hazards that can potentially affect the community. Additionally, during summer school holidays, a special two-week Civil Protection Summer School is organized for the children in the community.

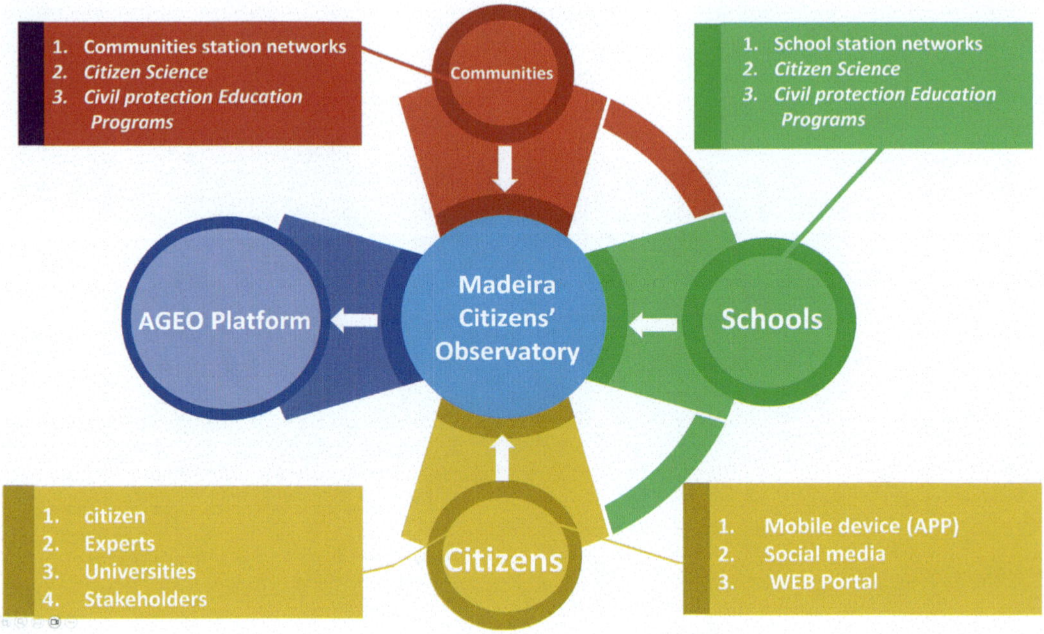

Fig. 13.8 Madeira Citizens Observatory

The initiatives outlined in the AGEO project were incorporated into ongoing initiatives of the Civil Protection Units, aligning with existing activities.

The data provided by weather stations and seismic stations implemented in certain schools, as part of the School Station Network, are accessible to the entire community and offer important real-time information, especially in high-vulnerability areas like Curral das Freiras and Ponta Delgada.

3. School level

During the AGEO project, most of the dissemination activities were conducted in schools that were integrated in the School Station Network. These activities were incorporated in the Funchal Municipality Civil Protection Service initiatives. Within the School Civil Protection Center, an annual program of activities is currently planned for the different school grades. Lectures, practical visits, and games are some of the strategies used to engage students.

Traditional games and dances from Madeira were introduced to encourage student participation in these activities, enhancing their understanding of geohazards and allowing them to identify essential aspects of Madeira's traditional culture. These games were easy to explain and adapt, making them simple and enjoyable while effectively addressing many types of hazards.

Students were also invited to use the AGEO mobile application, the AGEO Observatory platform and websites like Sentinel Playground, to understand the main geohazards that can affect the school and its surrounding.

Within the scope of the AGEO Project, books and videos were edited as didactic materials for teaching purposes. These cover themes such as the characterization of geohazards in Madeira, historical catastrophic events, self-protection measures and educational activities for the prevention of natural disasters.

Due to COVID-19 restrictions, schools and institutions were closed, impacting all interactions with citizens and students. Consequently, educational programs were transitioned to virtual platforms during that period. To facilitate online learning, new methodologies were developed, and students used platforms such as Sentinel playground, Google Earth, EO Browser, between others. These activities involved interpreting aerial photos, detecting landscape changes and associating them with increased vulnerability.

13.4.2 School Station Network and AGEO Platform

Since the pilot aims, as a long-term goal, to develop an Early Warning School Network, weather stations (Fig. 13.10) and seismic stations (Raspberry Shake RS1D) (Fig. 13.11) were made available at six schools. Five of them are located in Madeira Island (in Funchal, Câmara de Lobos, São Vicente and Machico) and one is in Porto Santo (Fig. 13.9). Schools were selected based on three criteria: (1) location in areas highly affected by natural hazards, (2) geographical distribution and (3) schools already involved in Civil Protection activities.

The data can easily be accessed in real-time through the AGEO platform or through the AGEO mobile application (Fig. 13.12), and there are monthly detailed data reports from the stations. This data is being used and analysed in schools, during various classes such as geography, mathematics, as well as at the Civil Protection Center. This real-time data is accessible to the entire community, and it has proven highly beneficial, particularly for communities located nearby, in high-risk areas.

13.5 Data Handled by the Pilot

The AGEO Madeira Pilot used and managed a significant amount of data and services, including:

- Historical data for the characterization of catastrophic events that occurred in the Madeira Archipelago in the past that resulted in severe damage and numerous fatalities.
- Aerial photos from the 1970s–1980s were used to evaluate the evolution and the rate of

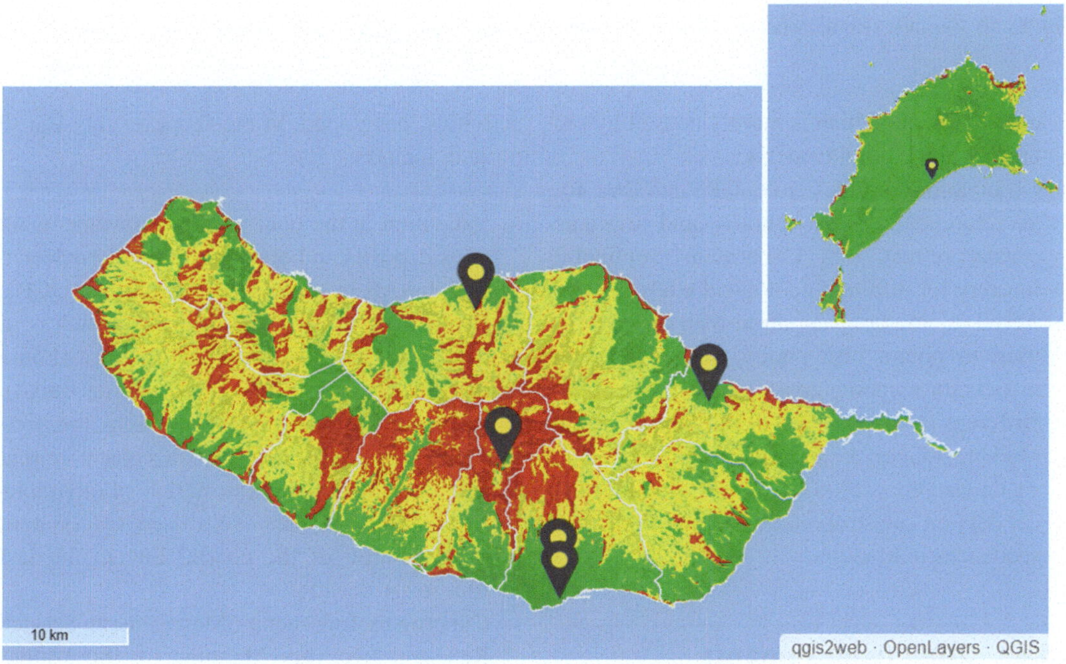

Fig. 13.9 Location of the six schools that integrate the School Station Network. The map also shows the susceptibility to mass movements (Red—High, Yellow—Moderate; Green—low). *Source* Online Platform (https://psprociv.github.io/ARRAM/): "Avaliação de Risco da Região Autónoma da Madeira (Abril, 2023)

Fig. 13.10 Weather station installed at Curral das Freiras School, as part of the AGEO School Station Network

coastal retreat, which is mainly caused by sea erosion and coastal landslides.
- Observation services, like GPS surveys for georeferencing aerial photos and evaluate coastal retreat, and Copernicus Sentinel-2 imagery for different geological analyses.
- Aerial photography captured with drones was used to obtain orthophotos and 3D terrain models for mapping coastal areas.
- Different databases were used, including the Madeira—Georisk Database (Rodrigues e Ayala-Carcedo, 2000), for collecting historical data on catastrophic events and their consequences in Madeira.

13.6 Other Achievements

Geological fieldwork was conducted for surveying and analysing the geological processes concerning coastal retreat and erosion. These activities, developed on the scope of the AGEO project, included:

- Validation in the field of creep movements on talus deposits and erosion previously detected with InSAR by a partner of the project (UCD).
- Evaluation of the coastal retreat index in three pilot areas of the Madeira coast (Ponta Delgada, Fajã da Areia, and Lugar de Baixo). This work aimed at understanding the geological processes involved in coastal retreat and its rate. Different aerial photographs from different years were used to compare the evolution of the coastal line in the last four to six decades.
- Copernicus Emergency Management Service Risk and Recovery Mapping for the evaluation of the impact and extent of the damage caused by mass movements that occurred on December 25, 2020, as a consequence of a storm with heavy rain on the northern coast

Fig. 13.11 Seismograph (Raspberry Shake 1D) installed at schools (AGEO School Station Network)

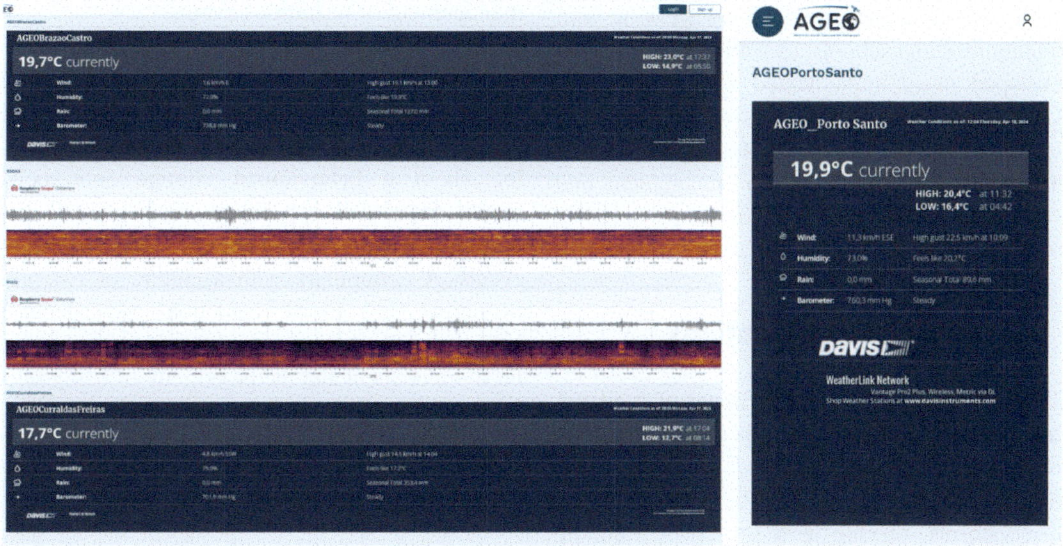

Fig. 13.12 Weather stations and seismographs real-time data. Source AGEO platform and mobile app (Madeira pilot)

of Madeira Island. Debris flow and landslides occurred in Ponta Delgada and Boaventura, both in the municipality of São Vicente. One house was destroyed, and several others were severely damaged; 20 people were in Ponta Delgada and 7 others in Boaventura.

13.7 Interaction with Stakeholders and Risk Management

AGEO Madeira Pilot includes, as stakeholders, the Municipal Civil Protection Service of Funchal, and the Municipal Civil Protection Service of Câmara de Lobos, as well as the six schools that integrate the AGEO School Station Network (Fig. 13.9).

The School Station Network, also referred to as School Early Warning Network, and its real-time data is accessible to all the citizens. Schools within the AGEO network and communities in those municipalities, where weather and seismic stations were installed, actively follow and analyse this data.

The tools and methodologies for working with communities and schools, as well as the activities developed, were implemented with the support of the Municipal Service of Civil Protection of Funchal. Therefore, meetings for the implementation of strategies for citizen engagement and evaluation of its progress were frequent.

13.8 Conclusion and Recommendations

The development and establishment of a citizen platform on Atlantic geohazards risk assessment, preparedness, mitigation, and prevention, is crucial. This platform should be fed by Citizen Observatories, serving as a source of valuable information to enhance risk preparedness and management.

The Portuguese Civil Protection Authority uses a sequential model of risk management for emergency response (ISO 31000), which is a model focused on instrumental and organizational aspects of the processes. It is not quite suitable for fostering more direct and active citizen participation in risk management through Citizen Observatories and platforms.

The implementation of other models of risk management, such as the circular model (IRGC 2017), and especially the **multilevel** model of risk management in Civil Protection, could enhance the levels of participation of different actors, such as universities and citizens.

However, at the local authority level (municipality and parish), especially with the new goals introduced by the Portuguese National Strategy for Preventive Civil Protection 2030 and the Civil Protection Basic Law—Article 43.° in the engagement of citizens and the creation of Local Civil Protection Units, Citizens Observatories can be important and valuable support for decision making.

At the local authority level, the municipalities or parishes that receive detailed information issued by citizens, Citizen Observatories and School Early Warning Systems can be an important, and sometimes the only, source of information, contributing decisively to improve the risk management of hazardous events.

The continuous engagement of citizens, communities, and schools through citizen observatories has the objective of improving **risk perception** and, especially and far more importantly, bringing about a **behavioural change** and social regulation that will increase the resilience of communities during the occurrence of hazardous events. A community with a culture of risk preparedness high social regulation, and adequate behaviour knowledge, especially during crises, is far more resilient and the probability of being seriously affected is significantly decreased.

References

Figueira, R., Peixoto, A., Pestana, J., & Quintal, S (2022). *A civil protection of proximity: making a resilient community.* Global Assessment Report on Disaster Risk Reduction—United Nations Office for Disaster Risk Reduction.

Nguyen, H., Wiatr, T., Fernández-Steeger, T., Reicherter, K., Rodrigues, D., & Azzam, R. (2013). *Landslide hazard and cascading effects following the extreme rainfall event on Madeira Island (February 2010).* Nat. Hazards 65, 635–652.

Peer review—report Portugal. (2019). 2018–2019 Programme for peer reviews in the framework of EU cooperation on civil protection and disaster risk management.

Madeira, J., da Silveira, A. B., Ramalho, R., Fonseca, P., & Prada, S. (2007). Carta geológica da Madeira na escala 1:50000—Relatório Técnico.

NP EN 1998-1. (2010). Eurocódigo 8—Projeto de Estruturas para Resistência aos Sismos—Parte 1: Regras gerais, ações sísmicas, e regras para edifícios. Instituto Português da Qualidade.

Reis, E, Bergonse, R., Simões, E., & Filipe, P. Impactos e vulnerabilidades às alterações climáticas. Riscos hidrogeomorfológicos, Relatório de vulnerabilidades e medidas de adaptação. Observatório Clima Madeira.

Rodrigues, D., & Ayala-Carcedo, F. (2000). Georisk: A natural hazard database of historic events in Madeira Island. 25th General Assembly EGS, Nice, p. 286.

14 Recommendations for Future Citizens' Observatories

Rui Carrilho Gomes

14.1 Citizens' Observatories Objectives

The AGEO project implemented five pilots of Citizens' Observatories to improve the monitoring and management of natural hazards in the Atlantic Arc Area.

The key actions undertaken addressed:

(i) improving and capitalizing on knowledge of hazards and exposure, by using long-term monitoring and data dissemination;
(ii) improving risk management and prevention by engaging diverse stakeholders and providing tools to help decision-making;
(iii) promoting the risk culture by involving citizens in monitoring and/or by providing them with information materials.

More specific objectives shared among all pilots included:

(i) Data collection from monitoring hazards and definition of measurement protocols;
(ii) Data standardization, to providing standardized data for public authorities responsible for risk management;
(iii) Dedicated web-portal, providing access to documentary resources and/or results;
(iv) Earth Observation product and services development and integration in the risk management system and use and dissemination of Copernicus emergency services;
(v) Stakeholders engagement, by building/maintaining partnerships with local authorities, local associations, and other local and regional stakeholders to support citizens engagement;
(vi) Dissemination action, including leading/animating groups of natural risk management professionals and citizens;
(vii) Delivering training sessions/workshops and awareness-raising measures for target groups, as students, scouts, local associations, among others;
(viii) Scientific expertise sharing, providing expertise in civil engineering, geology, computer science, citizen science among other to support risk management policies at local scale;
(ix) Setting up warning systems.

14.2 Technological Tools

The AGEO project brought together experts from several disciplines to create a platform and app that can be used by citizens, stakeholders,

R. C. Gomes (✉)
CERIS, Instituto Superior Técnico, Universidade de Lisboa, Lisboa, Portugal
e-mail: rui.carrilho.gomes@tecnico.ulisboa.pt

R. C. Gomes et al. (eds.), *Citizens' Observatories on Geohazards*, Geoenvironmental Disaster Reduction,
https://doi.org/10.1007/978-3-031-53371-6_14

researchers, and decision-makers to monitor, assess, and mitigate geohazards risks.

The AGEO Platform was designed to be the medium between citizens and responsible authorities, national or local, working as an ice-breaker to start interaction. This collaboration encourages authorities to see the opportunity in citizen science and attract regular citizens to voluntarily be more interested in their environment's status and what is happening around them.

Overall, the AGEO platform and app aims to serve a diverse range of stakeholders involved in geohazards risk management, leveraging citizen science and technology to create a more resilient and sustainable future.

The primary audience for the AGEO platform is citizens, particularly those living in areas at risk of geohazard events. The web-based Citizens' Observatory platform developed provides citizens with tools and resources for reporting geohazards events, including earthquakes, landslides, geotechnical risks, floods, among others. It also provides citizens with real-time access to data and information on geohazards risks in their area, as well as guidance on how to prepare for and response to emergency situations.

Researchers are also an important audience for the AGEO platform, as it provides them with access to wealth of data on geohazard events that can be used to better understand the causes and impacts of these responses. This data can also be used to develop new risk management strategies and emergency response plans.

Decision-makers, including local and national governments officials and emergency responders, to whom the platform can provide real-time information on geohazards events in their area, allowing them to make more informed decisions about risk management and emergency response strategies.

AGEO platform offers a range of tools and services to help citizens and decision-makers monitor and manage geohazard risks, including:

- Data collection and sharing—the platform for citizens to collect and share data on geohazard events, can help to inform risk management strategies and emergency response plans.
- Hazard mapping and risk assessment—the platform offers mapping and geohazard risks, for assessing the potential impact of these risks on communities and infrastructure. This can help-decision-makers to develop more effective risk reduction strategies, and to prioritize investments in disaster preparedness and response.
- Citizen engagement and education: engaging citizens in the monitoring and management of geohazard risks, helping to build community resilience and awareness. By involving citizens in the data collection and risk assessment, a culture of preparedness is promoted and helps communities to better understand the risks they face.
- Early-warning and emergency response: help decision-makers and citizens to respond more quickly and effectively to geohazard events, aiming to reduce the impact of these events on communities.

AGEO platform can partner with a wide range of organizations and stakeholders such as:

- Research institutions—to develop and implement new methods for monitoring and managing geohazards, aiming at ensuring the latest research and scientific findings, and that its tools and services are based on sound scientific principles.
- Local and national government agencies—to develop more effective risk reduction and emergency response strategies, by sharing data and expertise to improve the resilience and preparedness of communities at risk.
- Non-governmental organizations (NGOs) focused on disaster relief and recovery—to ensure that citizens affected by geohazard events have access to resources and support they need, mitigating the impact of these events on vulnerable communities.
- Private sector partners—to develop new technologies and services for monitoring and managing geohazard risks, such as

technology companies could develop new tools and platforms for collecting and analysing data. While insurance companies could use the collected data to develop more accurate risk models and insurance products.

The platform was designed to be accessible via mobile devices, is optimized for use on mobile browsers and through the mobile applications (ISO and Android).

The AGEO Citizens' Observatory platform and mobile applications can be accessed to report geohazards events, access real-time data on geohazards risks in their area, upload georeferenced pictures and videos uploaded for enhancing report, and receive guidance on how to prepare for and response to emergency situations. The platform's mobile-friendly design means that it can be accessed and used by citizens, researchers, and decision-makers on the go, providing users with real-time information and support for geohazard risk management.

An easy-use, open-source, multi-temporal InSAR processing tool for geohazard monitoring named EZ-InSAR has been released, providing satellite-derived multi-temporal Interferometry of Synthetic Aperture Radar (InSAR) data. This data can be added to the Citizens' Observatory, increasing the quality and diversity of information on the geohazard or their triggers (e.g. precipitation), but also to educate citizens on the Earth-Observation potential.

14.3 Recommendations from AGEO Pilots

The deployment of AGEO pilots in 5 areas with very diversified contexts, as described in Chap.7 , brought up the following recommendations:

1. The integration of the Citizens' Observatories into the risk management system has been addressed in the 5 AGEO pilots, through the involvement of citizens before, during and after the events, enabling data to be collected at numerous times and locations in a cost-efficient approach that has the added benefits of providing educational information to the user. It is recommended to promote the involvement of the citizens in data collection that should be tailored with respect to the characteristics of the geohazard, the legal framework, the stakeholders' culture and practice, to maximize the impact and overall acceptance and contribution of the Citizens' Observatory.
2. To assure the Citizens' Observatory sustainability, it is recommended to integrate it into an existing local/regional public institution related with the emergency and risk management structure and/or with direct interaction with citizens, as a municipality. For the Citizens' Observatory to work, collaborative links must be created between all relevant stakeholders (e.g. local associations, scouts, …), resulting in the increase of the Citizens' Observatory effectiveness.
3. Citizens' Observatories technological tools, namely the app to collect data and platform to organize and share data, revealed to be effective in collecting data in all pilots. The app and platform should be user-friendly and should return information of interest to users, e.g. educational contents and/or information on the event reported. It is also recommended to co-design the Citizens' Observatory technological tools with the local/regional institution, to assure:
   a. the effective connectivity between hardware/software systems of the host institution and of the app/platform;
   b. an effective workflow of information, i.e., the information reaches the right person or system and needed actions are taken.
   c. The platform's interface should adopt simple language and predominantly graphic, to enable any citizen to share information.
4. The implementation of a citizen platform on Atlantic geohazards risk assessment, preparedness, mitigation, and prevention fed by the Citizens' Observatories are a source of valuable information to improve the quality of risk preparedness and management. It is

recommended to use the open-access platform developed in AGEO project or another that can transmit the key data in a visual georeferenced manner, linked to an app, web-portal, or any other system to collect information.

5. Earth-Observation Systems, including Copernicus products and services, combined with detailed topography obtained with drones and field validation, are very effective in the identification of risk areas, to characterize in detail the environment, and to know both the conditioning characteristics and the triggering factors that can cause/contribute to the geohazard. It is recommended to create a common reference datasets of the geomorphological features, e.g. using LiDAR and photogrammetry techniques. The incorporation of citizen photos into future photogrammetry models will enable repeated assessments and change detection through the use of the common reference dataset.
6. The legal framework regarding geohazards, emergency response and civil protection vary significantly from country to country, e.g. and may not be suitable for the involvement of citizens and Citizens' Observatories as in the case of the sequential model of risk management for emergency response. The circular risk management model but especially the multilevel model, can better profit from Citizens' Observatories. It is recommended to identify the gaps in the legislation and propose a new legislative framework where Citizens' Observatories can be included and regulated.
7. Citizens' engagement can be limited in the beginning, but can grow over time. It is recommended to persist in awareness campaigns with support of local and regional stakeholders, promoting direct contact with citizens, involving them in the different phases of hazard assessment and management and include clear information on how they, as contributors, will benefit from the collective actions of participants, bringing behavioral change that will increase the resilience of communities during the occurrence of hazardous events, developing a culture of risk in the community, high social control and regulation, especially during crises.
8. A consistent communication strategy is recommended, for example with a mascot/character common to all educational contents and dissemination actions, bringing additional visibility and coherence to the overall message.
9. In high interest sites (e.g. touristic, monumental, geoparks), include the public in geohazard monitoring is an attractive asset to assist both understanding processes, morphological changes, managing the risk and the sustainability of the sites. The occurrence of geohazards like landslides at high interest sites are unlikely to decrease in the future. Options for engineered mitigation measures are also limited given the sites designations and importance of natural conservation, therefore, it is critical that the risk is managed in a way to minimize public risk and environmental damage.
10. Particularly in isolated areas, it is recommended to install acquisition equipment in schools, as seismograph and meteorological station. This equipment are an important asset to raise awareness to students, sometimes the fastest and/or only source of information available, and can work even as a rudimental early warning system, contributing decisively improve the risk management of a hazardous event.
11. Citizens' Observatory are recommended to rely on the co-construction and the close interaction between academics, local/regional authorities, and stakeholders, which requires in-depth involvement of all parties in joint actions (data collection, sharing expertise, ...), as well as political support to achieve partnership agreements and engage in long-term observation of the vulnerabilities.

14.4 Final Remarks

The global geohazard monitoring and management applications is expected to grow significantly in the coming years. The increasing frequency and intensity of geohazard events, coupled with growing population and infrastructure development in hazard-prone areas, are driving the demand for geohazard monitoring and management solutions, integrating real-time monitoring, early-warning systems, and crowd-source data.

A geohazard Citizens' Observatory can be a powerful tool for monitoring, rasing awareness and managing geohazard risk, as it enables the public to participate in geohazard monitoring and reporting.

To maximize the effectiveness of the Citizens' Observatory, a user-friendly platform is needed, as persistent training and support to citizens to ensure participation and data quality. Fostering collaboration between citizens and professionals is crucial for effective geohazard management, based on mutual trust and ensuring privacy and security of the data to be used for decision-making.

The Geohazard mobile application is a powerful tool for monitoring and managing geohazards, enabling users to receive real-time alerts and updates on geohazard events. The mobile application features should include user-friendly interface, incorporate real-time monitoring data, provide personal alerts, include safety tips and recommendations, enable report and feedback, collaborate with local authorities and ensure privacy and security.

The manufacturer's authorised representative in the EU is Springer Nature Customer Service Centre GmbH, Europaplatz 3, 69115 Heidelberg, Germany. If you have any concerns regarding our products, please contact ProductSafety@springernature.com

Printed and bound by CPI Group (UK) Ltd, Croydon, CR0 4YY
01/10/2026
02225855-0018